Year 3A

A Guide to Teaching for Mastery

Series Editor: Tony Staneff

Contents

Introduction

Foreword by the series editor and author, Tony Staneff

For far too long in the UK, maths has been feared by learners – and by many teachers, too. As a result, most learners consistently underachieve. More crucially, negative beliefs about ability, aptitude and the nature of maths are entrenched in children's thinking from an early age.

Yet, as someone who has loved maths all my life, I've always believed that every child has the capacity to succeed in maths. I've also had the great pleasure of leading teams and departments who share that belief and passion. Teaching for mastery, as practised in China and other South-East Asian jurisdictions since the 1980s, has confirmed my conviction that maths really is for everyone and not just those who have a special talent. In recent years, my team and I at Trinity Academy, Halifax, have had the privilege of researching with and working alongside some of the finest mastery practitioners from the UK and beyond, whose impact on learners' confidence, achievement and attitude is an inspiration.

The mastery approach recognises the value of developing the power to think rather than just do. It also recognises the value of making a coherent journey in which whole-class groups tackle concepts in very small steps, one by one. You cannot build securely on loose foundations – and it is just the same with maths: by creating a solid foundation of deep understanding, our children's skills and confidence will be strong and secure. What's more, the mindset of learner and teacher alike is fundamental: everyone can do maths … EVERYONE CAN!

I am proud to have been part of the extensive team responsible for turning the best of the world's practice, research, insights, and shared experiences into *Power Maths*, a unique teaching and learning resource developed especially for UK classrooms. *Power Maths* embodies our vision to help and support primary maths teachers to transform every child's mathematical and personal development. 'Everyone can!' has become our mantra and our passion, and we hope it will be yours, too.

Now, explore and enjoy all the resources you need to teach for mastery, and please get back to us with your *Power Maths* experiences and stories!

What is *Power Maths*?

Created especially for UK primary schools, and aligned with the new National Curriculum, *Power Maths* is a whole-class, textbook-based mastery resource that empowers every child to understand and succeed. *Power Maths* rejects the notion that some people simply 'can't' do maths. Instead, it develops growth mindsets and encourages hard work, practice and a willingness to see mistakes as learning tools.

Best practice consistently shows that mastery of small, cumulative steps builds a solid foundation of deep mathematical understanding. *Power Maths* combines interactive teaching tools, high-quality textbooks and continuing professional development (CPD) to help you equip children with a deep and long lasting understanding. Based on extensive evidence, and developed in partnership with practising teachers, *Power Maths* ensures that it meets the needs of children in the UK.

Power Maths and Mastery

Power Maths makes mastery practical and achievable by providing the structures, pathways, content, tools and support you need to make it happen in your classroom.

To develop mastery in maths children need to be enabled to acquire a deep understanding of maths concepts, structures and procedures, step by step. Complex mathematical concepts are built on simpler conceptual components and when children understand every step in the learning sequence, maths becomes transparent and makes logical sense. Interactive lessons establish deep understanding in small steps, as well as effortless fluency in key facts such as tables and number bonds. The whole class works on the same content and no child is left behind.

Power Maths

- Builds every concept in small, progressive steps.
- Is built with interactive, whole-class teaching in mind.
- Provides the tools you need to develop growth mindsets.
- Helps you check understanding and ensure that every child is keeping up.
- Establishes core elements such as intelligent practice and reflection.

The *Power Maths* approach

Everyone can!

Founded on the conviction that every child can achieve, *Power Maths* enables children to build number fluency, confidence and understanding, step by step.

Child-centred learning

Children master concepts one step at a time in lessons that embrace a Concrete-Pictorial-Abstract (C-P-A) approach, avoid overload, build on prior learning and help them see patterns and connections. Same-day intervention ensures sustained progress.

Continuing professional development

Embedded teacher support and development offer every teacher the opportunity to continually improve their subject knowledge and manage whole-class teaching for mastery.

Whole-class teaching

An interactive, whole-class teaching model encourages thinking and precise mathematical language and allows children to deepen their understanding as far as they can.

Introduction to the author team

Power Maths arises from the work of maths mastery experts who are committed to proving that, given the right mastery mindset and approach, **everyone can do maths**. Based on robust research and best practice from around the world, *Power Maths* was developed in partnership with a group of UK teachers to make sure that it not only meets our children's wide-ranging needs but also aligns with the National Curriculum in England.

Tony Staneff, Series Editor and author

Vice Principal at Trinity Academy, Halifax, Tony also leads a team of mastery experts who help schools across the UK to develop teaching for mastery via nationally recognised CPD courses, problem-solving and reasoning resources, schemes of work, assessment materials and other tools.

 A team of experienced authors, including:

- ⚡ **Josh Lury** – a specialist maths teacher, author and maths consultant with a passion for innovative and effective maths education

- ⚡ **Trinity Academy Halifax** (Michael Gosling CEO, Tony Staneff, Emily Fox, Kate Henshall, Rebecca Holland, Stephanie Kirk, Stephen Monaghan and Rachel Webster)

- ⚡ **David Board, Belle Cottingham, Jonathan East, Tim Handley, Derek Huby, Neil Jarrett, Stephen Monaghan, Beth Smith, Tim Weal, Paul Wrangles** – skilled maths teachers and mastery experts

- ⚡ **Cherri Moseley** – a maths author, former teacher and professional development provider

Professors Liu Jian and Zhang Dan, Series Consultants and authors, and their team of mastery expert authors:

- ⚡ **Wei Huinv, Huang Lihua, Zhu Dejiang, Zhu Yuhong, Hou Huiying, Yin Lili, Zhang Jing, Zhou Da and Liu Qimeng**

Used by over 20 million children, Professor Liu Jian's textbook programme is one of the most popular in China. He and his author team are highly experienced in intelligent practice and in embedding key maths concepts using a C-P-A approach.

 A group of 15 teachers and maths co-ordinators

We have consulted our teacher group throughout the development of *Power Maths* to ensure we are meeting their real needs in the classroom.

Your *Power Maths* resources

To help you teach for mastery, *Power Maths* comprises a variety of high-quality resources.

Pupil Textbooks

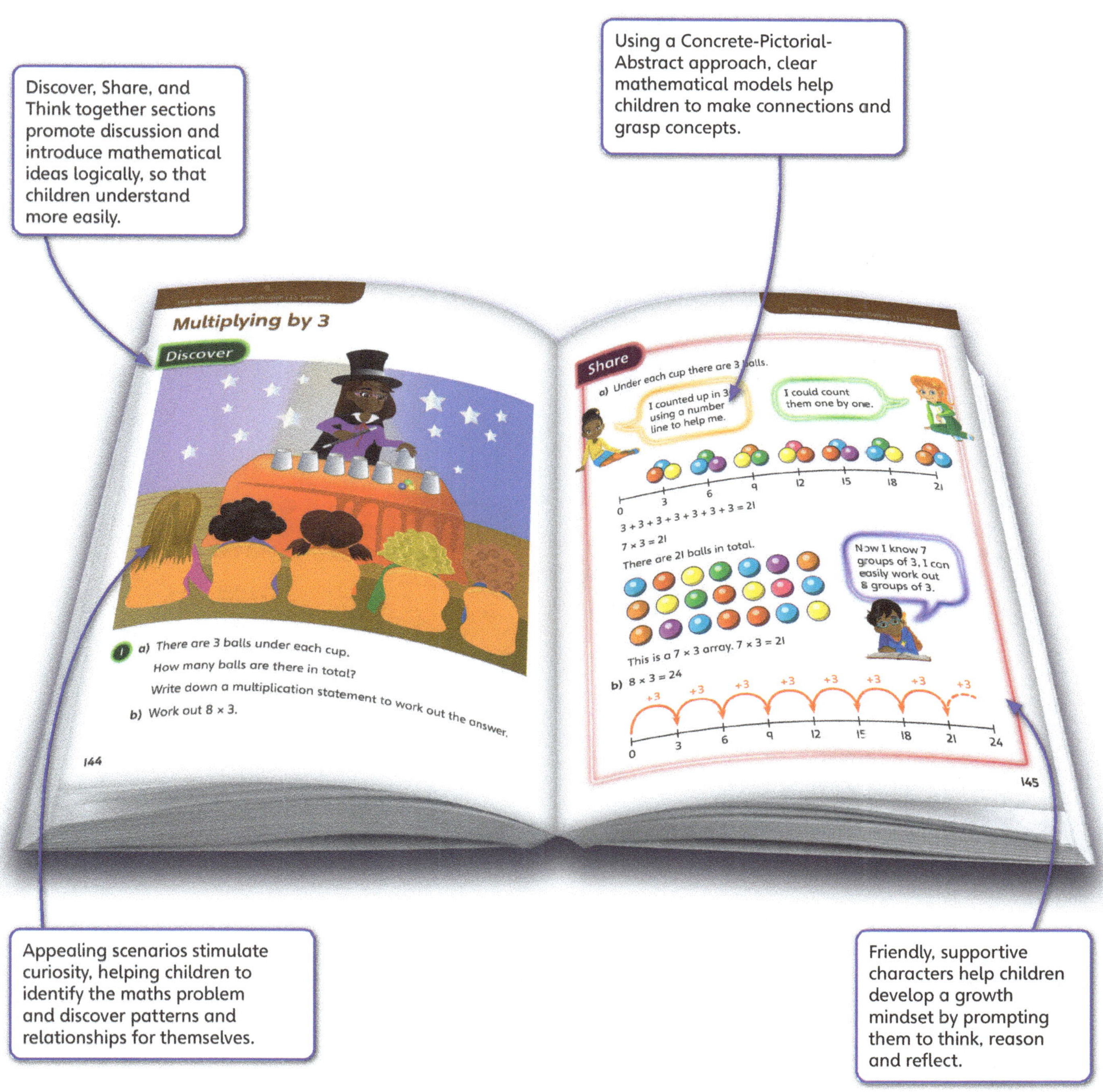

The coherent *Power Maths* lesson structure carries through into the vibrant, high-quality textbooks. Setting out the core learning objectives for each class, the lesson structure follows a carefully mapped journey through the curriculum and supports children on their journey to deeper understanding.

Pupil Practice Books

The Practice Books offer just the right amount of intelligent practice for children to complete independently in the final section of each lesson.

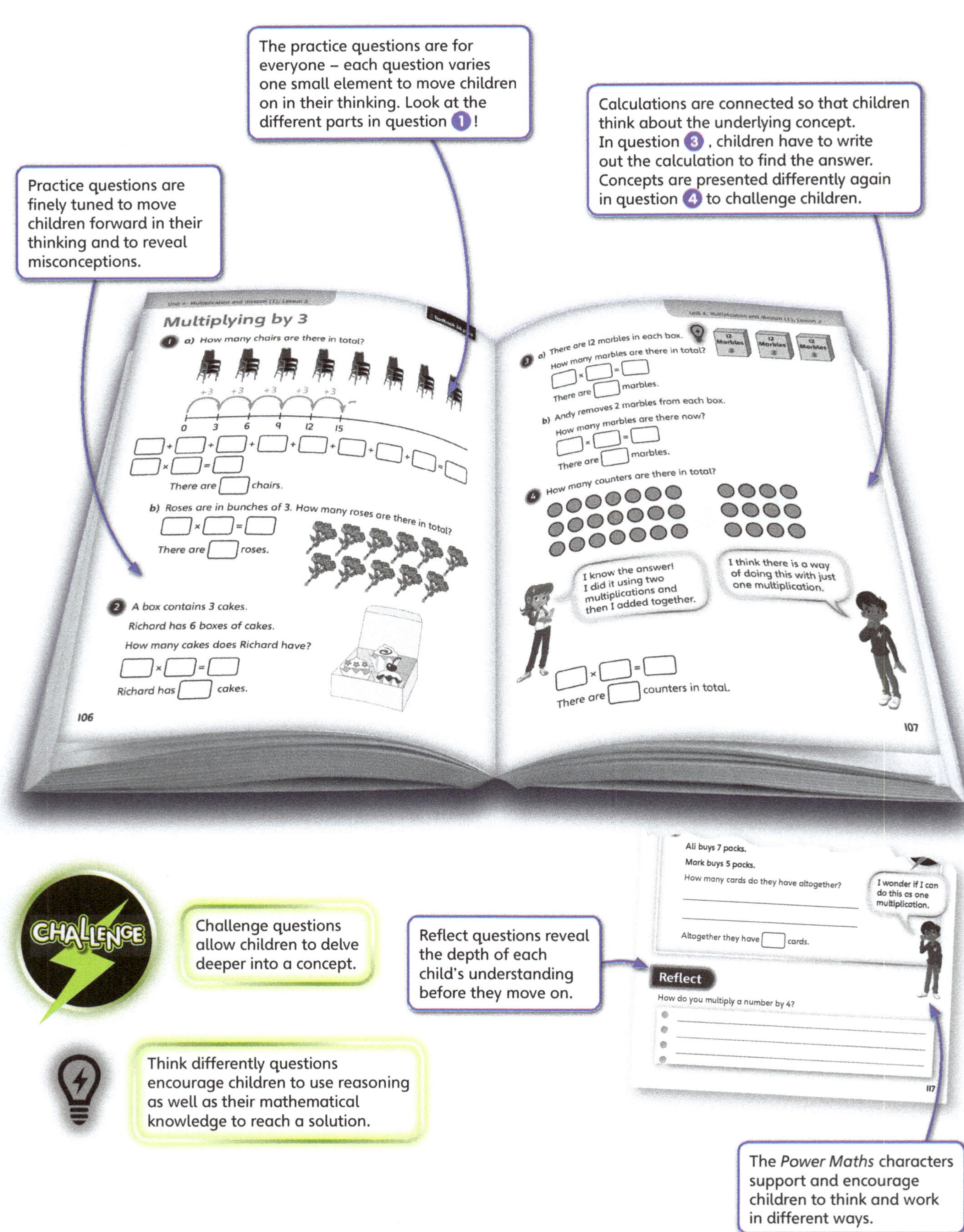

Challenge questions allow children to delve deeper into a concept.

Reflect questions reveal the depth of each child's understanding before they move on.

Think differently questions encourage children to use reasoning as well as their mathematical knowledge to reach a solution.

The *Power Maths* characters support and encourage children to think and work in different ways.

Online subscriptions

The online subscription will give you access to additional resources.

eTextbooks

Digital versions of *Power Maths* Textbooks allow class groups to share and discuss questions, solutions and strategies. They allow you to project key structures and representations at the front of the class, to ensure all children are focusing on the same concept.

Teaching tools

Here you will find interactive versions of key *Power Maths* structures and representations.

Power Ups

Use this series of daily activities to promote and check number fluency.

Online versions of Teacher Guide pages

PDF pages give support at both unit and lesson levels. You will also find help with key strategies and templates for tracking progress.

Unit videos

Watch the professional development videos at the start of each unit to help you teach with confidence. The videos explore common misconceptions in the unit, and include intervention suggestions as well as suggestions on what to look out for when assessing mastery in your children.

End of unit Strengthen and Deepen materials

Each Strengthen activity at the end of every unit addresses a key misconception and can be used to support children who need it. The Deepen activities are designed to be 'Low Threshold High Ceiling' and will challenge those children who can understand more deeply. These resources will help you ensure that every child understands and will help you keep the class moving forward together. These printable activities provide an optional resource bank for use after the assessment stage.

Underpinning all of these resources, *Power Maths* is infused throughout with continual professional development, supporting you at every step.

The *Power Maths* teaching model

At the heart of *Power Maths* is a clearly structured teaching and learning process that helps you make certain that every child masters each maths concept securely and deeply. For each year group, the curriculum is broken down into core concepts, taught in units. A unit divides into smaller learning steps – lessons. Step by step, strong foundations of cumulative knowledge and understanding are built.

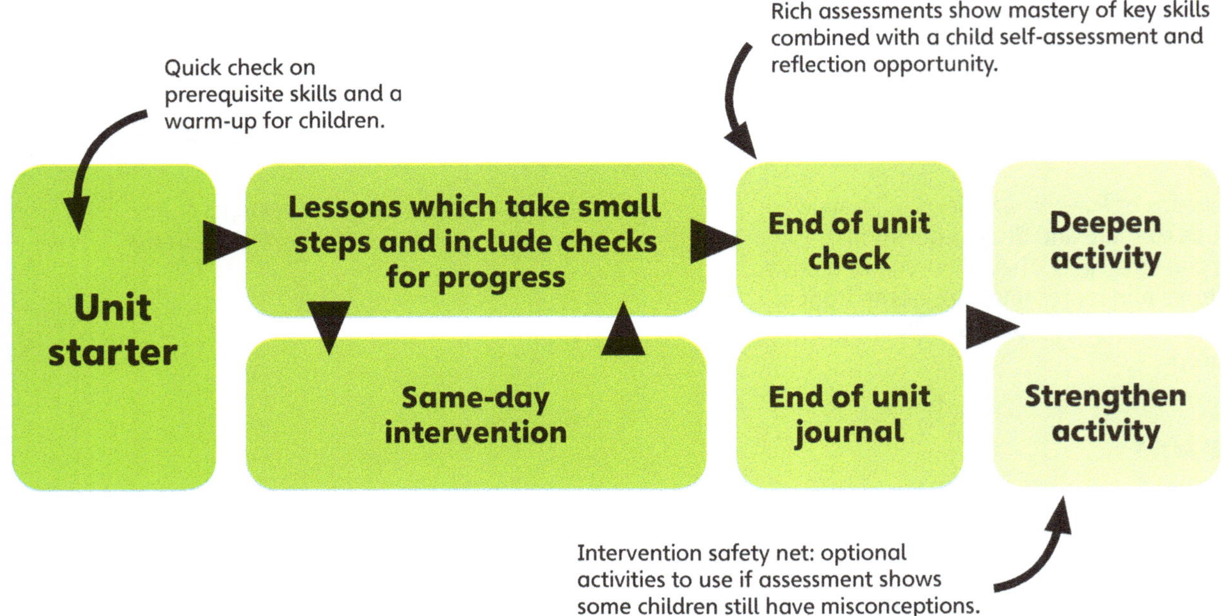

Unit starter

Each unit begins with a unit starter, which introduces the learning context along with key mathematical vocabulary, structures and representations.

- The Textbooks include a check on readiness and a warm-up task for children to complete.

- Your Teacher Guide gives support right from the start on important structures and representations, mathematical language, common misconceptions and intervention strategies.

- Unit-specific videos develop your subject knowledge and insights so you feel confident and fully equipped to teach each new unit. These are available via the online subscription.

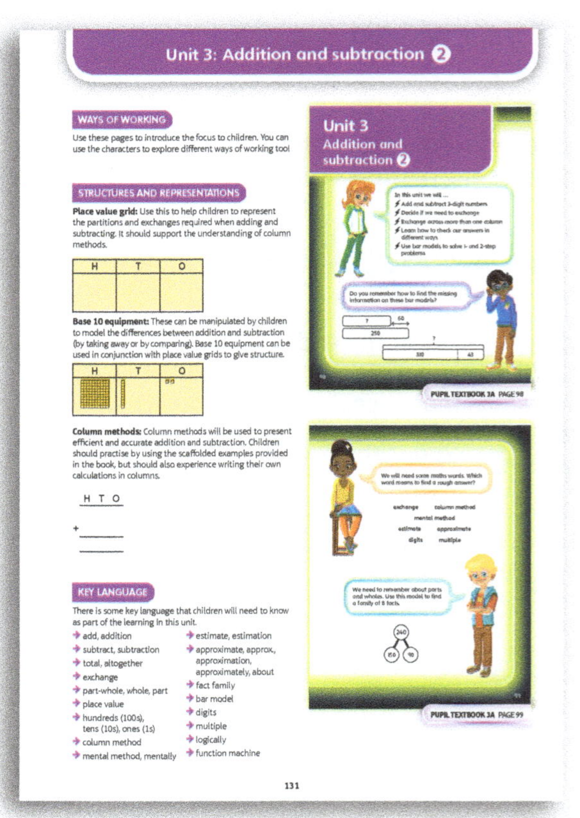

Lesson

Once a unit has been introduced, it is time to start teaching the series of lessons.

- Each lesson is scaffolded with Textbook and Practice Book activities and always begins with a Power Up activity (available via online subscription).

- *Power Maths* identifies lesson by lesson what concepts are to be taught.

- Your Teacher Guide offers lots of support for you to get the most from every child in every lesson. As well as highlighting key points, tricky areas and how to handle them, you will also find question prompts to check on understanding and clarification on why particular activities and questions are used.

Same-day intervention

Same-day interventions are vital in order to keep the class progressing together. Therefore, *Power Maths* provides plenty of support throughout the journey.

- Intervention is focused on keeping up now, not catching up later, so interventions should happen as soon as they are needed.

- Practice questions are designed to bring misconceptions to the surface, allowing you to identify these easily as you circulate during independent practice time.

- Child-friendly assessment questions in the Teacher Guide help you identify easily which children need to strengthen their understanding.

End of unit check and journal

At the end of a unit, summative assessment tasks reveal essential information on each child's understanding. An End of unit check in the Pupil Textbook lets you see which children have mastered the key concepts, which children have not and where their misconceptions lie. The Practice Book includes an End of unit journal in which children can reflect on what they have learned. Each unit also offers Strengthen and Deepen activities, available via the online subscription.

The Teacher Guide offers support with handling misconceptions.

The End of unit check presents six to nine multiple-choice questions. These questions are designed to reveal misconceptions and help you target areas that need strengthening.

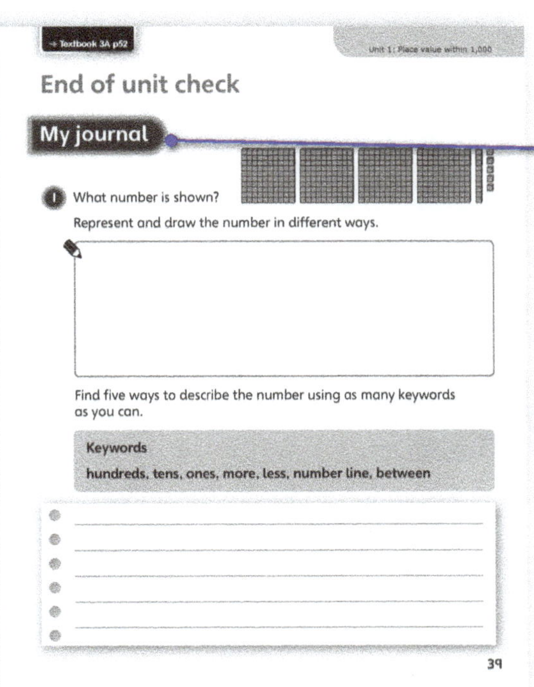

The End of unit journal is an opportunity for children to test out their learning and reflect on how they feel about it. Tackling the 'journal' problem reveals whether a child understands the concept deeply enough to move on to the next unit.

In KS2, the End of unit assessment will also include one SATs-style question.

The *Power Maths* lesson sequence

At the heart of *Power Maths* is a unique lesson sequence designed to empower children to understand core concepts and grow in confidence. Embracing the National Centre for Excellence in the Teaching of Mathematics' (NCETM's) definition of mastery, the sequence guides and shapes every *Power Maths* lesson you teach.

Flexibility is built into the *Power Maths* programme so there is no one-to-one mapping of lessons and concepts meaning you can pace your teaching according to your class. While some children will need to spend longer on a particular concept (through interventions or additional lessons), others will reach deeper levels of understanding. However, it is important that the class moves forward together through the termly schedules.

Power Up ⏱ 5 minutes

Each lesson begins with a Power Up activity (available via the online subscription) which supports fluency in key number facts.

The whole-class approach depends on fluency, so the Power Up is a powerful and essential activity.

TOP TIP

If the class is struggling with the task, revisit it later and check understanding.

Power Ups reinforce key skills such as times-tables, number bonds and working with place value.

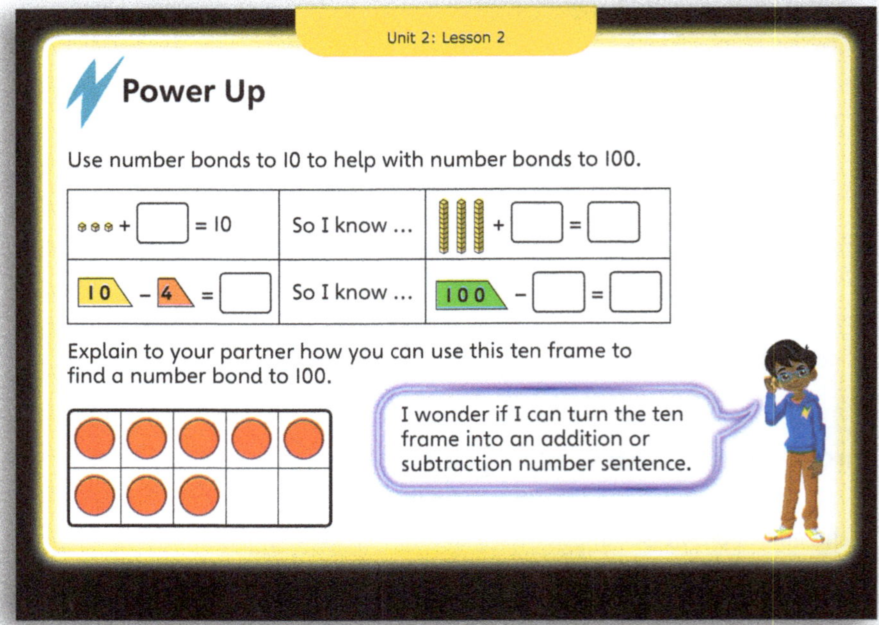

Discover ⏱ 10 minutes

A practical, real-life problem arouses curiosity. Children find the maths through story-telling.

A real-life scenario is provided for the Discover section but feel free to build upon these with your own examples that are more relevant to your class.

TOP TIP

Discover works best when run at tables, in pairs with concrete objects.

Question ❶ a) tackles the key concept and question ❶ b) digs a little deeper. Children have time to explore, play and discuss possible strategies.

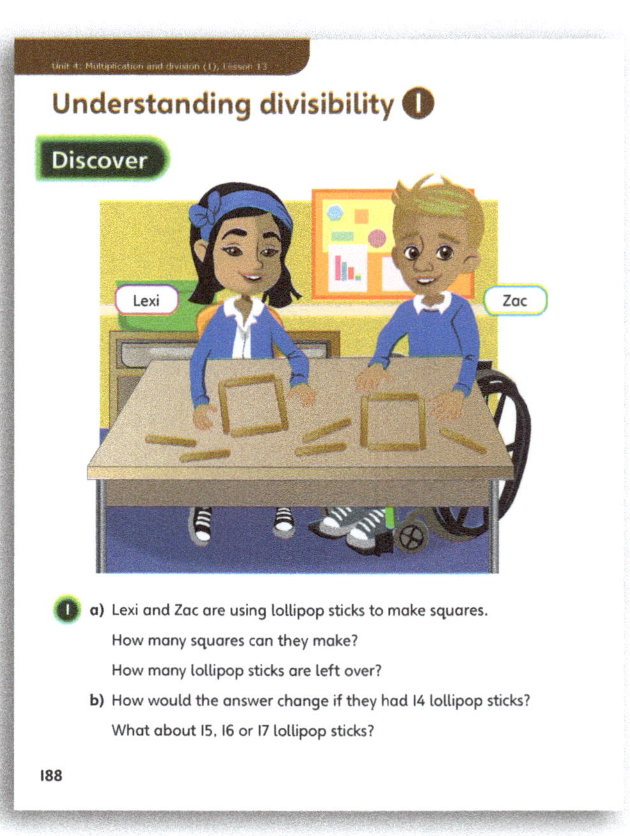

Share ⏱ 10 minutes

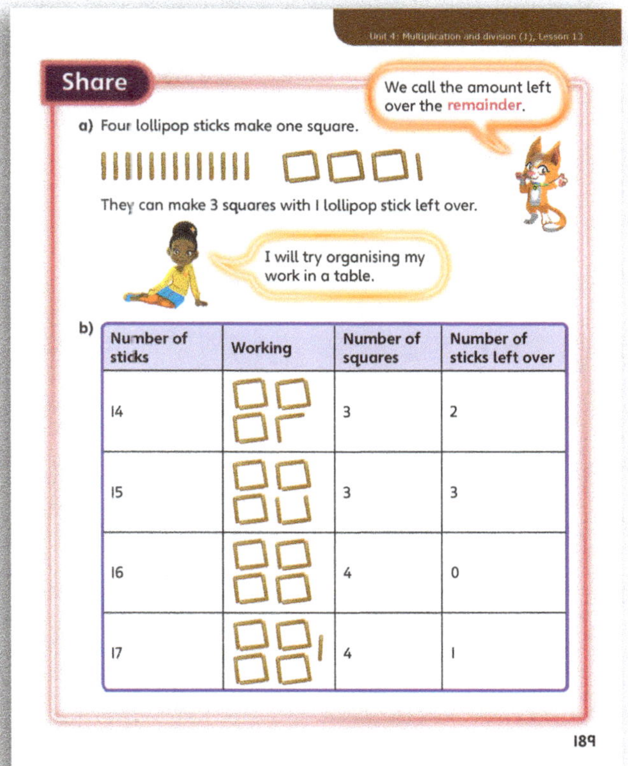

b)

Number of sticks	Working	Number of squares	Number of sticks left over
14		3	2
15		3	3
16		4	0
17		4	1

Think together

⏱ 10 minutes

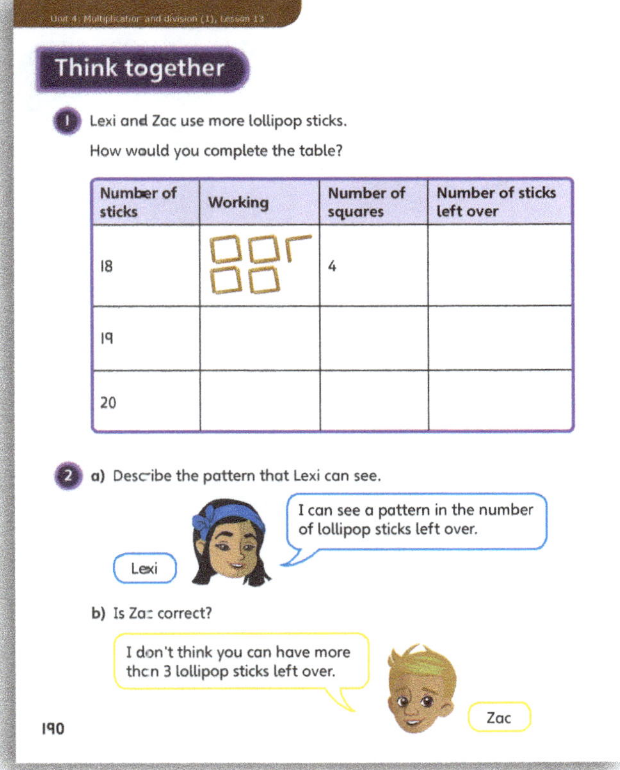

Number of sticks	Working	Number of squares	Number of sticks left over
18		4	
19			
20			

Practice ⏱ 15 minutes

Using their Practice Books, children work independently while you circulate and check on progress.

Questions follow small steps of progression to deepen learning.

TOP TIP
Some children could work separately with a teacher or assistant.

Are some children struggling? If so, work with them as a group, using mathematical structures and representations to support understanding as necessary.

There are no set routines: for real understanding, children need to think about the problem in different ways.

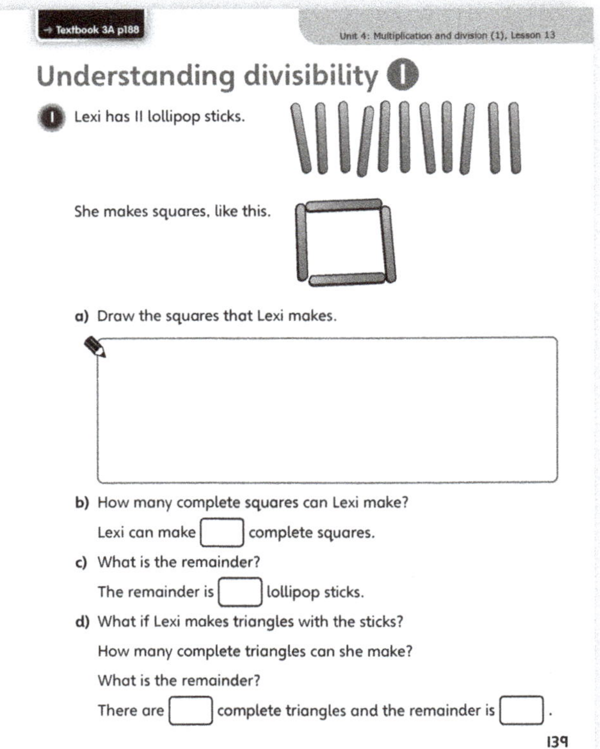

Reflect ⏱ 5 minutes

'Spot the mistake' questions are great for checking misconceptions.

The Reflect section is your opportunity to check how deeply children understand the target concept.

The Practice Books use various approaches to check that children have fully understood each concept.

Looking like they understand is not enough! It is essential that children can show they have grasped the concept.

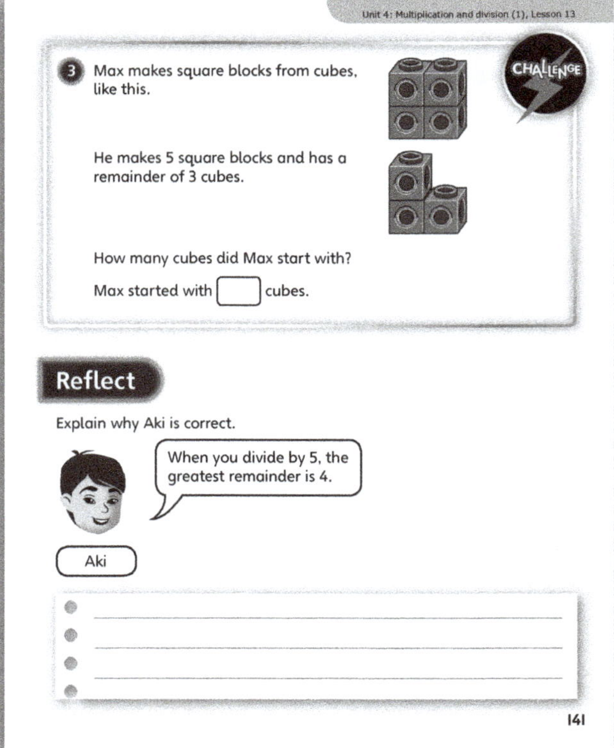

Using the *Power Maths* Teacher Guide

Think of your Teacher Guides as *Power Maths* handbooks that will guide, support and inspire your day-to-day teaching. Clear and concise, and illustrated with helpful examples, your Teacher Guides will help you make the best possible use of every individual lesson. They also provide wrap-around professional development, enhancing your own subject knowledge and helping you to grow in confidence about moving your children forward together.

There is a Teacher Guide per year group for every term with unit and lesson level guidance and support.

Tips and advice on key elements such as C-P-A approaches, misconceptions, language, modelling growth mindsets and same-day intervention.

Annotations for every Pupil Textbook and Practice Book page, providing prompts for key questions to ask to expose understanding and explanations as to why key questions have been chosen.

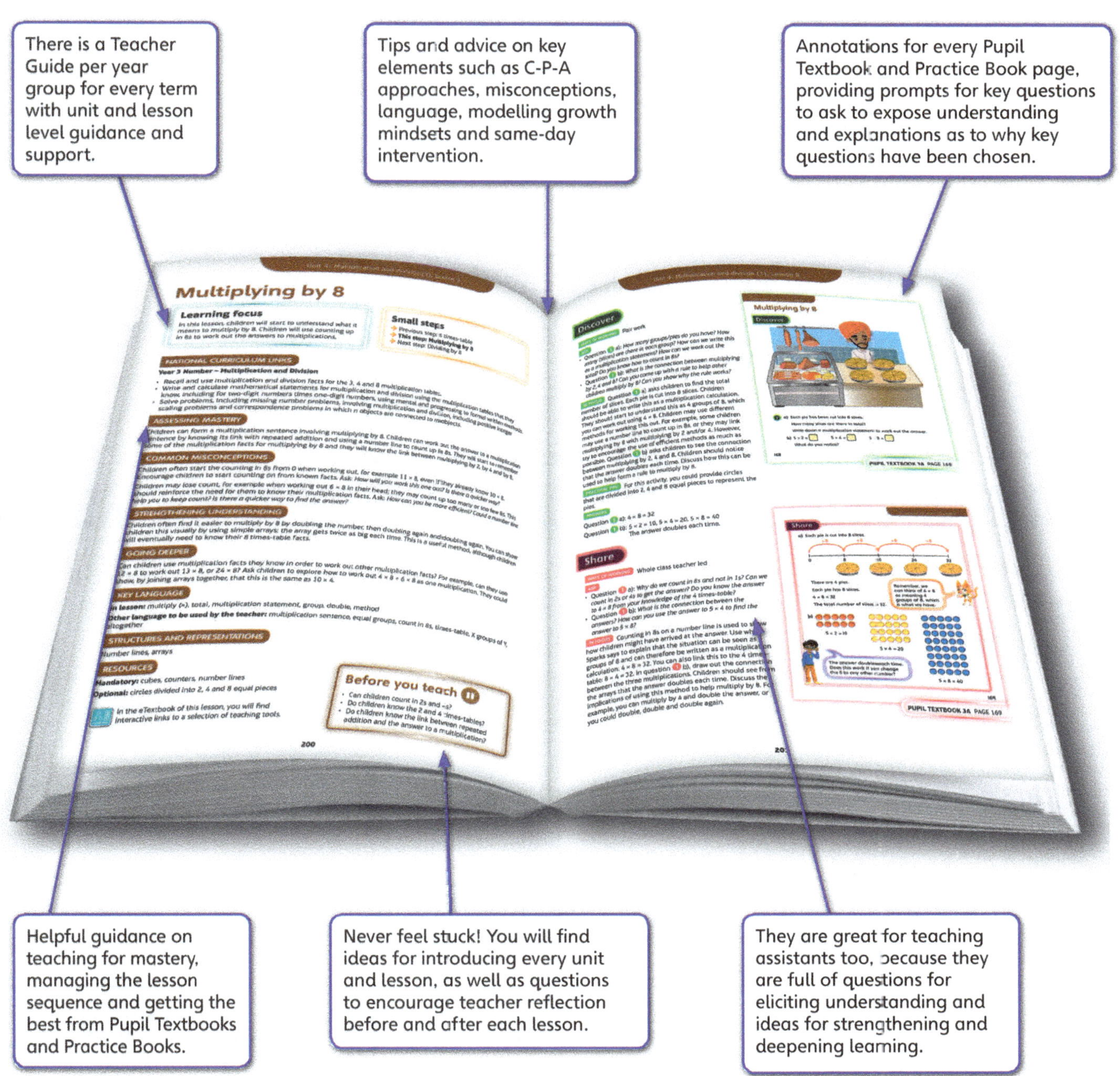

Helpful guidance on teaching for mastery, managing the lesson sequence and getting the best from Pupil Textbooks and Practice Books.

Never feel stuck! You will find ideas for introducing every unit and lesson, as well as questions to encourage teacher reflection before and after each lesson.

They are great for teaching assistants too, because they are full of questions for eliciting understanding and ideas for strengthening and deepening learning.

At the end of each unit, your Teacher Guide helps you identify who has fully grasped the concept, who has not and how to move every child forward. This is covered later in the Assessment strategies section.

Power Maths Year 3, yearly overview

Textbook	Strand	Unit		Number of Lessons
Textbook A / Practice Book A (Term 1)	Number – number and place value	1	Place value within 1,000	11
	Number – addition and subtraction	2	Addition and subtraction (1)	10
	Number – addition and subtraction	3	Addition and subtraction (2)	9
	Number – multiplication and division	4	Multiplication and division (1)	15
Textbook B / Practice Book B (Term 2)	Number – multiplication and division	5	Multiplication and division (2)	14
	Measurement	6	Money	5
	Statistics	7	Statistics	5
	Measurement	8	Length	11
	Number – fractions	9	Fractions (1)	11
Textbook C / Practice Book C (Term 3)	Number – fractions	10	Fractions (2)	9
	Measurement	11	Time	11
	Geometry – properties of shapes	12	Angles and properties of shapes	9
	Measurement	13	Mass	6

Power Maths Year 3, Textbook 3A (Term I) Overview

Strand 1	Strand 2	Unit		Lesson number	Lesson title	NC Objective 1	NC Objective 2	NC Objective 3
Number – number and place value		Unit 1	Place value within 1,000	1	Counting in 100s	Recognise the place value of each digit in a three-digit number (hundreds, tens, ones)	Read and write numbers up to 1,000 in numerals and in words	Identify, represent and estimate numbers using different representations
Number – number and place value		Unit 1	Place value within 1,000	2	Representing numbers to 1,000	Identify, represent and estimate numbers using different representations	Recognise the place value of each digit in a three-digit number (hundreds, tens, ones)	Read and write numbers up to 1,000 in numerals and in words
Number – number and place value		Unit 1	Place value within 1,000	3	100s, 10s and 1s (1)	Recognise the place value of each digit in a three-digit number (hundreds, tens, ones)	Identify, represent and estimate numbers using different representations	Read and write numbers up to 1,000 in numerals and in words
Number – number and place value		Unit 1	Place value within 1,000	4	100s, 10s and 1s (2)	Recognise the place value of each digit in a three-digit number (hundreds, tens, ones)	Identify, represent and estimate numbers using different representations	Read and write numbers up to 1,000 in numerals and in words
Number – number and place value		Unit 1	Place value within 1,000	5	The number line to 1,000 (1)	Recognise the place value of each digit in a three-digit number (hundreds, tens, ones)	Identify, represent and estimate numbers using different representations	Read and write numbers up to 1,000 in numerals and in words
Number – number and place value		Unit 1	Place value within 1,000	6	The number line to 1,000 (2)	Compare and order numbers up to 1,000	Read and write numbers up to 1,000 in numerals and in words	Recognise the place value of each digit in a three-digit number (hundreds, tens, ones)
Number – number and place value		Unit 1	Place value within 1,000	7	Finding 1, 10 and 100 more or less	Recognise the place value of each digit in a three-digit number (hundreds, tens, ones)	Count from 0 in multiples of 4, 8, 50 and 100; find 10 or 100 more or less than a given number	Identify, represent and estimate numbers using different representations
Number – number and place value		Unit 1	Place value within 1,000	8	Comparing numbers to 1,000 (1)	Compare and order numbers up to 1,000	Identify, represent and estimate numbers using different representations	Read and write numbers up to 1,000 in numerals and in words
Number – number and place value		Unit 1	Place value within 1,000	9	Comparing numbers to 1,000 (2)	Compare and order numbers up to 1,000	Solve number problems and practical problems involving these ideas	Recognise the place value of each digit in a three-digit number (hundreds, tens, ones)
Number – number and place value		Unit 1	Place value within 1,000	10	Ordering numbers to 1,000	Compare and order numbers up to 1,000	Recognise the place value of each digit in a three-digit number (100s, 10s, 1s)	Read and write numbers up to 1000 in numerals and in words
Number – number and place value		Unit 1	Place value within 1,000	11	Counting in 50s	Count from 0 in multiples of 4, 8, 50 and 100; find 10 or 100 more or less than a given number	Solve number problems and practical problems involving these ideas	

Strand 1	Strand 2	Unit		Lesson number	Lesson title	NC Objective 1	NC Objective 2	NC Objective 3
Number – addition and subtraction		Unit 2	Addition and subtraction (1)	1	Adding and subtracting 100s	Add and subtract numbers mentally, including: a three-digit number and ones, a three-digit number and tens, a three-digit number and hundreds		
Number – addition and subtraction		Unit 2	Addition and subtraction (1)	2	Adding and subtracting a 3-digit number and 1s	Add and subtract numbers mentally, including: a three-digit number and ones, a three-digit number and tens, a three-digit number and hundreds	Solve problems, including missing number problems, using number facts, place value, and more complex addition and subtraction	
Number – addition and subtraction		Unit 2	Addition and subtraction (1)	3	Adding a 3-digit number and 1s	Add and subtract numbers mentally, including: a three-digit number and ones, a three-digit number and tens, a three-digit number and hundreds	Solve problems, including missing number problems, using number facts, place value, and more complex addition and subtraction	
Number – addition and subtraction		Unit 2	Addition and subtraction (1)	4	Subtracting 1s from a 3-digit number	Add and subtract numbers mentally, including: a three-digit number and ones, a three-digit number and tens, a three-digit number and hundreds		
Number – addition and subtraction		Unit 2	Addition and subtraction (1)	5	Adding and subtracting a 3-digit number and 10s	Add and subtract numbers mentally, including: a three-digit number and ones, a three-digit number and tens, a three-digit number and hundreds	Solve problems, including missing number problems, using number facts, place value, and more complex addition and subtraction	
Number – addition and subtraction		Unit 2	Addition and subtraction (1)	6	Adding a 3-digit number and 10s	Add and subtract numbers mentally, including: a three-digit number and ones, a three-digit number and tens, a three-digit number and hundreds	Solve problems, including missing number problems, using number facts, place value, and more complex addition and subtraction	
Number – addition and subtraction		Unit 2	Addition and subtraction (1)	7	Subtracting 10s from a 3-digit number	Add and subtract numbers mentally, including: a three-digit number and ones, a three-digit number and tens, a three-digit number and hundreds	Solve problems, including missing number problems, using number facts, place value, and more complex addition and subtraction	
Number – addition and subtraction		Unit 2	Addition and subtraction (1)	8	Adding and subtracting a 3-digit and 2-digit number	Add and subtract numbers with up to three digits, using formal written methods of columnar addition and subtraction	Add and subtract numbers mentally, including: a three-digit number and ones, a three-digit number and tens, a three-digit number and hundreds	
Number – addition and subtraction		Unit 2	Addition and subtraction (1)	9	Adding a 3-digit and 2-digit number	Add and subtract numbers with up to three digits, using formal written methods of columnar addition and subtraction	Add and subtract numbers mentally, including: a three-digit number and ones, a three-digit number and tens, a three-digit number and hundreds	Solve problems, including missing number problems, using number facts, place value, and more complex addition and subtraction
Number – addition and subtraction		Unit 2	Addition and subtraction (1)	10	Subtracting a 2-digit number from a 3-digit number	Add and subtract numbers with up to three digits, using formal written methods of columnar addition and subtraction	Add and subtract numbers mentally, including: a three-digit number and ones, a three-digit number and tens, a three-digit number and hundreds	Solve problems, including missing number problems, using number facts, place value, and more complex addition and subtraction
Number – addition and subtraction		Unit 3	Addition and subtraction (2)	1	Addition and subtraction patterns	Add and subtract numbers with up to three digits, using formal written methods of columnar addition and subtraction	Add and subtract numbers mentally, including: a three-digit number and ones, a three-digit number and tens, a three-digit number and hundreds	Solve problems, including missing number problems, using number facts, place value, and more complex addition and subtraction
Number – addition and subtraction		Unit 3	Addition and subtraction (2)	2	Adding two 3-digit numbers (1)	Add and subtract numbers with up to three digits, using formal written methods of columnar addition and subtraction	Add and subtract numbers mentally, including: a three-digit number and ones, a three-digit number and tens, a three-digit number and hundreds	
Number – addition and subtraction		Unit 3	Addition and subtraction (2)	3	Adding two 3-digit numbers (2)	Add and subtract numbers with up to three digits, using formal written methods of columnar addition and subtraction	Add and subtract numbers mentally, including: a three-digit number and ones, a three-digit number and tens, a three-digit number and hundreds	Solve problems, including missing number problems, using number facts, place value, and more complex addition and subtraction
Number – addition and subtraction		Unit 3	Addition and subtraction (2)	4	Subtracting a 3-digit number from a 3-digit number (1)	Add and subtract numbers with up to three digits, using formal written methods of columnar addition and subtraction	Add and subtract numbers mentally, including: a three-digit number and ones, a three-digit number and tens, a three-digit number and hundreds	

Strand 1	Strand 2	Unit		Lesson number	Lesson title	NC Objective 1	NC Objective 2	NC Objective 3
Number – addition and subtraction		Unit 3	Addition and subtraction (2)	5	Subtracting a 3-digit number from a 3-digit number (2)	Add and subtract numbers with up to three digits, using formal written methods of columnar addition and subtraction	Add and subtract numbers mentally, including: a three-digit number and ones, a three-digit number and tens, a three-digit number and hundreds	Solve problems, including missing number problems, using number facts, place value, and more complex addition and subtraction
Number – addition and subtraction		Unit 3	Addition and subtraction (2)	6	Estimating answers to additions and subtractions	Estimate the answer to a calculation and use inverse operations to check answers		
Number – addition and subtraction		Unit 3	Addition and subtraction (2)	7	Checking strategies	Estimate the answer to a calculation and use inverse operations to check answers		
Number – addition and subtraction		Unit 3	Addition and subtraction (2)	8	Problem solving – addition and subtraction (1)	Solve problems, including missing number problems, using number facts, place value, and more complex addition and subtraction		
Number – addition and subtraction		Unit 3	Addition and subtraction (2)	9	Problem solving – addition and subtraction (2)	Solve problems, including missing number problems, using number facts, place value, and more complex addition and subtraction		
Number – multiplication and division		Unit 4	Multiplication and division (1)	1	Multiplication – equal grouping	Write and calculate mathematical statements for multiplication and division using the multiplication tables that they know, including for two-digit numbers times one-digit numbers, using mental and progressing to formal written methods	Recall and use multiplication and division facts for the 3, 4 and 8 multiplication tables	Solve problems, including missing number problems, involving multiplication and division, including positive integer scaling problems and correspondence problems in which n objects are connected to m objects
Number – multiplication and division		Unit 4	Multiplication and division (1)	2	Multiplying by 3	Recall and use multiplication and division facts for the 3, 4 and 8 multiplication tables	Solve problems, including missing number problems, involving multiplication and division, including positive integer scaling problems and correspondence problems in which n objects are connected to m objects	Write and calculate mathematical statements for multiplication and division using the multiplication tables that they know, including for two-digit numbers times one-digit numbers, using mental and progressing to formal written methods
Number – multiplication and division		Unit 4	Multiplication and division (1)	3	Dividing by 3	Recall and use multiplication and division facts for the 3, 4 and 8 multiplication tables	Write and calculate mathematical statements for multiplication and division using the multiplication tables that they know, including for two-digit numbers times one-digit numbers, using mental and progressing to formal written methods	Solve problems, including missing number problems, involving multiplication and division, including positive integer scaling problems and correspondence problems in which n objects are connected to m objects
Number – multiplication and division		Unit 4	Multiplication and division (1)	4	3 times-table	Recall and use multiplication and division facts for the 3, 4 and 8 multiplication tables	Write and calculate mathematical statements for multiplication and division using the multiplication tables that they know, including for two-digit numbers times one-digit numbers, using mental and progressing to formal written methods	Solve problems, including missing number problems, involving multiplication and division, including positive integer scaling problems and correspondence problems in which n objects are connected to m objects
Number – multiplication and division		Unit 4	Multiplication and division (1)	5	Multiplying by 4	Recall and use multiplication and division facts for the 3, 4 and 8 multiplication tables	Write and calculate mathematical statements for multiplication and division using the multiplication tables that they know, including for two-digit numbers times one-digit numbers, using mental and progressing to formal written methods	Solve problems, including missing number problems, involving multiplication and division, including positive integer scaling problems and correspondence problems in which n objects are connected to m objects
Number – multiplication and division		Unit 4	Multiplication and division (1)	6	Dividing by 4	Write and calculate mathematical statements for multiplication and division using the multiplication tables that they know, including for two-digit numbers times one-digit numbers, using mental and progressing to formal written methods	Recall and use multiplication and division facts for the 3, 4 and 8 multiplication tables	Solve problems, including missing number problems, involving multiplication and division, including positive integer scaling problems and correspondence problems in which n objects are connected to m objects

Strand 1	Strand 2	Unit		Lesson number	Lesson title	NC Objective 1	NC Objective 2	NC Objective 3
Number – multiplication and division		Unit 4	Multiplication and division (1)	7	4 times-table	Recall and use multiplication and division facts for the 3, 4 and 8 multiplication tables	Write and calculate mathematical statements for multiplication and division using the multiplication tables that they know, including for two-digit numbers times one-digit numbers, using mental and progressing to formal written methods	Solve problems, including missing number problems, involving multiplication and division, including positive integer scaling problems and correspondence problems in which n objects are connected to m objects
Number – multiplication and division		Unit 4	Multiplication and division (1)	8	Multiplying by 8	Recall and use multiplication and division facts for the 3, 4 and 8 multiplication tables	Write and calculate mathematical statements for multiplication and division using the multiplication tables that they know, including for two-digit numbers times one-digit numbers, using mental and progressing to formal written methods	Solve problems, including missing number problems, involving multiplication and division, including positive integer scaling problems and correspondence problems in which n objects are connected to m objects
Number – multiplication and division		Unit 4	Multiplication and division (1)	9	Dividing by 8	Recall and use multiplication and division facts for the 3, 4 and 8 multiplication tables	Write and calculate mathematical statements for multiplication and division using the multiplication tables that they know, including for two-digit numbers times one-digit numbers, using mental and progressing to formal written methods	Solve problems, including missing number problems, involving multiplication and division, including positive integer scaling problems and correspondence problems in which n objects are connected to m objects
Number – multiplication and division		Unit 4	Multiplication and division (1)	10	8 times-table	Recall and use multiplication and division facts for the 3, 4 and 8 multiplication tables	Write and calculate mathematical statements for multiplication and division using the multiplication tables that they know, including for two-digit numbers times one-digit numbers, using mental and progressing to formal written methods	Solve problems, including missing number problems, involving multiplication and division, including positive integer scaling problems and correspondence problems in which n objects are connected to m objects
Number – multiplication and division		Unit 4	Multiplication and division (1)	11	Problem solving – multiplication and division (1)	Solve problems, including missing number problems, involving multiplication and division, including positive integer scaling problems and correspondence problems in which n objects are connected to m objects	Write and calculate mathematical statements for multiplication and division using the multiplication tables that they know, including for two-digit numbers times one-digit numbers, using mental and progressing to formal written methods	Recall and use multiplication and division facts for the 3, 4 and 8 multiplication tables
Number – multiplication and division		Unit 4	Multiplication and division (1)	12	Problem solving – multiplication and division (2)	Solve problems, including missing number problems, involving multiplication and division, including positive integer scaling problems and correspondence problems in which n objects are connected to m objects	Write and calculate mathematical statements for multiplication and division using the multiplication tables that they know, including for two-digit numbers times one-digit numbers, using mental and progressing to formal written methods	Recall and use multiplication and division facts for the 3, 4 and 8 multiplication tables
Number – multiplication and division		Unit 4	Multiplication and division (1)	13	Understanding divisibility (1)	Solve problems, including missing number problems, involving multiplication and division, including positive integer scaling problems and correspondence problems in which n objects are connected to m objects		
Number – multiplication and division		Unit 4	Multiplication and division (1)	14	Understanding divisibility (2)	Solve problems, including missing number problems, involving multiplication and division, including positive integer scaling problems and correspondence problems in which n objects are connected to m objects	Write and calculate mathematical statements for multiplication and division using the multiplication tables that they know, including for two-digit numbers times one-digit numbers, using mental and progressing to formal written methods	Recall and use multiplication and division facts for the 3, 4 and 8 multiplication tables
Number – multiplication and division		Unit 4	Multiplication and division (1)	15	Related facts – multiplication and division	Write and calculate mathematical statements for multiplication and division using the multiplication tables that they know, including for two-digit numbers times one-digit numbers, using mental and progressing to formal written methods	Recall and use multiplication and division facts for the 3, 4 and 8 multiplication tables	Solve problems, including missing number problems, involving multiplication and division, including positive integer scaling problems and correspondence problems in which n objects are connected to m objects

Mindset: an introduction

Global research and best practice deliver the same message: learning is greatly affected by what learners perceive they can or cannot do. What is more, it is also shaped by what their parents, carers and teachers perceive they can do. Mindset – the thinking that determines our beliefs and behaviours – therefore has a fundamental impact on teaching and learning.

Everyone can!

Power Maths and mastery methods focus on the distinction between 'fixed' and 'growth' mindsets (Dweck, 2007).[1] Those with a fixed mindset believe that their basic qualities (for example, intelligence, talent and ability to learn) are pre-wired or fixed: 'If you have a talent for maths, you will succeed at it. If not, too bad!' By contrast, those with a growth mindset believe that hard work, effort and commitment drive success and that 'smart' is not something you are or are not, but something you become. In short, everyone can do maths!

Key mindset strategies

A growth mindset needs to be actively nurtured and developed. *Power Maths* offers some key strategies for fostering healthy growth mindsets in your classroom.

It is okay to get it wrong

Mistakes are valuable opportunities to re-think and understand more deeply. Learning is richer when children and teachers alike focus on spotting and sharing mistakes as well as solutions.

Praise hard work

Praise is a great motivator, and by focusing on praising effort and learning rather than success, children will be more willing to try harder, take risks and persist for longer.

Mind your language!

The language we use around learners has a profound effect on their mindsets. Make a habit of using growth phrases, such as, 'Everyone can!', 'Mistakes can help you learn' and 'Just try for a little longer'. The king of them all is one little word, 'yet … I cannot solve this … yet!' Encourage parents and carers to use the right language too.

Build in opportunities for success

The step-by-small-step approach enables children to enjoy the experience of success. In addition, avoid ability grouping and encourage every child to answer questions and explain or demonstrate their methods to others.

[1] Dweck, C (2007) *The New Psychology of Success*, Ballantine Books: New York

The *Power Maths* characters

The *Power Maths* characters model the traits of growth mindset learners and encourage resilience by prompting and questioning children as they work. Appearing frequently in the Textbooks and Practice Books, they are your allies in teaching and discussion, helping to model methods, alternatives and misconceptions, and to pose questions. They encourage and support your children, too: they are all hardworking, enthusiastic and unafraid of making and talking about mistakes.

Meet the team!

Creative Flo is open-minded and sometimes indecisive. She likes to think differently and come up with a variety of methods or ideas.

Determined Dexter is resolute, resilient and systematic. He concentrates hard, always tries his best and he'll never give up – even though he doesn't always choose the most efficient methods!

Curious Ash is eager, interested and inquisitive, and he loves solving puzzles and problems. Ash asks lots of questions but sometimes gets distracted.

Sparks the Cat

Brave Astrid is confident, willing to take risks and unafraid of failure. She is never scared to jump straight into a problem or question, and although she often makes simple mistakes she is happy to talk them through with others.

Mathematical language

Traditionally, we in the UK have tended to try simplifying mathematical language to make it easier for young children to understand. By contrast, evidence and experience show that by diluting the correct language, we actually mask concepts and meanings for children. We then wonder why they are confused by new and different terminology later down the line! *Power Maths* is not afraid of 'hard' words and avoids placing any barriers between children and their understanding of mathematical concepts. As a result, we need to be planned, precise and thorough in building every child's understanding of the language of maths. Throughout the Teacher Guides you will find support and guidance on how to deliver this, as well as individual explanations throughout the Pupil Textbooks.

Use the following key strategies to build children's mathematical vocabulary, understanding and confidence.

Precise and consistent

Everyone in the classroom should use the correct mathematical terms in full, every time. For example, refer to 'equal parts', not 'parts'. Used consistently, precise maths language will be a familiar and non-threatening part of children's everyday experience.

Full sentences

Teachers and children alike need to use full sentences to explain or respond. When children use complete sentences, it both reveals their understanding and embeds their knowledge.

Stem sentences

These important sentences help children express mathematical concepts accurately, and are used throughout the *Power Maths* books. Encourage children to repeat them frequently, whether working independently or with others. Examples of stem sentences are:

'4 is a part, 5 is a part, 9 is the whole.'

'There are … groups. There are … in each group.'

Key vocabulary

The unit starters highlight essential vocabulary for every lesson. In the Pupil Textbooks, characters flag new terminology and the Teacher Guide lists important mathematical language for every unit and lesson. New terms are never introduced without a clear explanation.

Mathematical signs

Mathematical signs are used early on so that children quickly become familiar with them and their meaning. Often, the *Power Maths* characters will highlight the connection between language and particular signs.

The role of talk and discussion

When children learn to talk purposefully together about maths, barriers of fear and anxiety are broken down and they grow in confidence, skills and understanding. Building a healthy culture of 'maths talk' empowers their learning from day one.

Explanation and discussion are integral to the *Power Maths* structure, so by simply following the books your lessons will stimulate structured talk. The following key 'maths talk' strategies will help you strengthen that culture and ensure that every child is included.

Sentences, not words

Encourage children to use full sentences when reasoning, explaining or discussing maths. This helps both speaker and listeners to clarify their own understanding. It also reveals whether or not the speaker truly understands, enabling you to address misconceptions as they arise.

Working together

Working with others in pairs, groups or as a whole class is a great way to support maths talk and discussion. Use different group structures to add variety and challenge. For example, children could take timed turns for talking, work independently alongside a 'discussion buddy', or perhaps play different *Power Maths* character roles within their group.

Think first – then talk

Provide clear opportunities within each lesson for children to think and reflect, so that their talk is purposeful, relevant and focused.

Give every child a voice

Where the 'hands up' model allows only the more confident child to shine, *Power Maths* involves everyone. Make sure that no child dominates and that even the shyest child is encouraged to contribute – and are praised when they do.

Assessment strategies

Teaching for mastery demands that you are confident about what each child knows and where their misconceptions lie: therefore, practical and effective assessment is vitally important.

Formative assessment within lessons

The Think together section will often reveal any confusions or insecurities: try ironing these out by doing the first Think together question as a class. For children who continue to struggle, you or your teaching assistant should provide support and enable them to move on.

Performance in Practice can be very revealing: check Practice Books and listen out both during and after practice to identify misconceptions.

The Reflect section is designed to check on the all-important depth of understanding. Be sure to review how children performed in this final stage before you teach the next lesson.

End of unit check – Textbook

Each unit concludes with a summative check to help you assess quickly and clearly each child's understanding, fluency, reasoning and problem-solving skills. In KS2 this check also contains a SATs-style question to help children become familiar with answering this type of question.

In KS2 we would suggest the End of unit check is completed independently in children's exercise books, but you can adapt this to suit the needs of your class.

End of unit check – Practice Book

The Practice Book contains further opportunities for assessment, and can be completed by children independently whilst you are carrying out diagnostic assessment with small groups. Your Teacher Guide will advise you on what to do if children struggle to articulate an explanation – or perhaps encourage you to write down something they have explained well. It will also offer insights into children's answers and their implications for the next learning steps. It is split into three main sections, outlined below.

My journal

My journal is designed to allow children to show their depth of understanding of the unit. It can also serve as a way of checking that children have grasped key mathematical vocabulary. Children should have some time to think about how they want to answer the question, and you could ask them to talk to a partner about their ideas. Then children should write their answer in their Practice Book.

Power check

The Power check allows children to self-assess their level of confidence on the topic by colouring in different smiley faces. You may want to introduce the faces as follows:

Power play or Power puzzle

Each unit ends with either a Power play or a Power puzzle. This is an activity, puzzle or game that allows children to use their new knowledge in a fun, informal way. In Key Stage 2 we have also included a deeper level to each game to help challenge those children who have grasped a concept quickly.

How to use diagnostic questions

The diagnostic questions provided in *Power Maths* Textbooks are carefully structured to identify both understanding and misconceptions (if children answer in a particular way, you will know why). The simple procedure below may be helpful:

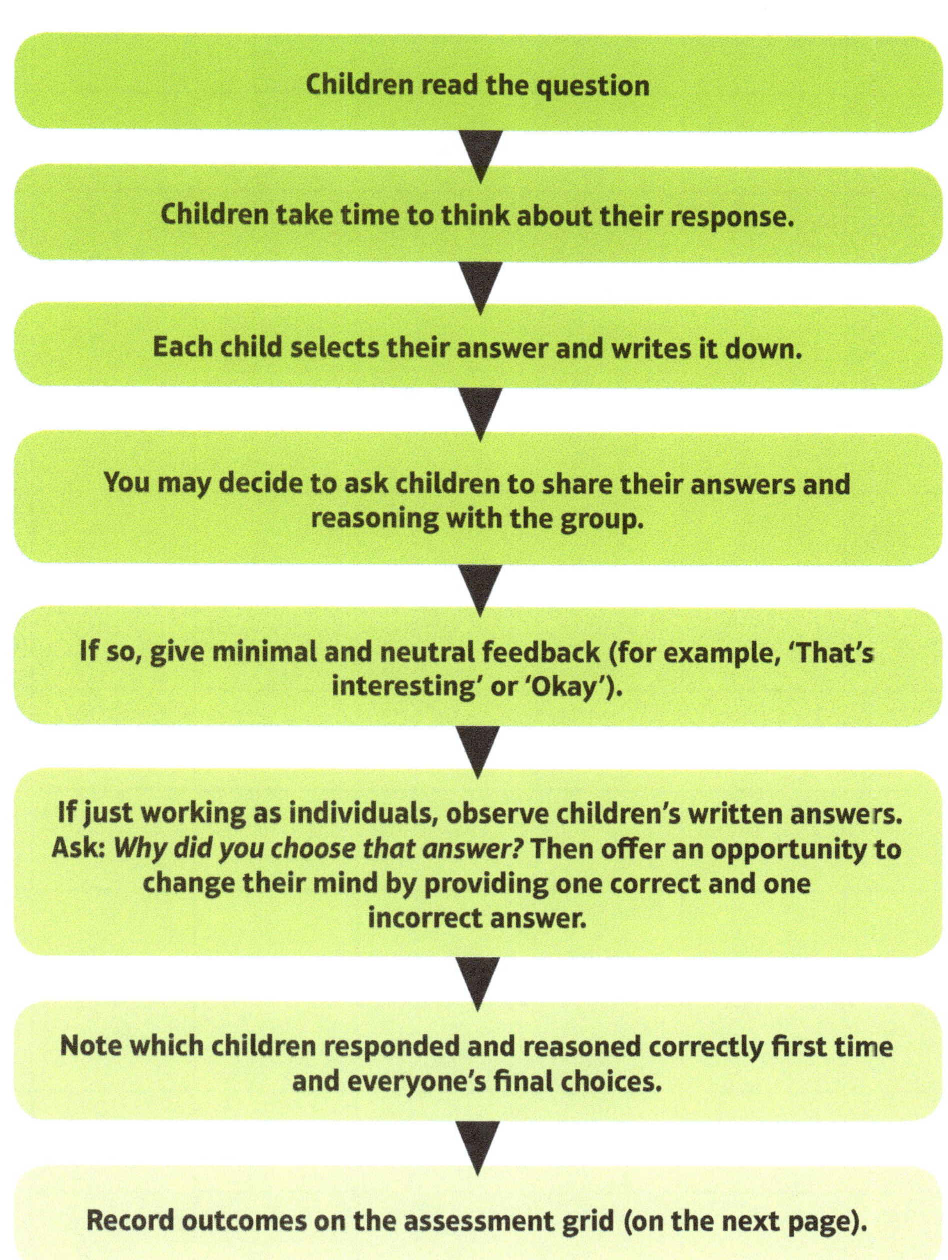

Power Maths unit assessment grid

Year ____   Unit ____ _______________________________________

Record only as much information as you judge appropriate for your assessment of each child's mastery of the unit and any steps needed for intervention.

Name	Diagnostic questions	SATs - style question	My journal	Power check	Power play/puzzle	Mastery	Intervention/ Strengthen

Keeping the class together

Traditionally, children who learn quickly have been accelerated through the curriculum. As a consequence, their learning may be superficial and will lack the many benefits of enabling children to learn with and from each other.

By contrast, *Power Maths'* mastery approach values real understanding and richer, deeper learning above speed. It sees all children learning the same concept in small, cumulative steps, each finding and mastering challenge at their own level. Remember that when you teach for mastery, EVERYONE can do maths! Those who grasp a concept easily have time to explore and understand that concept at a deeper level. The whole class therefore moves through the curriculum at broadly the same pace via individual learning journeys.

For some teachers, the idea that a whole class can move forward together is revolutionary and challenging. However, the evidence of global good practice clearly shows that this approach drives engagement, confidence, motivation and success for all learners, and not just the high flyers. The strategies below will help you keep your class together on their maths journey.

Mix it up

Do not stick to set groups at each table. Every child should be working on the same concept, and mixing up the groupings widens children's opportunities for exploring, discussing and sharing their understanding with others.

Recycling questions

Reuse the Pupil Textbook and Practice Book questions with concrete materials to allow children to explore concepts and relationships and deepen their understanding. This strategy is especially useful for reinforcing learning in same-day interventions.

Strengthen at every opportunity

The next lesson in a *Power Maths* sequence always revises and builds on the previous step to help embed learning. These activities provide golden opportunities for individual children to strengthen their learning with the support of teaching assistants.

Prepare to be surprised!

Children may grasp a concept quickly or more slowly. The 'fast graspers' won't always be the same individuals, nor does the speed at which a child understands a concept predict their success in maths. Are they struggling or just working more slowly?

Depth and breadth

Just as prescribed in the National Curriculum, the goal of *Power Maths* is never to accelerate through a topic but rather to gain a clear, deep and broad understanding.

"Pupils who grasp concepts rapidly should be challenged through being offered rich and sophisticated problems before any acceleration through new content. Those who are not sufficiently fluent with earlier material should consolidate their understanding, including through additional practice, before moving on."

National Curriculum: Mathematics programmes of study: KS1 & 2, 2013

The lesson sequence offers many opportunities for you to deepen and broaden children's learning, some of which are suggested below.

Discover

As well as using the questions in the Teacher Guide, check that children are really delving into why something is true. It is not enough to simply recite facts, such as '6 + 3 = 9'. They need to be able to see why, explain it, and to demonstrate the solution in several ways.

Share

Make sure that every child is given chances to offer answers and expand their knowledge and not just those with the greatest confidence.

Think together

Encourage children to think about how they found the solution and explain it to their partner. Be sure to make concrete materials available on group tables throughout the lesson to support and reinforce learning.

Practice

Avoid any temptation to select questions according to your assessment of ability: practice questions are presented in a logical sequence and it is important that each child works through every question.

Reflect

Open-ended questions allow children to deepen their understanding as far as they can by discovering new ways of finding answers. For example, *Give me another way of working out how high the wall is … And another way?*

Online materials

For each unit you will find additional strengthening activities to support those children who need it and to deepen the understanding of those who need the additional challenge.

Same-day intervention

Since maths competence depends on mastering concepts one-by-one in a logical progression, it is important that no gaps in understanding are ever left unfilled. Same-day interventions – either within or after a lesson – are a crucial safety net for any child who has not fully made the small step covered that day. In other words, intervention is always about keeping up, not catching up, so that every child has the skills and understanding they need to tackle the next lesson. That means presenting the same problems used in the lesson, with a variety of concrete materials to help children model their solutions.

We offer two intervention strategies below, but you should feel free to choose others if they work better for your class.

Within-lesson intervention

The Think together activity will reveal those who are struggling, so when it is time for Practice, bring these children together to work with you on the first Practice questions. Observe these children carefully, ask questions, encourage them to use concrete models and check that they reach and can demonstrate their understanding.

After-lesson intervention

You might like to use Think together before an assembly, giving you or teaching assistants time to recap and expand with slow graspers during assembly time. Teaching assistants could also work with strugglers at other convenient points in the school day.

The role of practice

Practice plays a pivotal role in the *Power Maths* approach. It takes place in class groups, smaller groups, pairs and independently, so that children always have the opportunities for thinking as well as the models and support they need to practise meaningfully and with understanding.

Intelligent practice

In *Power Maths*, practice never equates to the simple repetition of a process. Instead we embrace the concept of intelligent practice, in which all children become fluent in maths through varied, frequent and thoughtful practice that deepens and embeds conceptual understanding in a logical, planned sequence. To see the difference, take a look at the following examples.

Traditional practice

- Repetition can be rote – no need for a child to think hard about what they are doing.

- Praise may be misplaced.

- Does this prove understanding?

Intelligent practice

- Varied methods – concrete, pictorial and abstract.

- Calculations expressed in different ways, requiring thought and understanding.

- Constructive feedback.

All practice questions are designed to move children on and reveal misconceptions.

Simple, logical steps build onto earlier learning.

C-P-A runs throughout – different ways of modelling and understanding the same concept.

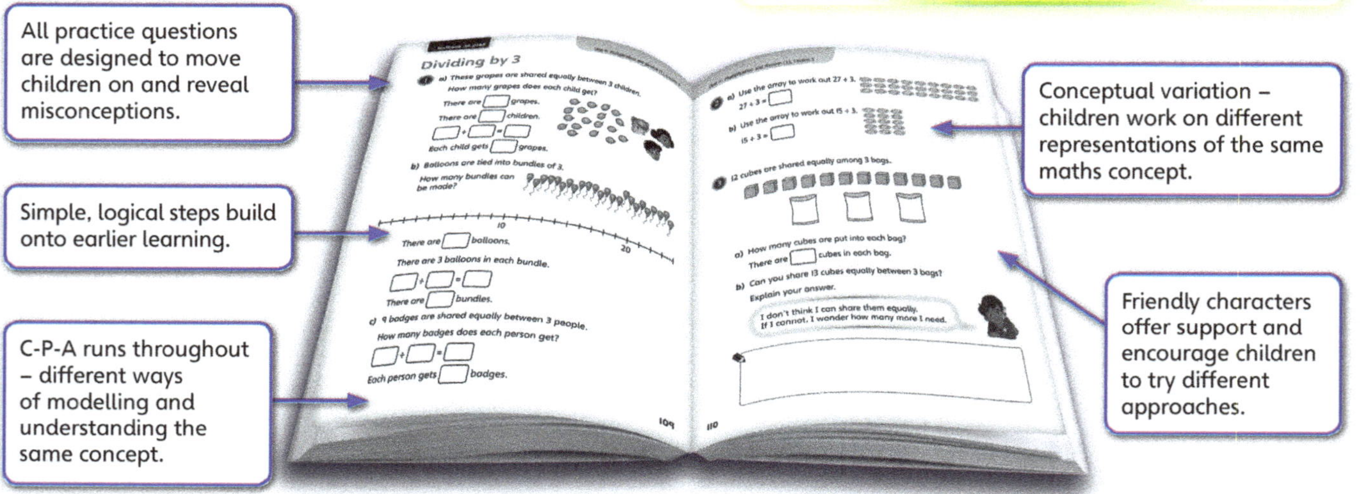

Conceptual variation – children work on different representations of the same maths concept.

Friendly characters offer support and encourage children to try different approaches.

A carefully designed progression

The Practice Books provide just the right amount of intelligent practice for children to complete independently in the final sections of each lesson. It is really important that all children are exposed to the Practice questions, and that children are not directed to complete different sections. That is because each question is different and has been designed to challenge children to think about the maths they are doing. The questions become more challenging so children grasping concepts more quickly will start to slow down as they progress. Meanwhile, you have the chance to circulate and spot any misconceptions before they become barriers to further learning.

Homework and the role of carers

While *Power Maths* does not prescribe any particular homework structure, we acknowledge the potential value of practice at home. For example, practising fluency in key facts, such as number bonds and times-tables, is an ideal homework task, and carers could work through uncompleted Practice Book questions with children at either primary stage.

However, it is important to recognise that many parents and carers may themselves lack confidence in maths, and few, if any, will be familiar with mastery methods. A Parents' and Carers' Evening that helps them understand the basics of mindsets, mastery and mathematical language is a great way to ensure that children benefit from their homework. It could be a fun opportunity for children to teach their families that everyone can do maths!

Structures and representations

Unlike most other subjects, maths comprises a wide array of abstract concepts – and that is why children and adults so often find it difficult. By taking a Concrete-Fictorial-Abstract (C-P-A) approach, *Power Maths* allows children to tackle concepts in a tangible and more comfortable way.

Non-linear stages

Concrete

Replacing the traditional approach of a teacher working through a problem in front of the class, the concrete stage introduces real objects that children can use to 'do' the maths – any familiar object that a child can manipulate and move to help bring the maths to life. It is important to appreciate, however, that children must always understand the link between models and the objects they represent. For example, children need to first understand that three cakes could be represented by three pretend cakes, and then by three counters or bricks. Frequent practice helps consolidate this essential insight. Although they can be used at any time, good concrete models are an essential first step in understanding.

Pictorial

This stage uses pictorial representations of objects to let children 'see' what particular maths problems look like. It helps them make connections between the concrete and pictorial representations and the abstract maths concept. Children can also create or view a pictorial representation together, enabling discussion and comparisons. The *Power Maths* teaching tools are fantastic for this learning stage, and bar modelling is invaluable for problem solving throughout the primary curriculum.

Abstract

Our ultimate goal is for children to understand abstract mathematical concepts, signs and notation and, of course, some children will reach this stage far more quickly than others. To work with abstract concepts, a child needs to be comfortable with the meaning of, and relationships between, concrete, pictorial and abstract models and representations. The C-P-A approach is not linear, and children may need different types of models at different times. However, when a child demonstrates with concrete models and pictorial representations that they have grasped a concept, we can be confident that they are ready to explore or model it with abstract signs such as numbers and notation.

Use at any time and with any age to support understanding.

Practical aspects of *Power Maths*

One of the key underlying elements of *Power Maths* is its practical approach, allowing you to make maths real and relevant to your children, no matter their age.

Manipulatives are essential resources for both key stages and *Power Maths* encourages teachers to use these at every opportunity, and to continue the Concrete-Pictorial-Abstract approach right through to Year 6.

The Textbooks and Teacher Guides include lots of opportunities for teaching in a practical way to show children what maths means in real life.

Discover and Share

The Discover and Share sections of the Textbook give you scope to turn a real-life scenario into a practical and hands-on section of the lesson. Use these sections as inspiration to get active in the classroom. Where appropriate, use the Discover contexts as a springboard for your own examples that have particular resonance for your children – and allow them to get their hands dirty trying out the mathematics for themselves.

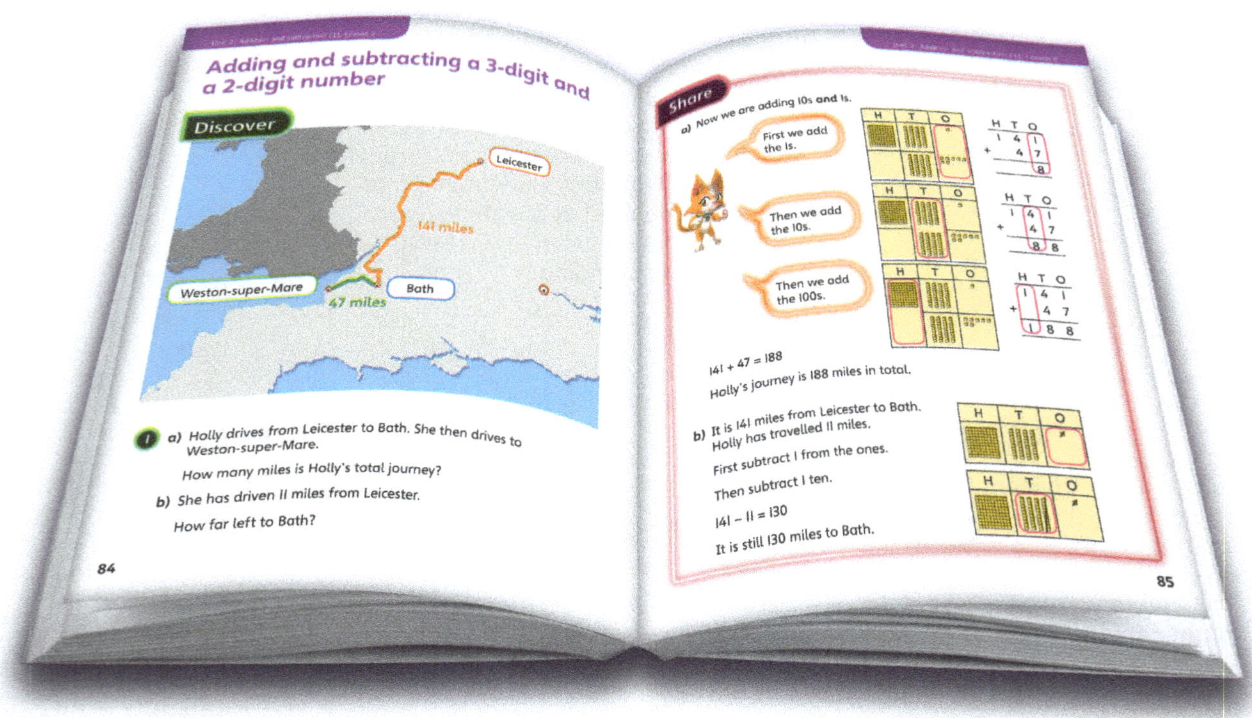

Unit videos

Every term has one unit video which incorporates real-life classroom sequences.

These videos show you how the reasoning behind mathematics can be carried out in a practical manner by showing real children using various concrete and pictorial methods to come to the solution. You can see how using these practical models, such as part-whole and bar models, helps them to find and articulate their answer.

Mastery tips

Mastery Experts give anecdotal advice on where they have used hands-on and real-life elements to inspire their children.

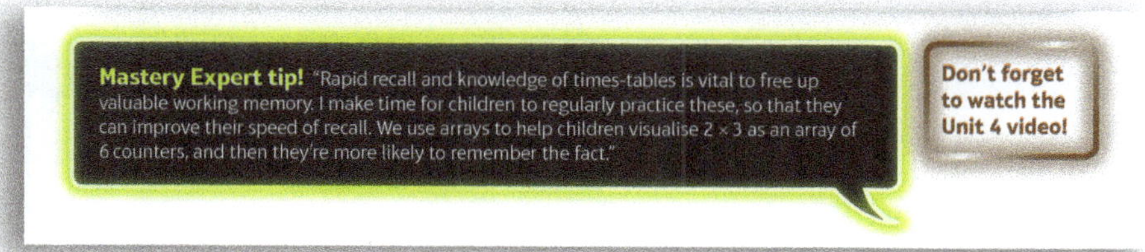

Concrete-Pictorial-Abstract (C-P-A) approach

Each Share section uses various methods to explain an answer, helping children to access abstract concepts by using concrete tools, such as counters. Remember this isn't a linear process, so even children who appear confident using the more abstract method can deepen their knowledge by exploring the concrete representations. Encourage children to use all three methods to really solidify their understanding of a concept.

Pictorial representation – drawing the problem in a logical way that helps children visualise the maths

Concrete representation – using manipulatives to represent the problem. Encourage children to physically use resources to explore the maths.

Abstract representation – using words and calculations to represent the problem.

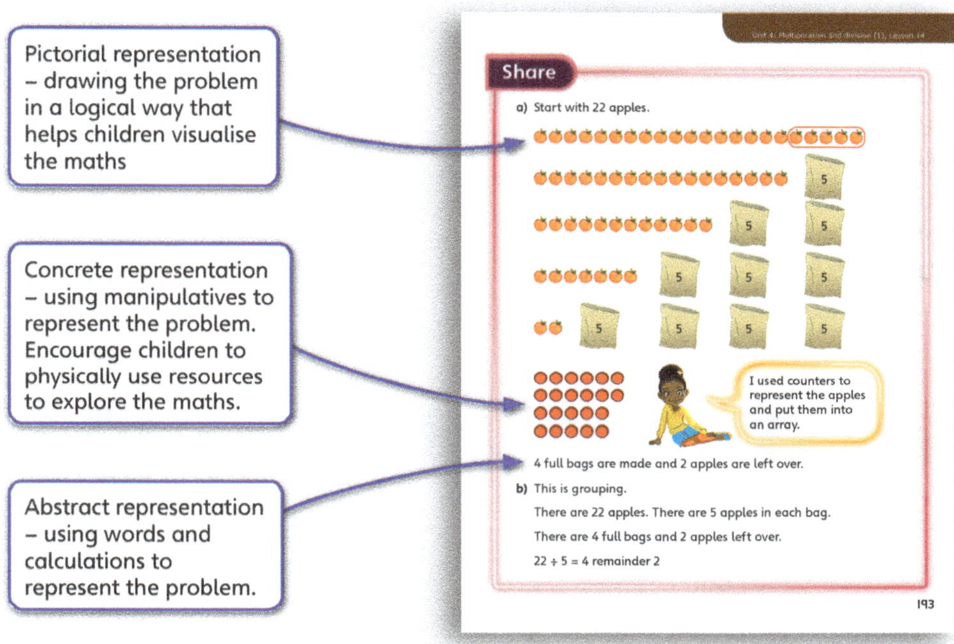

Practical tips

Every lesson suggests how to draw out the practical side of the Discover context.

You'll find these in the Discover section of the Teacher Guide for each lesson.

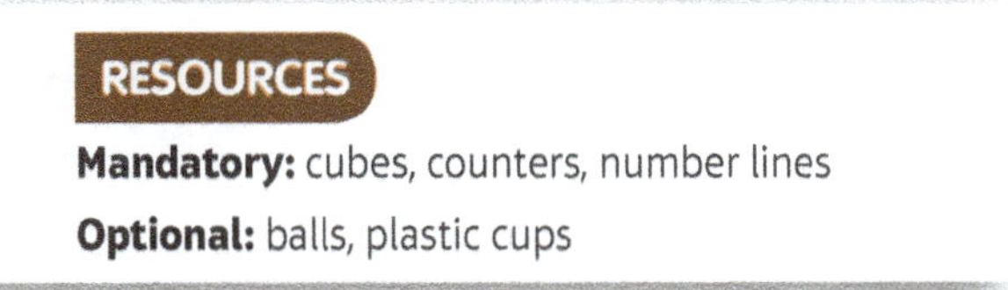

Resources

Every lesson lists the practical resources you will need or might want to use. There is also a summary of all of the resources used throughout the term on page 34 to help you be prepared.

RESOURCES

Mandatory: cubes, counters, number lines
Optional: balls, plastic cups

List of practical resources

Year 3A Mandatory resources

Resource	Lesson
Base 10 equipment	**Unit 1** lessons 1, 2, 3, 7, 8 **Unit 2** lessons 1, 2, 3, 4, 5, 6, 7, 8, 9, 10 **Unit 3** lessons 1, 2, 3, 4, 5
Counters	**Unit 4** lessons 1, 2, 3, 4, 5, 6, 7, 8, 9, 10, 14, 15
Number line	**Unit 1** lessons 5, 6, 11 **Unit 3** lesson 6, **Unit 4** lessons 2, 3, 4, 5, 6, 7, 8, 9, 10
Place value grids	**Unit 1** lessons 3, 4, 7, 9, 10 **Unit 2** lessons 2, 4, 5, 6, 7, 8, 9 **Unit 3** lessons 3, 4
Cubes	**Unit 4** lessons 1, 2, 3, 4, 5, 6, 7, 8, 9, 10, 14, 15
Lollipop sticks or strips of paper	**Unit 4** lesson 13
0–9 digit cards	**Unit 2** lessons 2, 3, 9, 10
Place value counters	**Unit 1** lesson 4, **Unit 2** lessons 6, 8, 9 **Unit 3** lessons 4, 5
Place value cards	**Unit 1** lesson 3, **Unit 2** lessons 2, 4, 5, 6, 7, 8, 9, 10
Bar model	**Unit 3** lesson 9

Year 3A Optional resources

Resource	Lesson
Large (laminated) place value grid	**Unit 1** lessons 3, 4 **Unit 2** lesson 1
Place value counters	**Unit 1** lessons 7, 8, 10, 11 **Unit 2** lessons 1, 2, 10
Bead strings	**Unit 1** lesson 1 **Unit 2** lesson 4
Place value abacus	**Unit 2** lesson 5
0–9 digit cards	**Unit 2** lessons 4, 6, 7, 8 **Unit 3** lesson 2
100 square	**Unit 1** lesson 1 **Unit 2** lesson 5
Place value cards	**Unit 1** lesson 8 **Unit 2** lesson 1
Bags and boxes of objects in 100s	**Unit 1** lesson 1
Large (laminated) part-whole model	**Unit 1** lesson 2 **Unit 3** lessons 1, 5, 7
Number lines	**Unit 1** lesson 7 **Unit 3** lessons 4, 5 **Unit 4** lessons 1, 11, 12
Ten frame	**Unit 2** lesson 3
Paper clips	**Unit 3** lesson 6
Wooden blocks	**Unit 4** lesson 12
Base 10 equipment	**Unit 1** lessons 1, 4, 5, 6, 9, 10, 11

Resource	Lesson
Paper circles (and paper circles divided into pieces)	**Unit 4** lessons 8, 9
Scrap paper	**Unit 3** lesson 7
Paper clips	**Unit 3** lessons 4, 6
50p coins	**Unit 1** lesson 11
Fruit juice	**Unit 4** lesson 9
Coloured rods	**Unit 3** lesson 9
Pegs	**Unit 3** lesson 6
Dice	**Unit 3** lesson 4
100 square	**Unit 1** lesson 1 **Unit 2** lesson 5
Place value cards	**Unit 1** lesson 8 **Unit 2** lesson 1
Counting sticks	**Unit 1** lessons 5, 6
Cubes	**Unit 3** lesson 6 **Unit 4** lessons 11, 12, 13
Ice lolly moulds	**Unit 4** lesson 9
4 times-table flashcards	**Unit 4** lesson 7
48 playing cards	**Unit 4** lesson 6
Plastic animals	**Unit 4** lesson 5
Times-tables written on stairs	**Unit 4** lesson 4
Balls	**Unit 4** lesson 2
Strips of paper	**Unit 3** lessons 8, 9
Washing line	**Unit 3** lesson 6
Number cards to 100	**Unit 3** lesson 6
Spinners	**Unit 3** lesson 4
Lengths of string	**Unit 3** lesson 9

Variation helps visualisation

Children find it much easier to visualise and grasp concepts if they see them presented in a number of ways, so be prepared to offer and encourage many different representations.

For example, the number six could be represented in various ways:

Getting started with *Power Maths*

As you prepare to put *Power Maths* into action, you might find the tips and advice below helpful.

STEP 1: Train up!

A practical, up-front full-day professional development course will give you and your team a brilliant head-start as you begin your *Power Maths* journey. You will learn more about the ethos, how it works and why.

STEP 2: Check out the progression

Take a look at the yearly and termly overviews. Next take a look at the unit overview for the unit you are about to teach in your Teacher Guide, remembering that you can match your lessons and pacing to your class.

STEP 3: Explore the context

Take a little time to look at the context for this unit: what are the implications for the unit ahead? (Think about key language, common misunderstandings and intervention strategies, for example.) If you have the online subscription, don't forget to watch the corresponding unit video.

STEP 4: Prepare for your first lesson

Familiarise yourself with the objectives, essential questions to ask and the resources you will need. The Teacher Guide offers tips, ideas and guidance on individual lessons to help you anticipate children's misconceptions and challenge those who are ready to think more deeply.

STEP 5: Teach and reflect

Deliver your lesson – and enjoy!

Afterwards, reflect on how it went… Did you cover all five stages?
Does the lesson need more time? How could you improve it?
What percentage of your class do you think mastered the concept?
How can you help those that didn't?

Unit 1
Place value within 1,000

Don't forget to watch the Unit 1 video!

WHY THIS UNIT IS IMPORTANT

This unit is important as it explores 3-digit numbers in depth. For many children, it will be the first time they have met these numbers. This work builds on the place value work that they did in Year 2 and they will extend many of the models and images that they have used previously.

Children begin with learning how to count in 100s. They will learn that a 3-digit number is made up of some 100s, 10s and 1s and they will be able to represent this in many ways (for example, on a place value grid with counters or in a part-whole model). They will extend the number line to 1,000 and know where different numbers lie. They will compare and order 3-digit numbers as well as count in 50s.

This unit underpins a lot of the subsequent work this year and it is essential that children gain a solid understanding of the key concepts within this unit.

WHERE THIS UNIT FITS

→ **Unit 1: Place value within 1,000**

→ Unit 2: Addition and subtraction (1)

This unit builds on children's work in Year 2 on 2-digit numbers. The work in this unit is essential for the work in the rest of this year when they look at the four rules of number, fractions and measure. In the next unit, children move on to adding and subtracting 3-digit numbers.

Before they start this unit, it is expected that children:
- know that a 2-digit number is made up of 10s and 1s
- can represent 2-digit numbers in different ways, such as base 10 equipment, place value grids and counters, part-whole models and number lines
- can find 1 and 10 more and less than a 2-digit number
- can compare and order 2-digit numbers
- know where a 2-digit number lies on a number line.

ASSESSING MASTERY

Children will know that a number is made up of some 100s, 10s and 1s and will be able to represent numbers in multiple ways. They will be able to find 100, 10 and 1 more or less than a 3-digit number. They will be able to compare and order 3-digit numbers by looking at the digits in each place value. They will understand the number line to 1,000 and start to know where numbers lie on the number line.

COMMON MISCONCEPTIONS	STRENGTHENING UNDERSTANDING	GOING DEEPER
Children may think that 2 tens + 5 hundreds + 7 ones is 257 as opposed to 527.	Use base 10 equipment and place value counters to secure understanding of 3-digit numbers.	Ask children to partition numbers in different ways. For example, 226 is 1 × 100, 12 × 10 and 6 × 1.
Children may compare a 3-digit and a 2-digit number by looking at the first digit rather than the numbers of 100s.	Use base 10 equipment or place value counters in a place value grid to emphasise that they need to compare numbers that have the same place value.	Challenge children to make as many numbers as possible using eight blank counters in an HTO place value grid. Ask them to order their numbers on a number line.

WAYS OF WORKING

Use these pages to introduce the unit focus to children. You can use the characters to explore different ways of working too!

STRUCTURES AND REPRESENTATIONS

Part-whole: This model will help children to see how numbers can be partitioned into 100s, 10s and 1s.

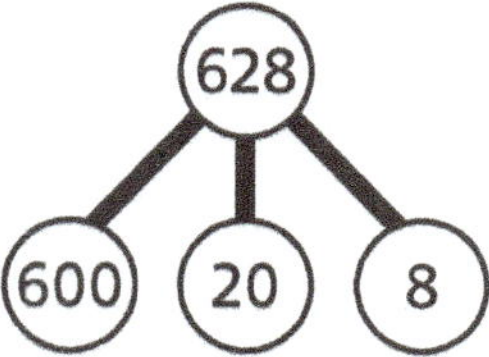

Place value grid, including using base 10 equipment and place value counters: This model will help children organise 3-digit numbers into 100s, 10s and 1s, with both concrete representations and abstract numbers.

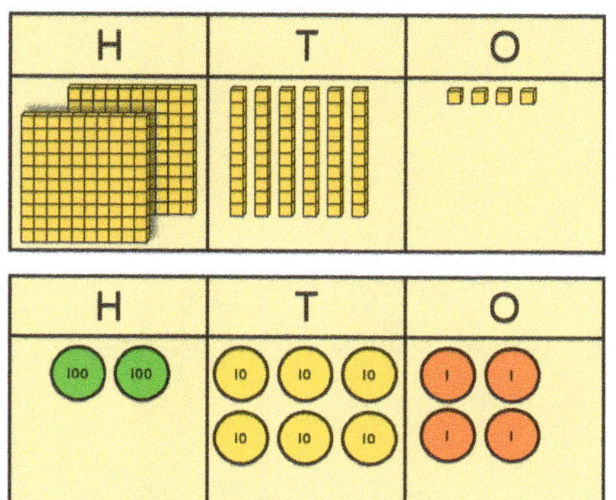

Number line to 1,000: This model will help children to visualise the order of numbers, and can help them to compare numbers.

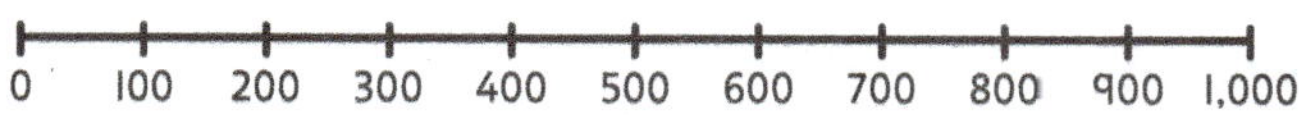

KEY LANGUAGE

There is some key language that children will need to know as part of the learning in this unit.

➜ hundreds (100s), tens (10s), ones (1s)

➜ place value

➜ more, less

➜ greater than (>), less than (<), equal to (=)

➜ order, compare

➜ digit, one thousand

➜ part-whole model, place value grid, number line

➜ estimate, halfway, exchange

➜ taller, tallest, longest, shortest, greatest, smallest, most, least, fewest

PUPIL TEXTBOOK 3A PAGE 6

PUPIL TEXTBOOK 3A PAGE 7

Counting in 100s

Learning focus

In this lesson, children will learn how to count in 100s from 0 to 1,000. They will write the numbers in both numerals and words.

Small steps

→ **This step: Counting in 100s**
→ Next step: Representing numbers to 1,000

NATIONAL CURRICULUM LINKS

Year 3 Number – Number and Place Value

- Recognise the place value of each digit in a three-digit number (hundreds, tens, ones).
- Read and write numbers up to 1,000 in numerals and in words.
- Identify, represent and estimate numbers using different representations.

ASSESSING MASTERY

Children can count in 100s from 0 to 1,000 and back again. They should understand what 100 is and the different ways of representing it. They will write the numbers in both numerals and words.

COMMON MISCONCEPTIONS

Often when children are counting from 0 to 1,000 and they get to 900, they say 'ten hundred' next. Explain that although they have 10 hundreds, we say this as one thousand. Ask:
- *Continue the count 700, 800, 900, … What comes after 900?*

STRENGTHENING UNDERSTANDING

Children who are struggling to count in 100s first need to understand what a 100 is. Give them a jar of 100 dice or other objects or get them to count out 100 cubes. Once they have done this give them multiple packs of 100. Use a base 10 hundred block for each pack of 100. The connection between the hundred block and the 100 items should be made clear.

GOING DEEPER

Ask children to count on 500 from 300. They need to keep track of how many 100s they have counted on as well as the number that they reach.

KEY LANGUAGE

In lesson: one thousand, hundreds (100s)

Other language to be used by the teacher: count forwards, count backwards, number track

STRUCTURES AND REPRESENTATIONS

Base 10 equipment

RESOURCES

Mandatory: base 10 equipment (100s)

Optional: bags and boxes of objects in 100s, bead strings, 100 square, base 10 equipment (1s and 10s)

 In the eTextbook of this lesson, you will find interactive links to a selection of teaching tools.

Before you teach 

- Can children count from 0 to 100?
- Do they know when to go up to the next 10? Do they know what happens when they get to 99?
- Do children already have an understanding of what 100 is?

Discover

ASK

• Question **1** a): *What happens if you lose count? Is there a way to make sure you can start again easily?*

IN FOCUS Question **1** a) is used to ensure that children recall from Year 2 the count from 0 to 100. It also allows the teacher to look at how children count. For example, do they count in 10s, just in case they lose count? Do they know what to do when they go from 9 to the next 10?

PRACTICAL TIPS Make sure each pair or group has 100 objects (for example, cubes) that they can use to represent the dice in the picture.

ANSWERS

Question **1** a): Yes, there are 100 dice.

Question **1** b): There are 300 in total.

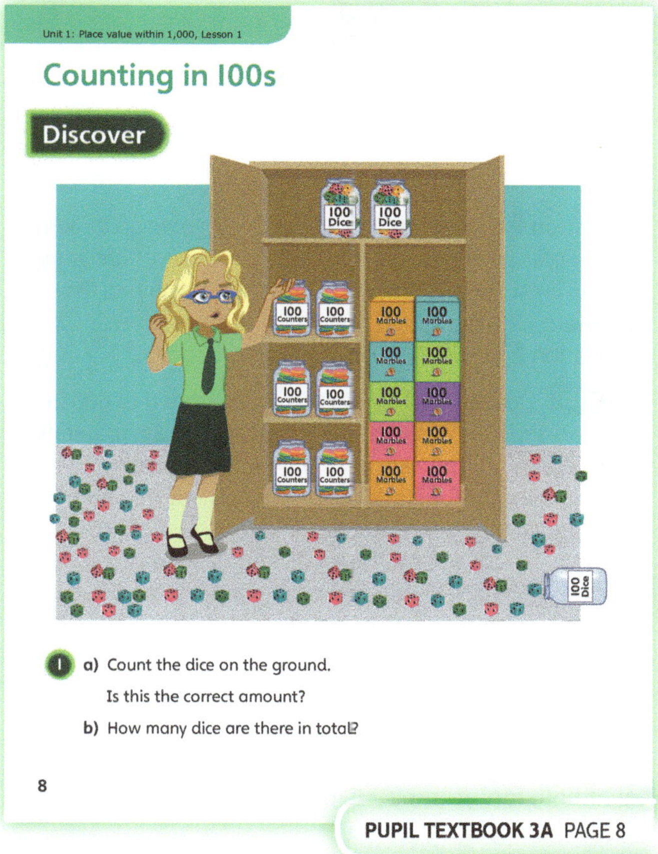

PUPIL TEXTBOOK 3A PAGE 8

Share

ASK

• Question **1** a): *How can you represent 100? What different ways do you know?*
• Question **1** b): *Do you need to count in 1s to count all the objects? Is there a quicker way?*

IN FOCUS Question **1** b) gets children to think about counting in 100s and checks whether, beyond 100, they can count in 100s instead of in 1s or 10s.

STRENGTHEN Count with children from 0 to 100 in 10s and pretend to lose count. Draw out how it helps to put the objects into groups of 10, in case they lose count. Draw the connection between counting in 1s and counting in 100s. Use a base 10 100 block to represent the 100 dice and then count up: 'One hundred, two hundred, …'.

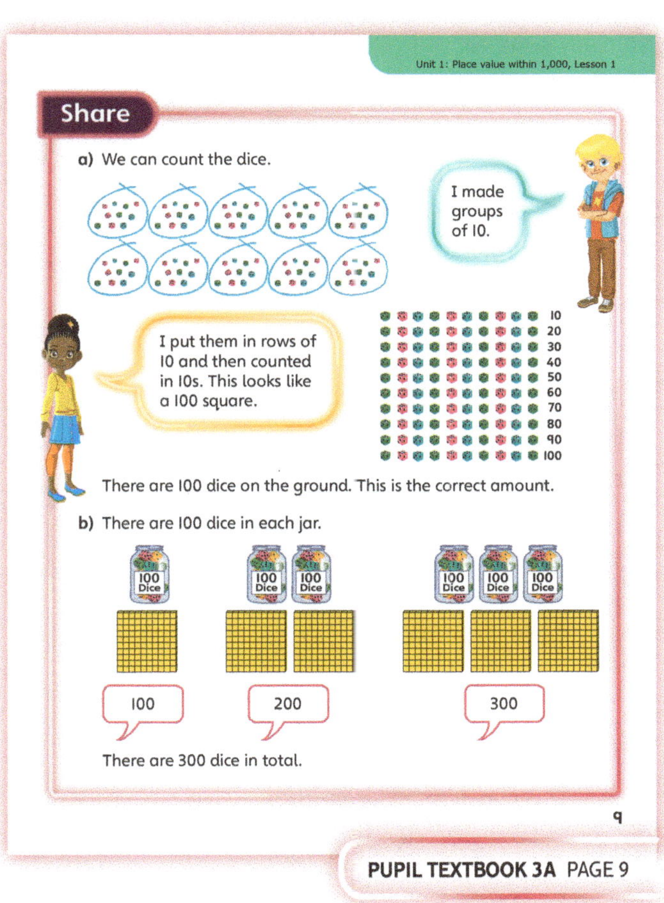

PUPIL TEXTBOOK 3A PAGE 9

Think together

 Whole class teacher led (I do, We do, You do)

ASK

- Question **2** : *What are these sequences going up in? Which sequences are going forwards? Which sequences are going backwards? How can counting in 100s help you find the answers?*
- Question **3** : *700, 800, 900 … What comes next?*

IN FOCUS Question **1** shows the count from 0 to 600 in 100s and shows all the numbers represented as numerals and words. It is important that children know these numbers. Question **2** applies their knowledge to number tracks and sequences. There is a mix of sequences to check that they can count both forwards and backwards. Question **3** encourages children to think what word comes after 900. Some children may think that the next 100 after 900 is ten hundred. Introduce the term 'one thousand'.

STRENGTHEN Use bags of 100 objects and show them alongside base 10 100 blocks. Ask children to count the 100 blocks and show them this alongside the numbers. Count both forwards (by adding blocks) and backwards (by removing blocks).

DEEPEN Ask children to count on 200 from 500, or 600 less than 800. This requires them to keep track of where they are and how many 100s they have counted forwards or back.

ASSESSMENT CHECKPOINT In questions **2** and **3** , assess whether children can count forwards and backwards in 100s from any number.

ANSWERS

Question **1** : 400, four hundred

500, five hundred

600, six hundred

700, seven hundred

800, eight hundred

Question **2** a): 400, 500

Question **2** b): 300, 100

Question **2** c): 500, 700, 900

Question **3** : There are 1,000 marbles.

There are one thousand marbles.

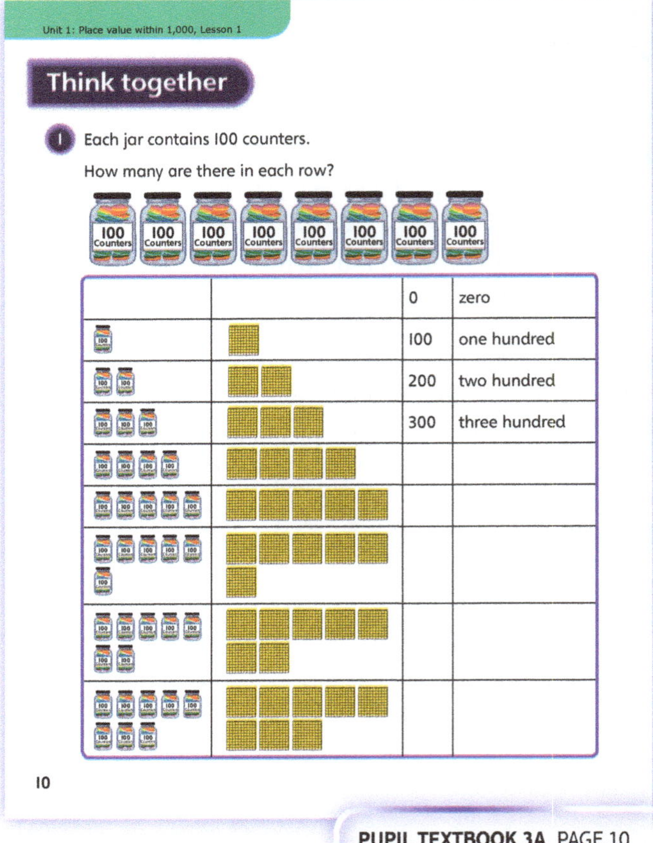

PUPIL TEXTBOOK 3A PAGE 10

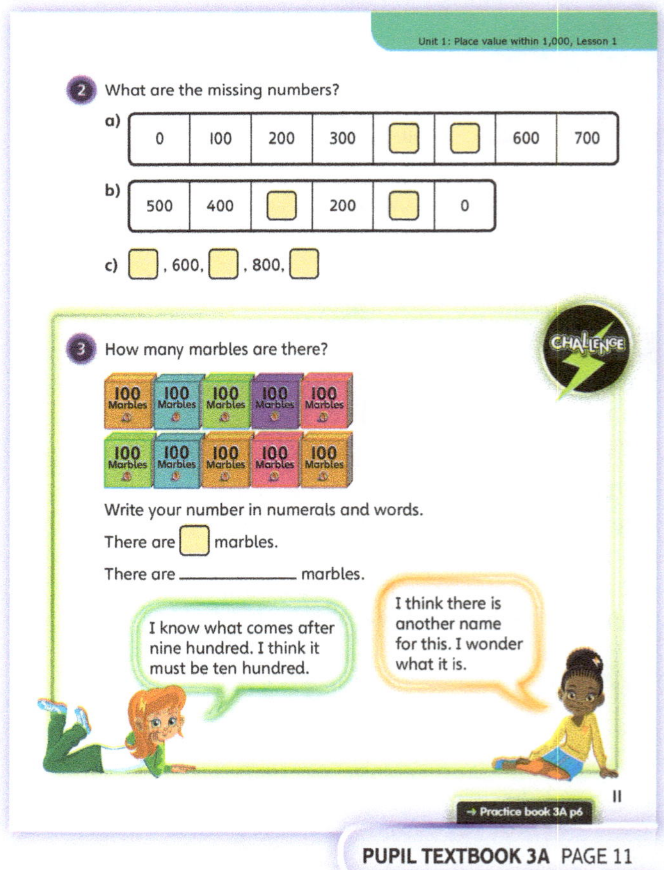

PUPIL TEXTBOOK 3A PAGE 11

Practice

WAYS OF WORKING Independent thinking

IN FOCUS Question ❶ checks whether children can count in 100s from 0 to 600. Question ❷ is more abstract and checks whether children know the forwards and backwards count from 0 to 1,000. Children have to work out the missing 100s. Question ❹ starts with the total number and asks how many 100s. Children need to count up in 100s until they reach 700. They also need to keep track of how many 100s they have counted.

STRENGTHEN Use bags of 100 objects to give children practice in counting in 100s. Then introduce 100 base 10 equipment, explaining that they represent the bags of objects. Ask children to count the 100 blocks and show them this alongside the numbers in both numerals and words. Count both forwards (by adding blocks) and backwards (by removing blocks).

DEEPEN Ask children to work out 300 more than 700 or 400 less than 900. They need to understand that 300 more than 700 means counting on 300 from 700, so they need to count on 3 hundreds. They need to keep track of where they are and how many 100s they have counted.

ASSESSMENT CHECKPOINT Use question ❶ to assess whether children know the numbers in words and numerals for 100 to 1,000.

Use question ❷ to assess whether children can count forwards and backwards in 100s from any number.

ANSWERS Answers for the **Practice** part of the lesson appear in the separate **Practice and Reflect answer guide**.

Reflect

WAYS OF WORKING Pair work

IN FOCUS The questions check whether children can count in 100s both forwards and backwards. They also check their understanding of how and when to say 1,000.

ASSESSMENT CHECKPOINT Check whether children are counting in 100s. Do this by getting children to do the counts individually first, then in pairs and then as a whole class. Look for children who are counting backwards instead of forwards, and vice versa.

ANSWERS Answers for the **Reflect** part of the lesson appear in the separate **Practice and Reflect answer guide**.

After the lesson ⏸

- Do children know the count from 0 to 1,000 in 100s?
- Do children know that 1,000 is one thousand and not said as ten hundred?
- Can children represent a number of 100s in base 10 equipment and in other ways?

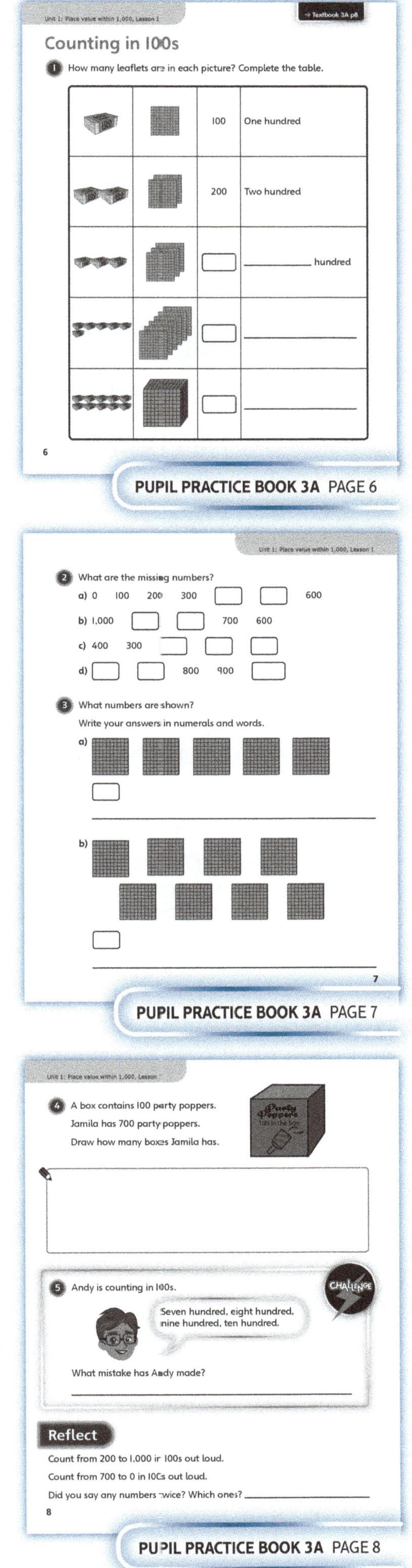

PUPIL PRACTICE BOOK 3A PAGE 6

PUPIL PRACTICE BOOK 3A PAGE 7

PUPIL PRACTICE BOOK 3A PAGE 8

Representing numbers to 1,000

Learning focus

In this lesson, children will understand that a number up to 1,000 is made up of some 100s, some 10s and some 1s. They will use base 10 equipment and part-whole models to represent numbers.

Small steps

→ Previous step: Counting in 100s
→ **This step: Representing numbers to 1,000**
→ Next step: 100s, 10s, 1s (1)

NATIONAL CURRICULUM LINKS

Year 3 Number – Number and Place Value

- Identify, represent and estimate numbers using different representations.
- Recognise the place value of each digit in a three-digit number (hundreds, tens, ones).
- Read and write numbers up to 1,000 in numerals and in words.

ASSESSING MASTERY

Children can represent 3-digit numbers using base 10 equipment and write them onto a part-whole model. Children can recognise that a 3-digit number is made up of some 100s, some 10s and some 1s.

COMMON MISCONCEPTIONS

A common mistake that children make is that they do not understand the place value and size. They may, for example, see 4 hundred blocks and 6 ones and think this is 460 as they are just going left to right. Show children representations not in order. Ask:

- *Can you represent 248 using base 10 equipment? What happens if I move the base 10 equipment around? What is my number now?*

STRENGTHENING UNDERSTANDING

First recap representing 2-digit numbers. Ask children to show, for example, 58. Explain that they can now add some 100s to this and this gives a 3-digit number. Encourage children to write out the numbers (for example, 348 is 3 hundreds, 4 tens and 8 ones) and to make each part with base 10 equipment.

GOING DEEPER

Ask if all 3-digit numbers are made up of three parts. Ask children to make numbers that can be represented on a part-whole model that just has two parts. Ask children if they can represent the number 228 in different ways using base 10 equipment (for example, 1 hundred, 12 tens and 8 ones) and to explain why these are equal.

KEY LANGUAGE

In lesson: hundreds (100s), tens (10s), ones (1s), part-whole model, digit

Other language to be used by the teacher: partition

STRUCTURES AND REPRESENTATIONS

Base 10 equipment, part-whole model

RESOURCES

Mandatory: base 10 equipment (including 100s)

Optional: large laminated part-whole model

 In the eTextbook of this lesson, you will find interactive links to a selection of teaching tools.

Before you teach 

- Can children represent 2-digit numbers in base 10 equipment? On a part-whole model?
- Given a number of 10s and 1s, do children know what number it is?
- Do children know that the number made from 7 tens and 3 ones is the same as the number made from 3 ones and 7 tens?

Discover

WAYS OF WORKING Pair work

ASK

- Question ❶ a): *What do the different base 10 blocks represent? What is a number between 100 and 999 made up of? How can you tell how many 100s, how many 10s and how many 1s?*
- Question ❶ b): *How can you use a part-whole model to show this? Why do you think that there are three parts?*

IN FOCUS Question ❶ a) introduces two different representations of 235. Firstly, children make 235 with concrete base 10 equipment and then represent it on a part-whole model. They should start to see that numbers between 100 and 999 are made up of some 100s, some 10s and some 1s. Question ❶ b) asks children to work the other way, using the representation to determine the number.

PRACTICAL TIPS Make sure children have all the necessary base 10 equipment. Say that there are 235 children in the school and ask children to represent this using the equipment.

ANSWERS

Question ❶ a): 235 using base 10 equipment

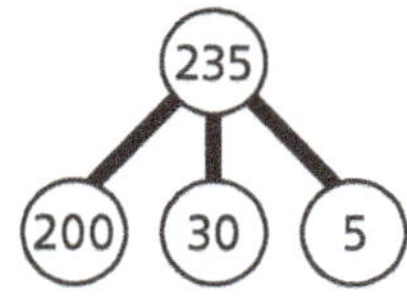

Question ❶ b): 124, 476

Share

WAYS OF WORKING Whole class teacher led

ASK

- Question ❶ a): *How many children are in the school? How can you represent these using base 10 equipment? Which blocks do you use for each representation? Why do you use 2 hundred blocks? How can you show this using a part-whole model? Why do you need to use three parts? Why do you not just use two parts?*
- Question ❶ b): *How can you work out what the number is? What information does the diagram show?*

IN FOCUS Question ❶ a) encourages children to represent objects using base 10 equipment. It is important they recognise which base 10 equipment represents each part of the number. By the end of **Share** they should have an understanding of how to represent 3-digit numbers in base 10 equipment and on part-whole models and should recognise numbers written in this form.

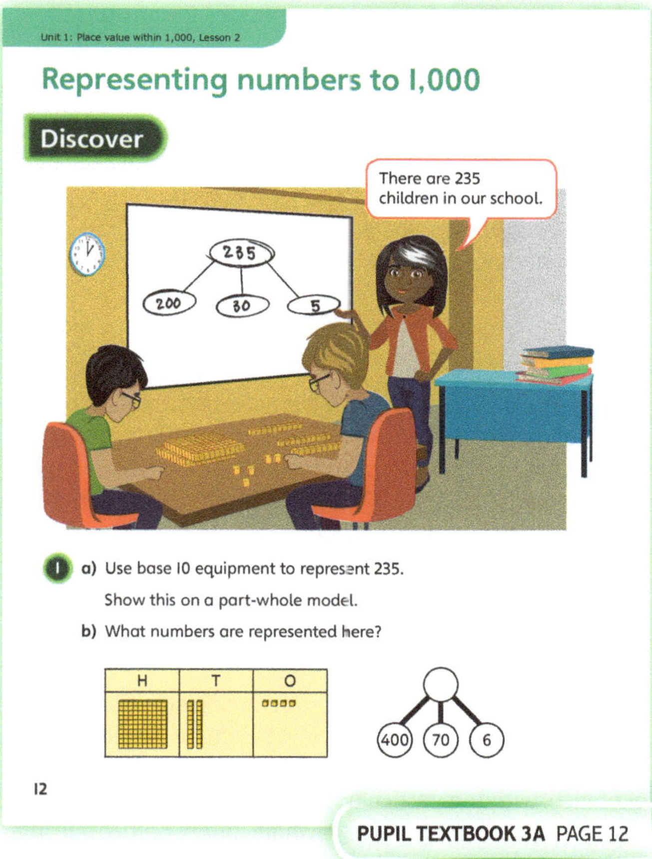

PUPIL TEXTBOOK 3A PAGE 12

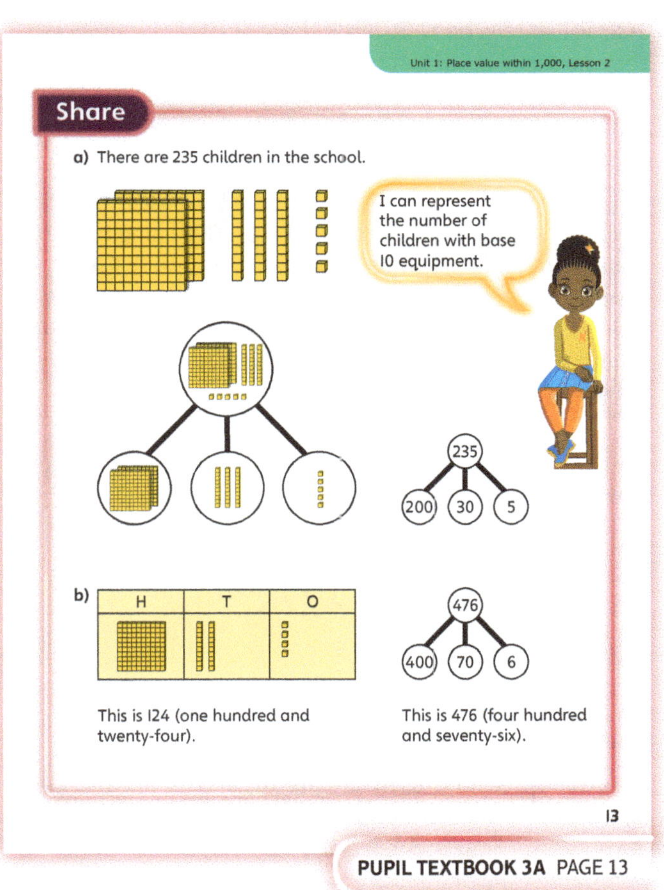

PUPIL TEXTBOOK 3A PAGE 13

Think together

WAYS OF WORKING Whole class teacher led (I do, We do, You do)

ASK

- Question **1**: *How can you use base 10 equipment to represent these numbers? Is there any other equipment you can use?*
- Question **3**: *Does it matter what order you put the blocks in? Will you still get the same number?*

IN FOCUS Questions **1** and **2** practise what children have been doing in **Discover**. They look at different ways of representing 3-digit numbers: base 10 equipment and part-whole models. Question **3** looks at common mistakes that are made with the place value of numbers. For example, children often think that the order of the blocks changes the numbers. They need to understand that the 100s will always represent the hundreds digit regardless of where they are in the representation.

STRENGTHEN Represent actual concrete objects (for example, 245 toys) using base 10 equipment. Separate the 100s from the 10s from the 1s. Clearly associate each group of objects with the particular base 10 equipment. Put the base 10 equipment into a part-whole model before putting in the numbers. This will help children understand the numbers.

DEEPEN Provide numbers in base 10 equipment that are not given in the order of 100s, 10s and 1s. This will help children understand that the order does not matter – 100s will always be 100s. Ask if there are 3-digit numbers that are made up of just two parts. Ask for examples and establish what is the same and what is different about these numbers. Ask whether there any 3-digit numbers that have only one part.

ASSESSMENT CHECKPOINT Use question **1** to assess whether children can represent numbers up to 1,000 using base 10 equipment and part-whole models. Use question **2** to assess whether children can identify the number from its representation. Children need to understand that a 3-digit number is made up of some 100s, some 10s and some 1s.

ANSWERS

Question **1** a): 365 and 130 in base 10 equipment

Question **1** b):

Question **2** a): 364

Question **2** b): 137

Question **2** c): 748

Question **3** : Andy has mixed the 10s and 1s. The number should be 306. The order of the blocks is not important, it is the size that is important.

Aki has made 562 and not 652. He has misunderstood place value and put the 10s before the 100s.

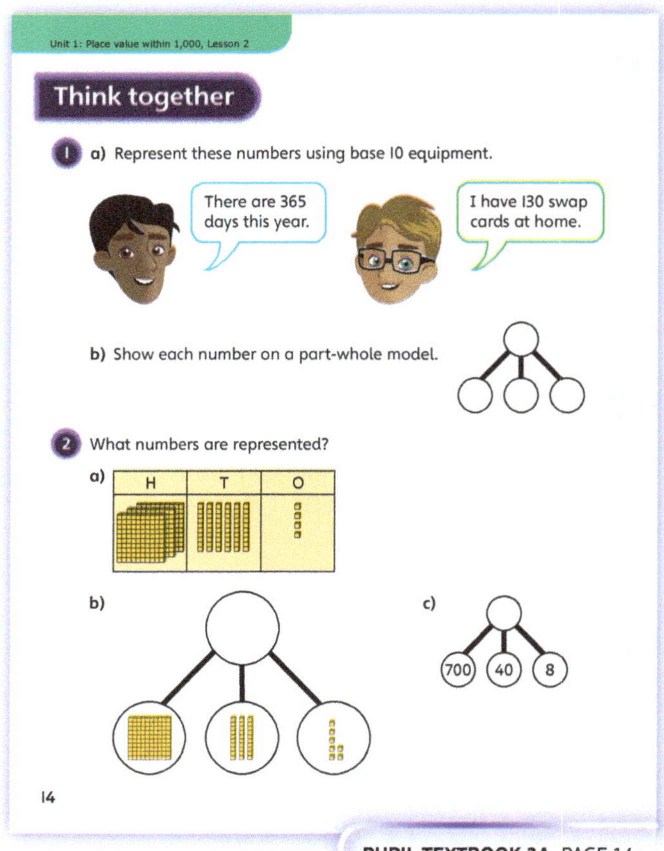

PUPIL TEXTBOOK 3A PAGE 14

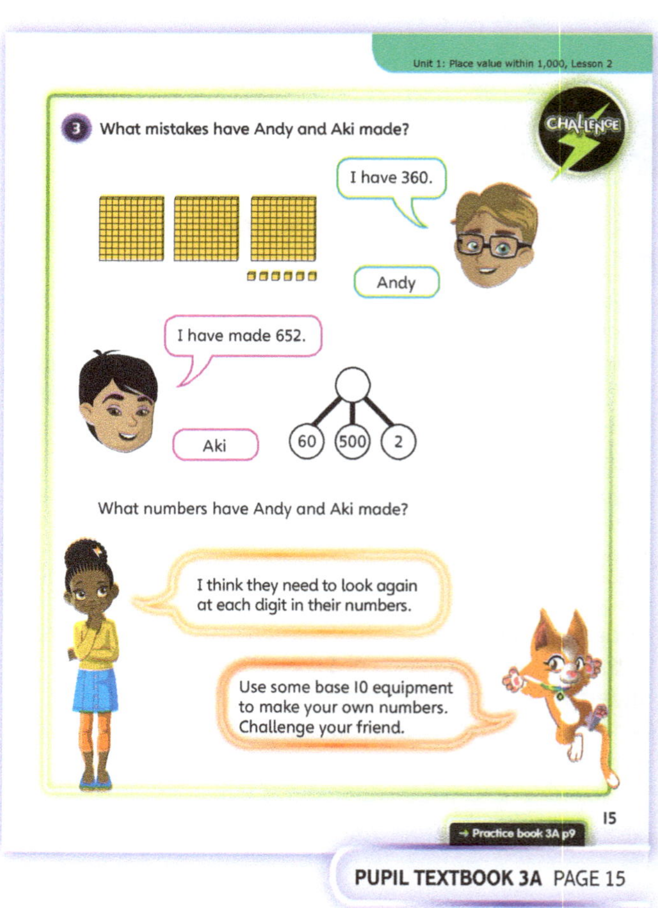

PUPIL TEXTBOOK 3A PAGE 15

Practice

WAYS OF WORKING Independent thinking

IN FOCUS Questions ❶ and ❷ aim to consolidate children's understanding of base 10 representations of 3-digit numbers, reinforcing understanding that a number is made up of 100s, 10s and 1s. Question ❹ develops the idea that there are not always three parts in a part-whole model for 3-digit numbers.

STRENGTHEN In question ❷ children can use base 10 equipment to replace the leaflets. This will help them understand the place value. When blocks are not given in 100s, 10s and 1s order, encourage children to put them in this order.

DEEPEN Question ❺ can be explored further by giving children another card and asking how many different 3-digit numbers they can make. Ask them how they know that they have made all the possible numbers.

Ask whether there are other ways they could represent a number such as 328 (for example, 2 hundreds, 12 tens and 8 ones). Ask them to explain why this is equal to 3 hundreds, 2 tens and 8 ones.

ASSESSMENT CHECKPOINT Use questions ❶ and ❸ to assess whether children can write 3-digit numbers represented by base 10 equipment or part-whole models. Use question ❹ to assess whether children can represent 3-digit numbers using part-whole models.

ANSWERS Answers for the **Practice** part of the lesson appear in the separate **Practice and Reflect answer guide**.

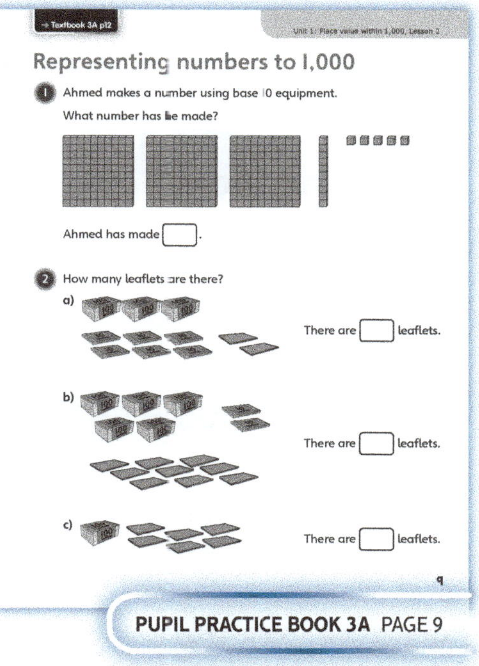

PUPIL PRACTICE BOOK 3A PAGE 9

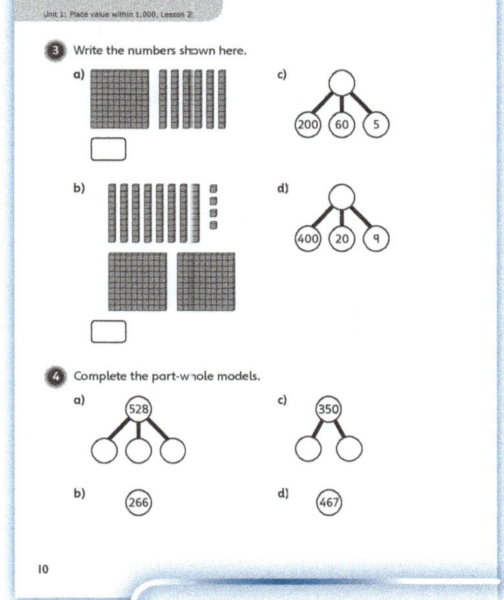

PUPIL PRACTICE BOOK 3A PAGE 10

Reflect

WAYS OF WORKING Pair work

IN FOCUS This question checks understanding of place value. The 10s are given before the 100s and so Ebo has made a mistake.

ASSESSMENT CHECKPOINT Children explain the mistake that has been made. They need to realise that the largest place value comes first – in this case it is the 100s.

ANSWERS Answers for the **Reflect** part of the lesson appear in the separate **Practice and Reflect answer guide**.

After the lesson 

- Can children represent a 3-digit number using base 10 equipment and on a part-whole model?
- Can children work out what numbers are represented by these representations?
- Do children understand that in 3-digit numbers the 100s come first and then the 10s and then the 1s?

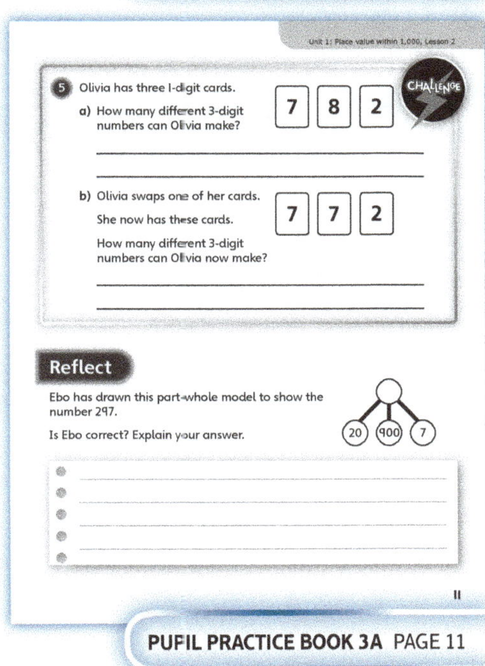

PUPIL PRACTICE BOOK 3A PAGE 11

100s, 10s and 1s ➊

Learning focus

In this lesson, children will represent numbers in place value grids and write 3-digit numbers in multiple ways.

Small steps

→ Previous step: Representing numbers to 1,000
→ **This step: 100s,10s and 1s (1)**
→ Next step: 100s, 10s and 1s (2)

NATIONAL CURRICULUM LINKS

Year 3 Number – Number and Place Value

- Recognise the place value of each digit in a three-digit number (hundreds, tens, ones).
- Identify, represent and estimate numbers using different representations.
- Read and write numbers up to 1,000 in numerals and in words.

ASSESSING MASTERY

Children can organise their work on a place value grid and are able to write the numbers that they represent. Children can show multiple ways of writing 3-digit numbers.

COMMON MISCONCEPTIONS

Children may believe that the order that numbers or base 10 equipment is provided in is the order that the digits will appear in the number. For example, children will think 400 + 8 + 60 is 486. Ask:
- *What does the 8 represent in 486? Which column does it appear in? What does this mean?*
- *How can you write 400 + 8 + 60 differently to show it is 468 and not 486?*

STRENGTHENING UNDERSTANDING

Encourage children to organise their work in a clearly labelled place value grid. Children should place the base 10 equipment in the grid and write down the amount of each that they have. They should then read the number. Ask children to write this as ___ hundreds, ___ tens and ___ ones.

GOING DEEPER

Ask children to write numbers in different ways. For example, can they work out 362 = 300 + 50 + ___? Ask them to explain how they can work out what is missing. Children can use their knowledge from Year 2 that 10 tens make 100, and 10 ones make 10, to help them make up and solve number sentences like this.

KEY LANGUAGE

In lesson: hundreds (100s), tens (10s), ones (1s), place value grid, digit

STRUCTURES AND REPRESENTATIONS

Place value grid, base 10 equipment

RESOURCES

Mandatory: base 10 equipment, place value grids, place value cards

Optional: large laminated place value grid

 In the eTextbook of this lesson, you will find interactive links to a selection of teaching tools.

Before you teach 

- Can children make a 3-digit number using base 10 equipment? Do they know what each digit represents?

Discover

 Pair work

- Question **1** a): *What can you do with Luis's base 10 equipment? Whose number is easier to work out from the picture? What has Lexi done with her base 10 equipment? How does grouping the 100s, 10s and 1s together help?*
- Question **1** b): *What are the headings on a place value grid? What comes first? Does it always come first? Does every column have to have a number or some base 10 equipment in it? What happens if it does not?*

 Question **1** a) will help children realise that Lexi's number is easier to work out as the blocks are ready grouped in 100s, 10s and 1s. This encourages them to group together the 100s, 10s and 1s to work out what number Luis has made, and to see that this is more effective than how Luis has presented it. Question **1** b) encourages children to start to order their work using a place value grid.

 Ensure that each pair has enough base 10 equipment to represent the numbers in the picture. Do not give them the exact amount required to make 241 and 262, so they need to select the correct amount.

Question **1** a): Luis has made 241. Luis has made two hundred and forty one. Lexi has made 262. Lexi has made two hundred and sixty two. Lexi's number is easier to work out as she has organised the base 10 equipment into columns.

Question **1** b): 358 in base 10 equipment on a place value grid: 3 hundreds, 5 tens and 8 ones
430 in base 10 equipment on a place value grid: 4 hundreds, 3 ten and 0 ones

Share

 Whole class teacher led

- Question **1** a): *How has Dexter counted Luis's blocks? Which blocks has he counted first? Why? Could you count the other blocks first?*
- Question **1** b): *Why do the columns help to organise the base 10? Which column does each base 10 block go into?*

 Question **1** a) develops understanding that it is easier to count the 100s, then the 10s and then the 1s, rather than counting the blocks as they come. Question **1** a) also introduces the place value grid as a way of organising the 100s, 10s and 1s in a number, making it easier to read the number.

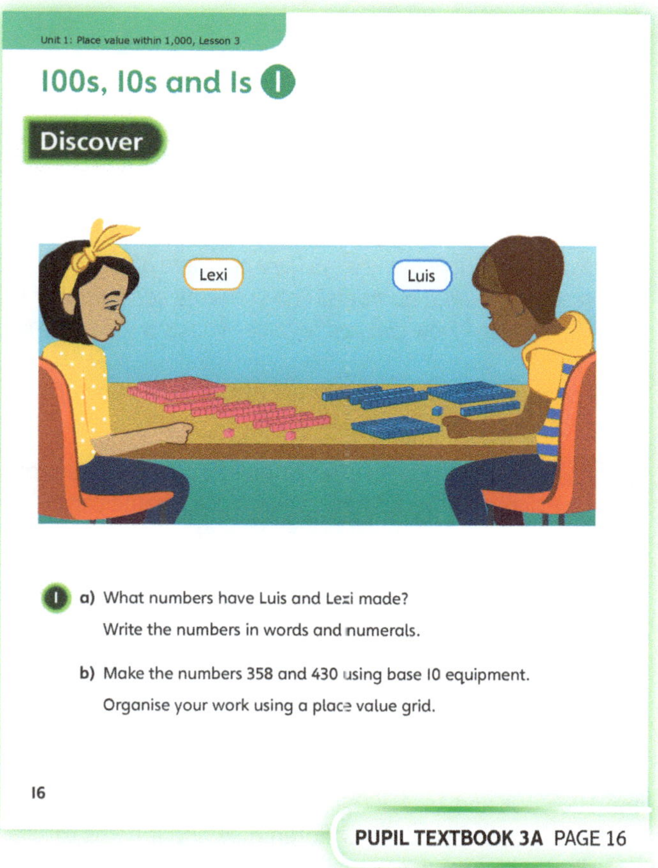

100s, 10s and Is **I**

Discover

I a) What numbers have Luis and Lexi made?
Write the numbers in words and numerals.

b) Make the numbers 358 and 430 using base 10 equipment.
Organise your work using a place value grid.

16

PUPIL TEXTBOOK 3A PAGE 16

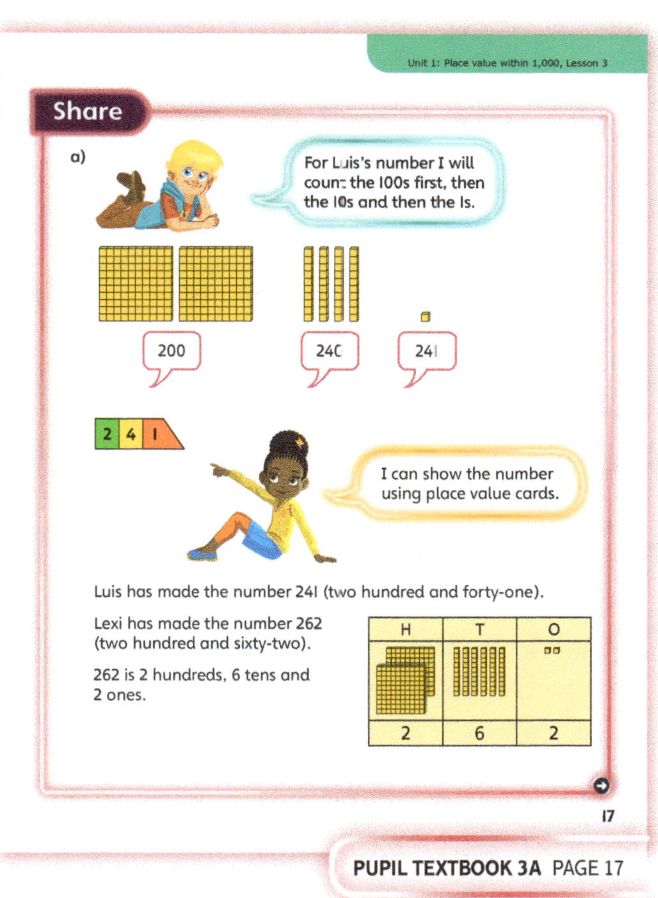

Share

a)

For Luis's number I will count the 100s first, then the 10s and then the Is.

200

240

241

2 4 I

I can show the number using place value cards.

Luis has made the number 241 (two hundred and forty-one).

Lexi has made the number 262 (two hundred and sixty-two).

262 is 2 hundreds, 6 tens and 2 ones.

H	T	O
2	6	2

I7

PUPIL TEXTBOOK 3A PAGE 17

Think together

WAYS OF WORKING Whole class teacher led (I do, We do, You do)

ASK

- Question ❶ : *How can you organise Luis's work?*
- Question ❷ : *How many ways here are there of representing 3-digit numbers? Can you write each number in a different way? How can you organise your work to make it clear?*
- Question ❸ : *How many different ways can you find of representing these numbers?*

IN FOCUS Question ❷ gives examples of different ways of writing 3-digit numbers. These include place value cards, which were not discussed extensively in **Share**. Question ❸ extends the ways given in question ❷ to include use of the addition sign. Encourage children to move between the different ways of writing numbers. Question ❸ also encourages children to see different ways of writing the same number using the addition sign by putting the numbers in a different order.

STRENGTHEN Encourage children to first group the objects into 100s, 10s and 1s. Ask them to count in 100s first and then in 10s and then in 1s. Children may find it helps to write it as an addition statement (such as 238 = 200 + 30 + 8) or on a part-whole model, to enable them to see the make-up of a number. Encourage them to use base 10 equipment to represent the number.

DEEPEN Challenge children to represent numbers to 1,000 in as many different ways as they can. For example, ask how many ways they can show 492. They may start with 400 + 90 + 2 and then move on to 3 hundreds, 19 tens and 2 ones. Ask them to explain why these representations are identical.

ASSESSMENT CHECKPOINT Use question ❶ to assess whether children can use a place value grid to organise their working and show 3-digit numbers. Watch for children who do not group the 100s, 10s and 1s together. Use question ❸ to assess whether they can write a 3-digit number from different ways of expressing the number.

ANSWERS

Question ❶ a): Luis used 2 hundreds, 5 tens and 3 ones. Luis made the number 253.

Question ❶ b): Lexi used 3 hundreds, 3 tens and 4 ones. Lexi made the number 334.

Question ❷: Each number in base 10 equipment:

Question ❷ a): 317: 3 hundreds, 1 ten and 7 ones

Question ❷ b): 650: 6 hundreds, 5 tens and 0 ones

Question ❷ c): 472: 4 hundreds, 7 tens and 2 ones

Question ❷ d): 702: 7 hundreds, 0 tens and 2 ones

Question ❸: Emma – 635

Bella – 870

Danny – 457

Mo – 457

Astrid is correct: Danny and Mo have made the same number.

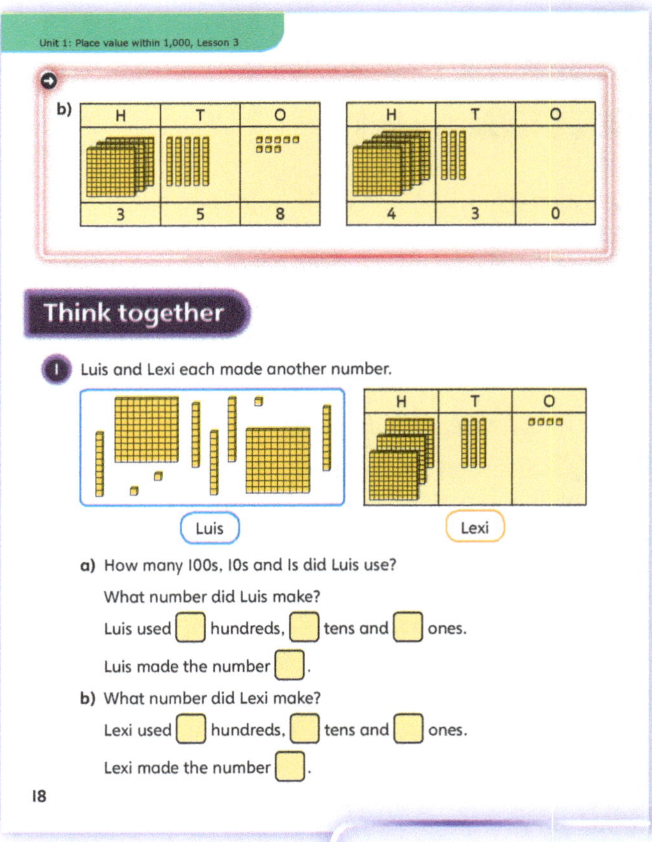

PUPIL TEXTBOOK 3A PAGE 18

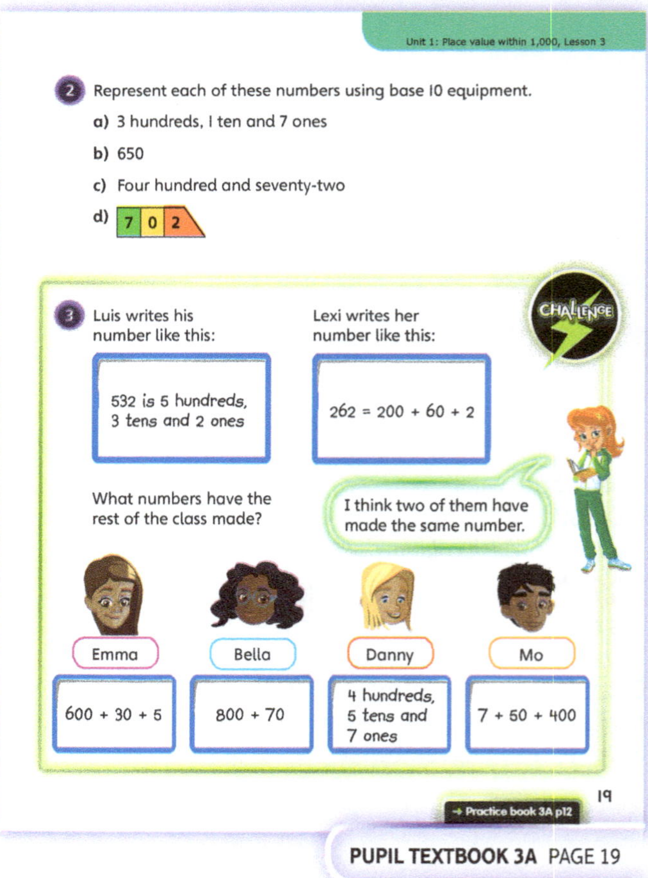

PUPIL TEXTBOOK 3A PAGE 19

Practice

WAYS OF WORKING Independent thinking

IN FOCUS Questions **1** to **3** reinforce the concept that numbers are made up of some 100s, some 10s and some 1s and develop the idea of organising numbers in a place value grid. Question **4** is more abstract, requiring children to complete number sentences. They need to be able to both write the elements that make the number and write the number from its components.

STRENGTHEN To support understanding, children should make the numbers using base 10 equipment and organise them into columns. In question **4**, some children may find it easier to first write the number in a part-whole model before writing it as an addition statement.

DEEPEN Question **6** helps children understand that it is not the order of the numbers that is important but the size of each number. Give children a number, for example 853, and ask them to write this number as an addition in as many ways as they can. When they have written the 100s, 10s and 1s (800, 50 and 3) in all possible orders, encourage them to find new ways of writing the number as an addition (similar to question **6** e).

THINK DIFFERENTLY Question **5** addresses the common misconception that 4 hundreds, 6 ones and 8 tens is equal to 468. Children need to understand that 400 + 6 + 80 is the same as 400 + 80 + 6, which is equal to 486.

ASSESSMENT CHECKPOINT Use questions **2** and **3** to assess whether children can write numbers presented as base 10 in place value grids. Use question **4** to assess whether children can write a 3-digit number in multiple ways: as an addition and as a number of 100s, 10s and 1s. Check that children know the different ways and can swap between them.

ANSWERS Answers for the **Practice** part of the lesson appear in the separate **Practice and Reflect answer guide**.

Reflect

WAYS OF WORKING Pair work

IN FOCUS This question encourages children to find as many ways as possible of writing a 3-digit number. Encourage children to choose a number where all the digits are different. They should make the number using base 10 equipment and show this on a place value grid. They can then think about the different ways they can write their number.

ASSESSMENT CHECKPOINT Look for children organising their work in columns and then look at the different ways that they represent the number. The different ways they represent the numbers should help you determine the depth of their understanding.

ANSWERS Answers for the **Reflect** part of the lesson appear in the separate **Practice and Reflect answer guide**.

After the lesson 

- Can children organise their work in columns?
- In what different ways do children write the numbers?
- Do children know that the order the numbers or base 10 equipment are presented in does not make a difference?

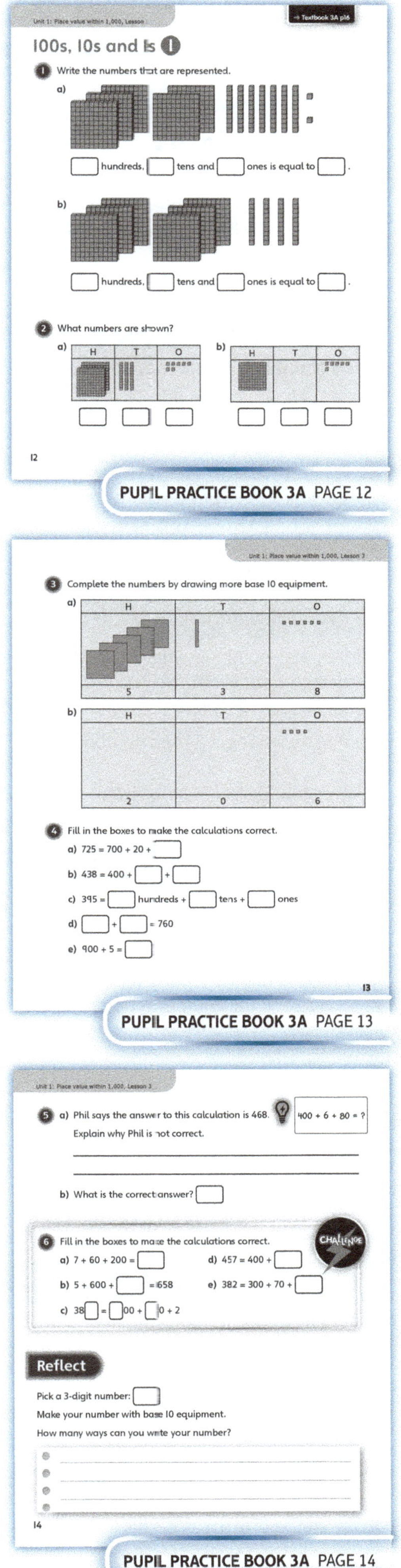

PUPIL PRACTICE BOOK 3A PAGE 12

PUPIL PRACTICE BOOK 3A PAGE 13

PUPIL PRACTICE BOOK 3A PAGE 14

100s, 10s and 1s ②

Learning focus

In this lesson, children will represent numbers in place value grids using counters. They will write numbers represented with counters in a place value grid.

Small steps

→ Previous step: 100s, 10s and 1s (1)
→ **This step: 100s,10s and 1s (2)**
→ Next step: The number line to 1,000 (1)

NATIONAL CURRICULUM LINKS

Year 3 Number – Number and Place Value

- Recognise the place value of each digit in a three-digit number (hundreds, tens, ones).
- Identify, represent and estimate numbers using different representations.
- Read and write numbers up to 1,000 in numerals and in words.

ASSESSING MASTERY

Children can represent numbers on a place value grid using labelled and blank counters. They can identify a number from its representation with counters.

COMMON MISCONCEPTIONS

Children may not know what to do when there is 0 in one of the columns. Ask:
- *How many tens in 605? How do you show this on a place value grid?*
- *What happens if a place value grid has 0 in the ones column? What does this tell you about the number?*

When representing a 2-digit number on an HTO place value grid, a common mistake is to fill in the columns from left to right. Ask:
- *How many 10s and 1s in 76? Which column do these go in on the place value grid?*
- *What goes in the hundreds column?*

STRENGTHENING UNDERSTANDING

Encourage children to organise their work in a clearly labelled place value grid. To strengthen understanding, get children to count aloud as they place counters on the grid. Provide further scaffolding by representing a number with base 10 equipment before using place value counters, so that they understand that a 100 counter represents a 100 block of base 10 equipment and a 10 counter represents a 10 stick of base 10 equipment and so on. This helps children understand that the physical size of the counter is not significant.

GOING DEEPER

Ask children to make as many numbers as they can using five counters. Ask how many numbers have at least one counter in every column of the HTO place value grid. Children should start to develop systematic approaches to answering this type of question.

KEY LANGUAGE

In lesson: hundreds (100s), tens (10s), ones (1s), place value grid

STRUCTURES AND REPRESENTATIONS

Place value grid, part-whole model

RESOURCES

Mandatory: place value counters, blank counters, place value grids

Optional: base 10 equipment, large laminated place value grid

 In the eTextbook of this lesson, you will find interactive links to a selection of teaching tools.

Before you teach 

- Can children make a 3-digit number using base 10 equipment?
- Do they know what each digit in a 3-digit number represents?

Discover

WAYS OF WORKING Pair work

ASK

• Question **1** a): *Which boxes should Toshi use first? Why? What can you use to record your counting?*

IN FOCUS Question **1** a) checks whether children can work out how many 100s, 10s and 1s make up a number, as in the previous lesson. Some children may try to start with boxes of 10 instead of 100. Discuss why it is better to start with the 100s (for example, it is quicker and you need fewer boxes). Question **1** b) further develops the concept of working out the number of 100s, 10s and 1s by introducing place value counters on a place value grid.

PRACTICAL TIPS Ensure that children have sufficient counters to be able to make the numbers. Encourage them to draw a place value grid for question **1** b). If possible, use place value counters numbered 1, 10 and 100, although blank counters can be used if necessary, with different colours for 1, 10 and 100.

ANSWERS

Question **1** a): Toshi needs 2 boxes of 100 bulbs, 1 box of 10 bulbs and 5 single bulbs.

Question **1** b):

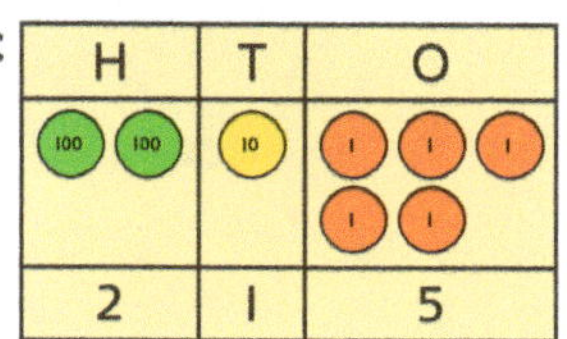

H	T	O
100 100	10	1 1 1 1 1
2	1	5

Share

WAYS OF WORKING Whole class teacher led

ASK

• Question **1** a): *Which size box did Ash start with? Why? How has Astrid recorded her thinking? How is Astrid's method different?*
• Question **1** b): *How does the number of counters relate to the number of boxes?*

IN FOCUS The two parts of the question develop from what children did in the previous lesson into use of place value counters. Use Ash and Astrid's methods to draw out the key points from the lesson. Children should realise that they need to start with the boxes of 100s, then the 10s and then the 1s. Ensure that the connection between the number of boxes and number of counters in the place value grid in question **1** b) is made explicit. For example, there are five boxes of 100 and five 100 counters on the place value grid.

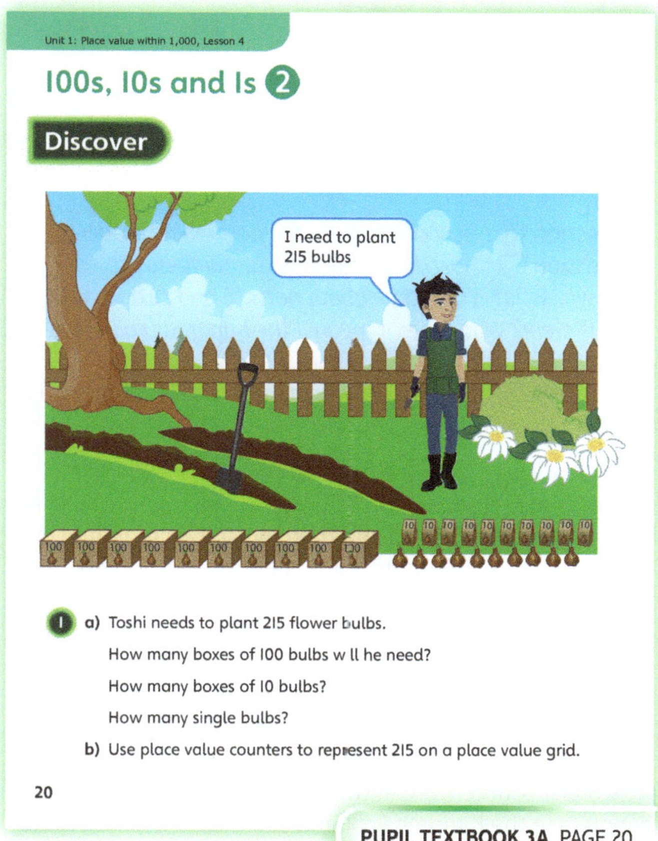

PUPIL TEXTBOOK 3A PAGE 20

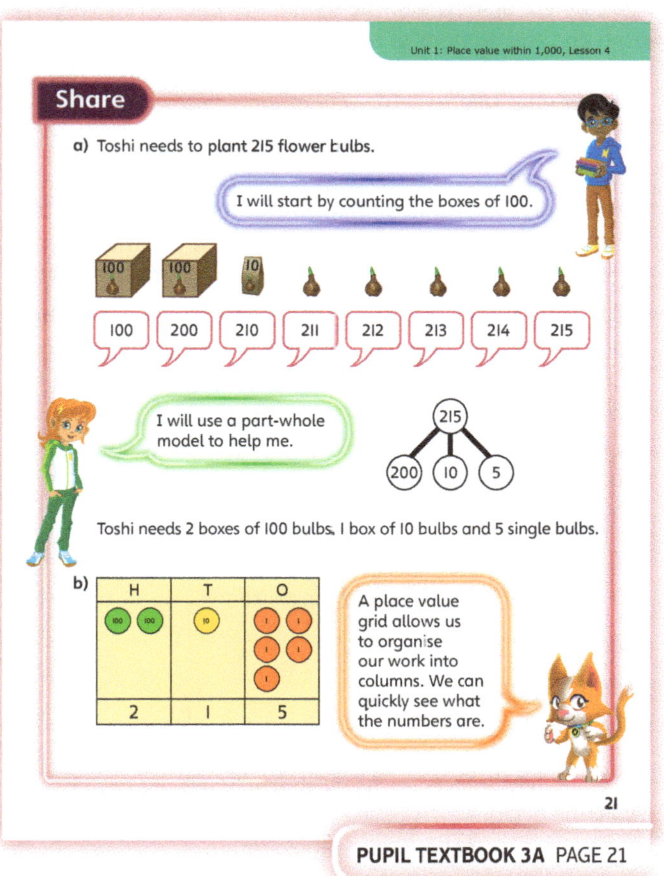

PUPIL TEXTBOOK 3A PAGE 21

Think together

WAYS OF WORKING Whole class teacher led (I do, We do, You do)

ASK

- Question **1** : *How can you use the part-whole model?*
- Question **2** : *How do you know what numbers are shown? What is the same about parts b) and c), and what is different? What does it mean if the hundreds column is blank?*
- Question **3** a): *Which of these numbers have counters in all the columns? How can you tell?*

IN FOCUS Question **2** asks children to work out what numbers are represented, starting with a number where all three digits are non-zero and then moving on to numbers where there are no counters in a column. Question **3** includes all the representations of 3-digit numbers that children have encountered in this unit and shows how they can all be represented on a place value grid. Question **3** b) moves from place value counters, which are labelled with the counter value, to using plain counters in a place value grid. It is important for children to see that what matters is how many counters are in the column, for example 6 blank counters in the hundreds column still means 600.

STRENGTHEN In question **1**, represent the seeds with base 10 equipment. Ask children to count aloud in 100s first until they get to the correct number of 100s and then count in 10s and then count in 1s. As they count ask them to select the correct base 10 equipment. Show explicitly the connection between the number of base 10 equipment used and the number of counters on the place value grid. Then ask children to count the counters in the same way as they counted the base 10 equipment.

DEEPEN Ask children to make as many numbers as they can using eight counters. Ask whether they can make a number with the same digit in each column, explaining their answer. Ask how many numbers they can make with 4 tens. Encourage them to investigate the numbers they can make.

ASSESSMENT CHECKPOINT Use question **2** to assess whether children can write numbers represented with counters in a place value grid. Use question **3** to assess whether children can represent numbers with counters on a place value grid. Check that they understand what to do when there are no counters in a particular column.

ANSWERS

Question **1** : Faisal needs 3 packs of 100 seeds, 9 packs of 10 seeds and 2 single seeds.

Question **2** a): 415

Question **2** b): 630

Question **2** c): 503

Question **2** d): 23

Question **3** a): 627, 143, 150, 606, 85 represented on place value grids with counters

Question **3** b): Meg made 627

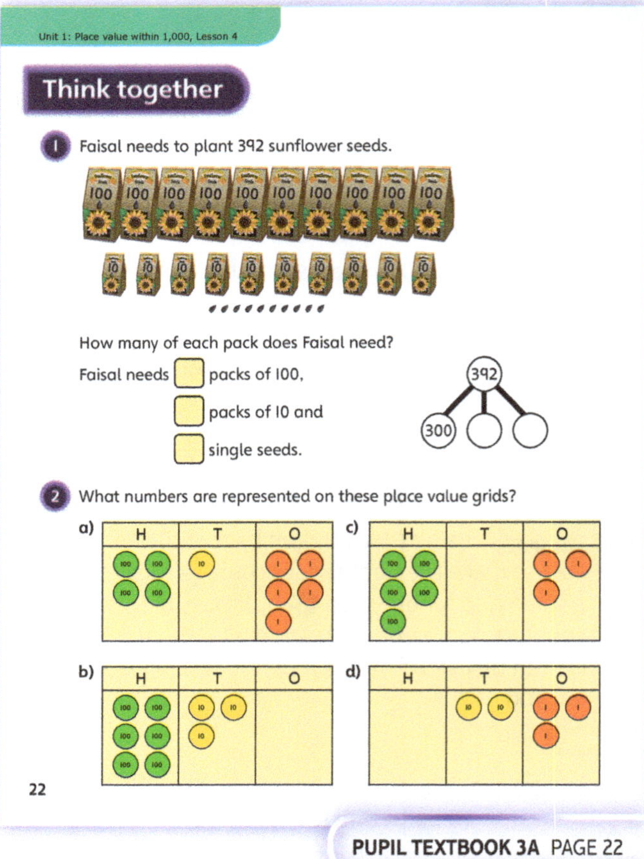

PUPIL TEXTBOOK 3A PAGE 22

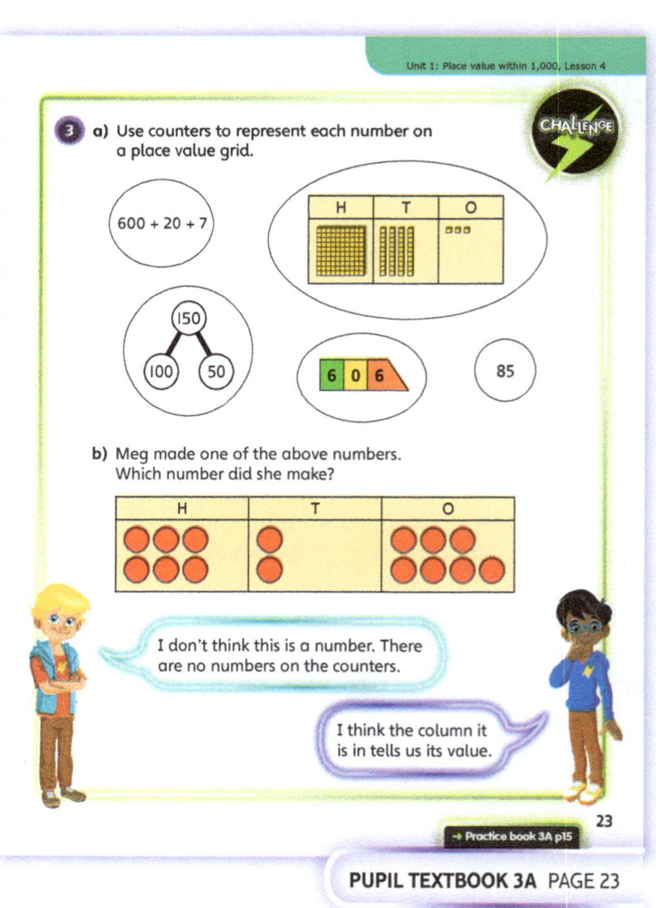

PUPIL TEXTBOOK 3A PAGE 23

Practice

WAYS OF WORKING Independent thinking

IN FOCUS In question **1** children are given place value counters in place value grids, including examples where there are 0 counters in certain columns. Question **2** then turns the question round, giving children the number and asking them to make it using place value counters. This also includes a number where children need to leave one column empty. Question **3** develops this further by using different representations of numbers. In question **5** children have to start using the knowledge that ten 10s is equal to 100 and use this to explain why the two numbers are identical.

STRENGTHEN To strengthen understanding, use base 10 equipment alongside the counters. In question **1**, ask children to create the place value grids using base 10 equipment instead of counters. Then arrange the base 10 equipment from the place value grid in a row, 100s then 10s then 1s. Ask children to count aloud in 100s, then 10s, then 1s. Now ask children to count the counters in the same way.

DEEPEN Ask children to make numbers using seven blank counters. Ask how they can be confident that they have found all the possible numbers. They should be able to develop a strategy for finding all the numbers.

THINK DIFFERENTLY In question **4**, children start to use blank counters instead of place value counters. For children who struggle with this concept, create Tim's place value grid using place value counters. Then replace each place value counter with a blank counter, explaining that its value stays the same because the counter is still in the same column.

ASSESSMENT CHECKPOINT Use questions **1** and **2** to assess whether children can work out what numbers are represented by place value counters on a place value grid, and whether they can represent numbers on a place value grid with place value counters. Assess whether children understand what to do when there is 0 in a particular column. Use question **4** to assess whether children can use blank counters in a place value grid.

ANSWERS Answers for the **Practice** part of the lesson appear in the separate **Practice and Reflect answer guide**.

Reflect

WAYS OF WORKING Pair work

IN FOCUS Children record the numbers they can make using six blank counters and reflect on any strategy that they use to ensure they get all the numbers. For example, do they just do trial and error, or do they develop a systematic approach?

ASSESSMENT CHECKPOINT Check whether children understand what happens when there are no counters in a particular column, particularly the first column.

ANSWERS Answers for the **Reflect** part of the lesson appear in the separate **Practice and Reflect answer guide**.

After the lesson 

- Can children represent numbers (written in different forms) with counters on a place value grid?
- Do children know how many 100s, 10s and 1s there are in numbers represented on a place value grid? Can they write the numbers?
- Do children understand what it means to have 0 in one of the columns?
- Do they know when they need to record the 0 and when they do not?

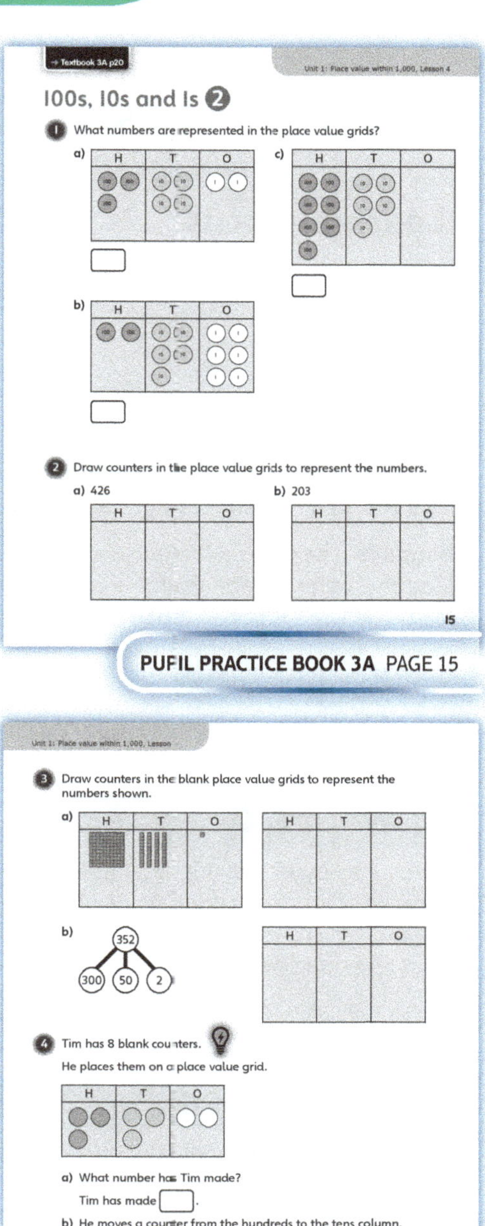

PUPIL PRACTICE BOOK 3A PAGE 15

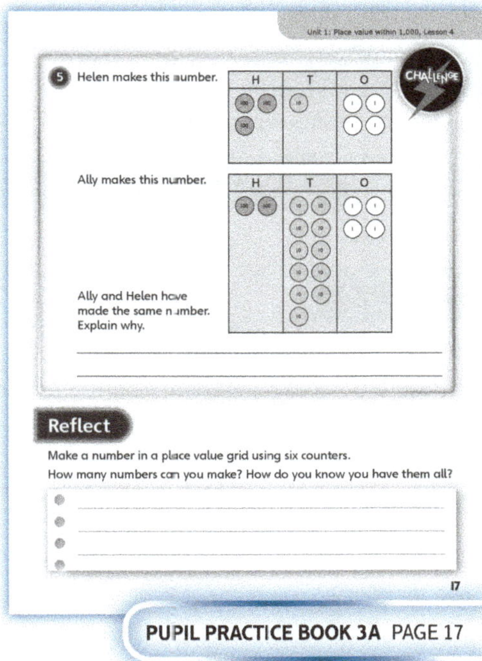

PUPIL PRACTICE BOOK 3A PAGE 16

PUPIL PRACTICE BOOK 3A PAGE 17

The number line to 1,000 ①

Learning focus

In this lesson, children will identify values and mark points on number lines that go up in 100s, 10s and 1s.

Small steps

→ Previous step: 100s, 10s and 1s (2)
→ **This step: The number line to 1,000 (1)**
→ Next step: The number line to 1,000 (2)

NATIONAL CURRICULUM LINKS

Year 3 Number – Number and Place Value

- Recognise the place value of each digit in a three-digit number (hundreds, tens, ones).
- Identify, represent and estimate numbers using different representations.
- Read and write numbers up to 1,000 in numerals and in words.

ASSESSING MASTERY

Children can work out whether a number line goes up in 100s, 10s or 1s. They can identify values on a number line and mark on given values.

COMMON MISCONCEPTIONS

Children may expect all number lines to go up in 100s or fail to identify correctly what the number line does go up in. Encourage children to count aloud while pointing to each mark. Ask:

- *What will you count up in? Have you said the correct end point?*
- *What else could you count up in?*

Children may also mark numbers such as 980 half-way between 900 and 1,000. Ask:

- *Is 980 closer to 900 or 1,000? How could you check?*

STRENGTHENING UNDERSTANDING

Encourage children to label numbers at every mark, to reinforce counting in 100s, 10s and 1s. Counting aloud to check whether a number goes up in 100s, 10s and 1s will also help children. Consider also using base 10 equipment next to the numbers on a large number line, to help children understand the meaning of the numbers.

GOING DEEPER

Using a number line going up in 100s from 0 to 1,000, ask children to mark points such as 50, 110, 295, 500 and 770. Ask what happens if the number line does not have any markings on. Discuss whether they can still mark the positions of these numbers.

KEY LANGUAGE

In lesson: estimate, hundreds (100s), tens (10s), ones (1s), number line

STRUCTURES AND REPRESENTATIONS

Number line

RESOURCES

Mandatory: number lines

Optional: base 10 equipment, counting stick

 In the eTextbook of this lesson, you will find interactive links to a selection of teaching tools.

Before you teach 

- Can children work out missing values on number lines that go in 10s and 1s within 100?
- Can children count up in 100s, 10s and 1s within 1,000?

Discover

 Pair work

- Question **1** a): *How long is the course? What is the distance between the markings? How can you work out how far each boat has travelled?*

IN FOCUS In this question, children extend the number line to 1,000, going up in 100s in the context of a boat race. Question **1** a) asks children to work out where each boat is and how far it has travelled. This involves their knowledge of counting up in 100s. Boat C also involves them estimating a position half-way between two known marks.

PRACTICAL TIPS The number line is extended to 1,000. Consider laying out the classroom, hall or playground so that there are 11 posts or chairs, labelled with every 100 m from 0 m to 1,000 m. Get children to stand in pairs where they think the boats are. Do not give them any additional information that is not in the picture. A counting stick could also be used, where the markings go up in 100s.

ANSWERS

Question **1** a): Boat A has travelled 300 metres.
Boat B has travelled 600 metres.
Boat C has travelled about 750 metres.

Question **1** b): Point to 900 metres for boat D.

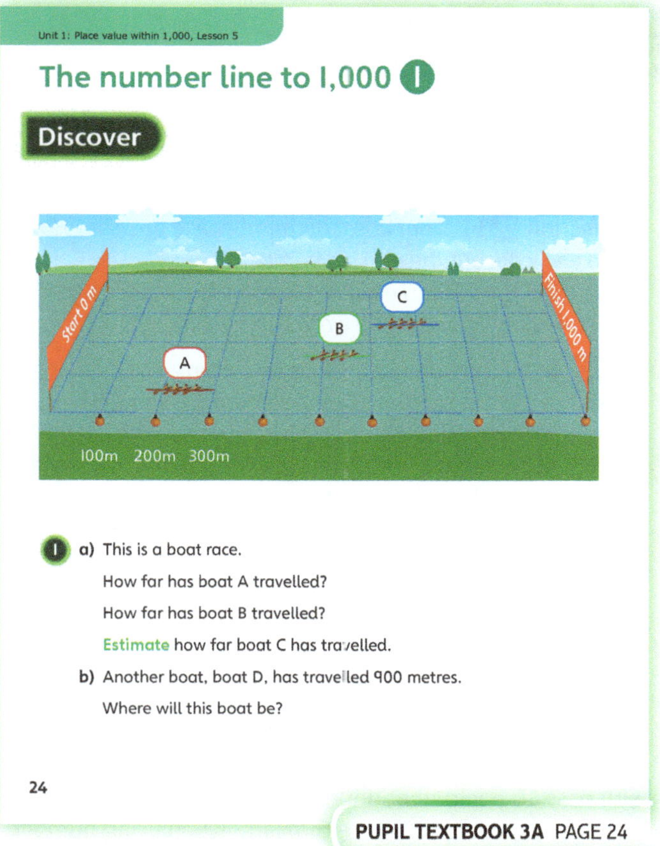

PUPIL TEXTBOOK 3A PAGE 24

Share

 Whole class teacher led

- Question **1** a): *Can you see how the number line matches the boat course? What does the number line go up in? What is half-way between each pair of marks on the number line?*
- Question **1** b): *How can you work out where boat D is? Did you count from the left? Can you work it out by counting from the right?*

IN FOCUS Question **1** b) encourages children to identify steps of 100 and to use them to solve a problem. Ensure that children understand that the marks are every 100 metres. Get them to check by counting up in 100s from 0 to 1,000, recapping the work that they did in the first lesson. Discuss Sparks's comments about 750 lying half-way between 700 and 800 and why 750 is an estimate. Ask children to find half-way points between any other two marks.

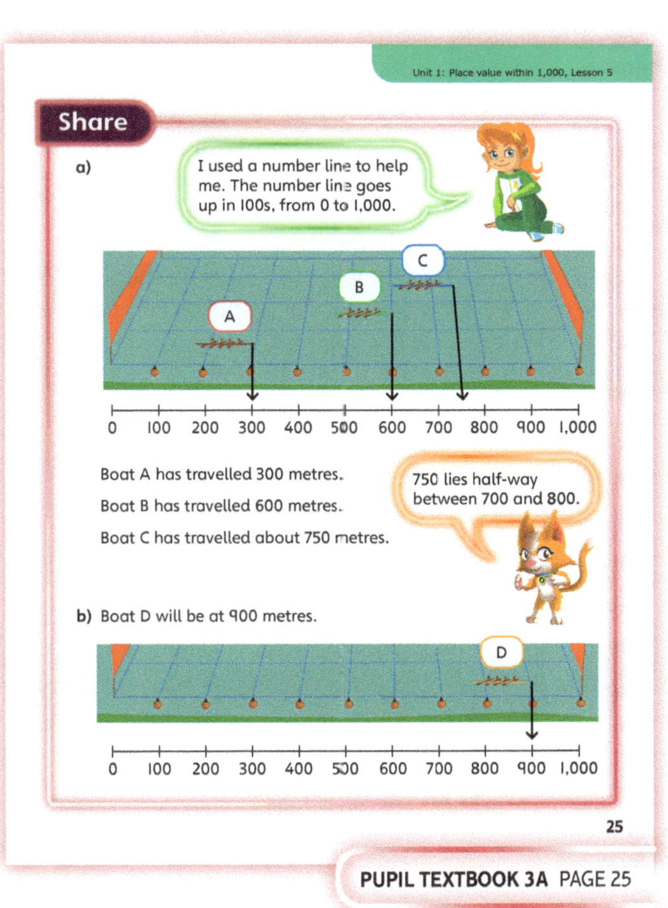

PUPIL TEXTBOOK 3A PAGE 25

Think together

WAYS OF WORKING Whole class teacher led (I do, We do, You do)

ASK

• Question **2** b): *What does this number line go up in? How can you work out the missing numbers?*
• Question **2** a): *Does each arrow point to a mark on the number line? How can you work out the number?*

IN FOCUS All the questions focus on finding and marking 3-digit values on number lines. Question **1** and **2** both use number lines that go up in 100s. Question **1** uses the familiar context of the boat race from **Discover**, while question **2** includes values that do not lie on marks on the number line. Encourage children to check first that the number line does increase in 100s. Question **3** extends work from Year 2, where children met number lines that go up in 10s and 1s, to include 3-digit numbers. Children should try and work out what the number lines go up in by counting and checking that the start and end points match.

STRENGTHEN Count aloud with children from 0 to 1,000 in 100s, pointing to each mark. You may additionally support with base 10 equipment underneath each number to show that each time it is 100 more. Similarly, when using number lines that go up in 10s or 1s, count aloud while pointing to each mark.

DEEPEN Ask children to draw their own number line from 0 to 1,000 and place numbers such as 10, 375, 402, 798 and 925 on it. Ask them to describe where the numbers will be before marking them on the number line. For example, do they know that 375 is three-quarters of the way between 300 and 400?

ASSESSMENT CHECKPOINT Use question **2** a) to assess whether children can identify the position of arrows on 100s and at half-way points between two numbers. Use question **3** to assess whether children can work out what the lines go up in (for example, 10s). Children should also be able to draw an arrow to particular points on number lines that go up in 100s, 10s and 1s.

ANSWERS

Question **1** a): 100, 200, 500, 700, 900

Question **1** b): Boat A has travelled 100 metres.

Question **1** c): Boat C has travelled 800 metres.

Question **1** d): Boat B has travelled about 450 metres.

Question **2** a): 100, 550, 800

Question **2** b): Children point to 300, 500 and 990.

Question **3** a): 270, 271, 272, 274, 275, 276, 277, 278, 279,
210, 220, 230, 240, 260, 270, 280, 290, 300

Question **3** b): Children point to 275 on each line.

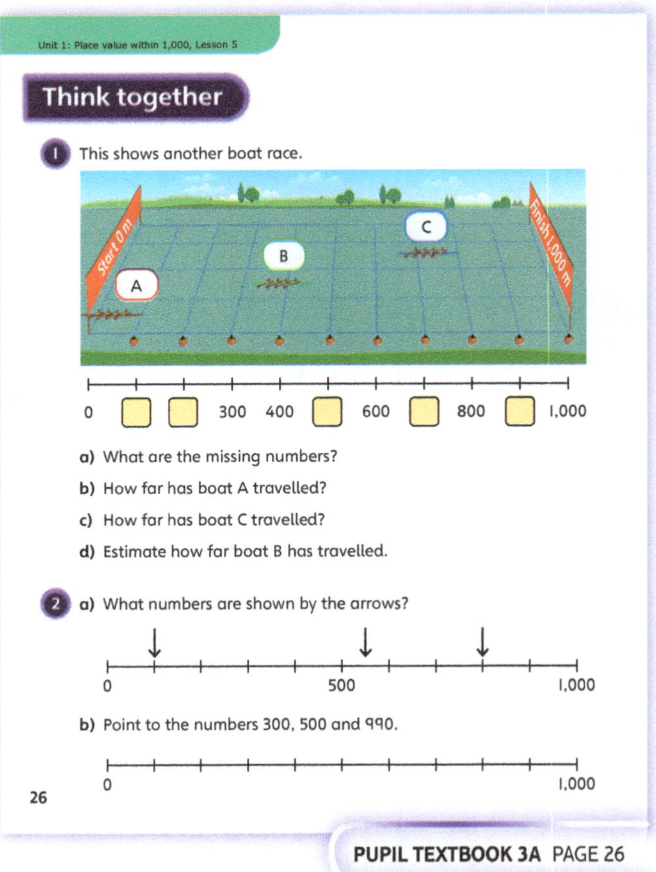

PUPIL TEXTBOOK 3A PAGE 26

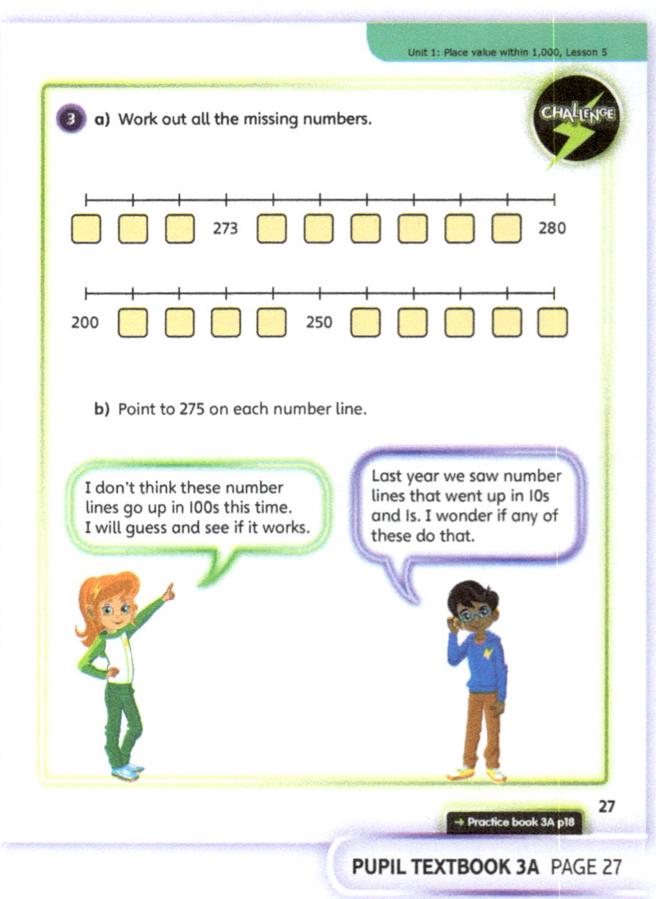

PUPIL TEXTBOOK 3A PAGE 27

Practice

WAYS OF WORKING Independent thinking

IN FOCUS In questions **2** and **3** children need to reason with number lines, finding missing values and putting on values. Children first need to work out what the number lines go up in, so that they can work out the correct position of the number. Question **6** requires an alternative way of thinking, as children have to think about the different possible number lines that could exist with 500 in the middle.

STRENGTHEN Ask children what the number lines could go up in. Ask how they could check whether they are correct. Count aloud with children to help them check. Encourage children to write on the values at each mark to help their understanding. A counting stick may be useful in questions **2** and **3**.

DEEPEN Ask children to add their own arrows to the number line in question **4** and challenge their friend. This encourages them to think about what makes a number easy or difficult to mark.

THINK DIFFERENTLY Question **5** provides a blank number line for the first time, where only one value is marked. Children need to reason about what a missing number could be. Children should start to realise that the missing number will depend on where the number line ends or what it goes up in. So Isla could be right or she might not be; it will depend on what it goes up in.

ASSESSMENT CHECKPOINT Use question **2** to assess whether children can identify missing numbers on number lines that go up in 100s and 10s. Questions **3** and **4** assess whether they can identify numbers on number lines that go up in 100s, 10s and 1s, including half-way numbers on lines going up in 100s or 10s, and draw arrows to numbers.

ANSWERS Answers for the **Practice** part of the lesson appear in the separate **Practice and Reflect answer guide**.

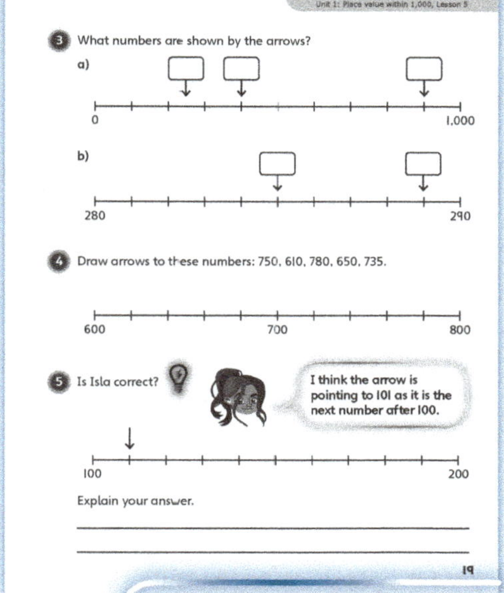

PUPIL PRACTICE BOOK 3A PAGE 18

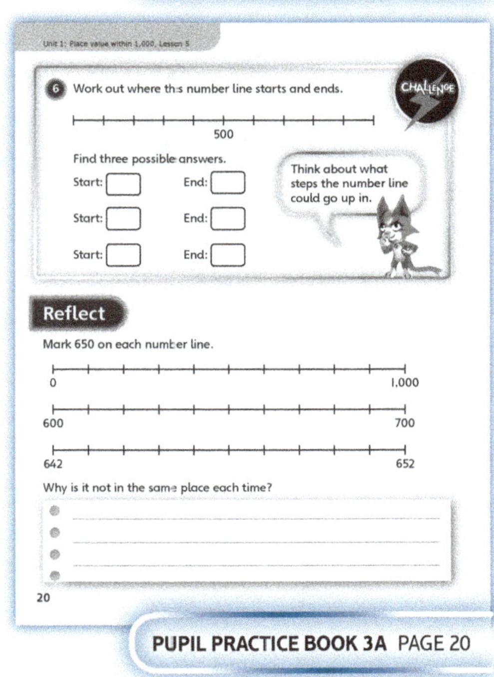

PUPIL PRACTICE BOOK 3A PAGE 19

Reflect

WAYS OF WORKING Pair work

IN FOCUS Marking 650 on each number line practises the main part of the lesson. However, children then have to reason why the number will not be in the same position each time, showing their understanding of number lines that go up in different steps and with different start and end points.

ASSESSMENT CHECKPOINT Check that children can identify what each number line goes up in and can then place 650 accurately on each line.

ANSWERS Answers for the **Reflect** part of the lesson appear in the separate **Practice and Reflect answer guide**.

After the lesson 

- Can children identify whether a number line goes up in 100s, 10s or 1s?
- Can they mark points on the number line or identify the values that arrows are pointing to?
- Do children know how to find numbers like 550 on a number line that goes up in 100s?

PUPIL PRACTICE BOOK 3A PAGE 20

The number line to 1,000 ②

Learning focus

In this lesson, children will begin to understand where numbers lie on a number line. They will identify numbers that lie between two points.

Small steps

→ Previous step: The number line to 1,000 (1)
→ **This step: The number line to 1,000 (2)**
→ Next step: Finding 1, 10 and 100 more or less

NATIONAL CURRICULUM LINKS

Year 3 Number – Number and Place Value

- Compare and order numbers up to 1,000.
- Read and write numbers up to 1,000 in numerals and in words.
- Recognise the place value of each digit in a three-digit number (hundreds, tens, ones).

ASSESSING MASTERY

Children can work out points that lie between two points on a number line. They can also identify start and/or end points on a number line given a set of points in between.

COMMON MISCONCEPTIONS

Children may often just look at the first digit to work out if a number lies between two points on a line. For example, they identify 72 as lying between 700 and 800 because it starts with the digit 7. Ask:
- *What does the 7 represent in 72? What does the 7 represent in 700?*

Making the numbers with base 10 equipment may also help them see that 72 does not lie between 700 and 800.

STRENGTHENING UNDERSTANDING

Ask children to make the start and end numbers using base 10 equipment. They can then make each of the given numbers and compare to work out whether the new number lies between the two points.

GOING DEEPER

Ask children to draw their own number line between two points and identify numbers that lie between the two points. Alternatively, give children a set of numbers for them to work out two possible numbers that the numbers could lie between. Ask how they can work out where the points would go.

KEY LANGUAGE

In lesson: number line, halfway

Other language to be used by the teacher: hundreds (100s), tens (10s), ones (1s)

STRUCTURES AND REPRESENTATIONS

Number line

RESOURCES

Mandatory: number lines

Optional: base 10 equipment, counting sticks

 In the eTextbook of this lesson, you will find interactive links to a selection of teaching tools.

Before you teach 

- Can children work out missing values on number lines that go up in 100s, 10s or 1s within 1,000?
- Do children know the place value of each digit in a number?

Discover

 Pair work

ASK

- Question ① a): *What number is on the card? Is 500 greater than 0? Is 500 less than 1,000?*
- Question ① b): *Where does the number line start and end? What numbers could go in between?*

IN FOCUS Each number line in the **Discover** is different. The aim of both questions ① a) and ① b) is for children to start to get a feel for where numbers lie on a number line.

PRACTICAL TIPS This lesson builds on the previous lesson by looking more at open number lines. An open number line gives children more freedom, as it can be blank or just have the start or end point. There are no markings indicating in-between points. You might find it useful to have the four lines from the picture as string around the classroom with numbers pegged on the ends. You could encourage children to come and peg numbers to these lines during the lesson.

ANSWERS

Question ① a): You can peg 500 to lines A and C.

Question ① b): You can peg any three numbers from 550 to 650 inclusive.

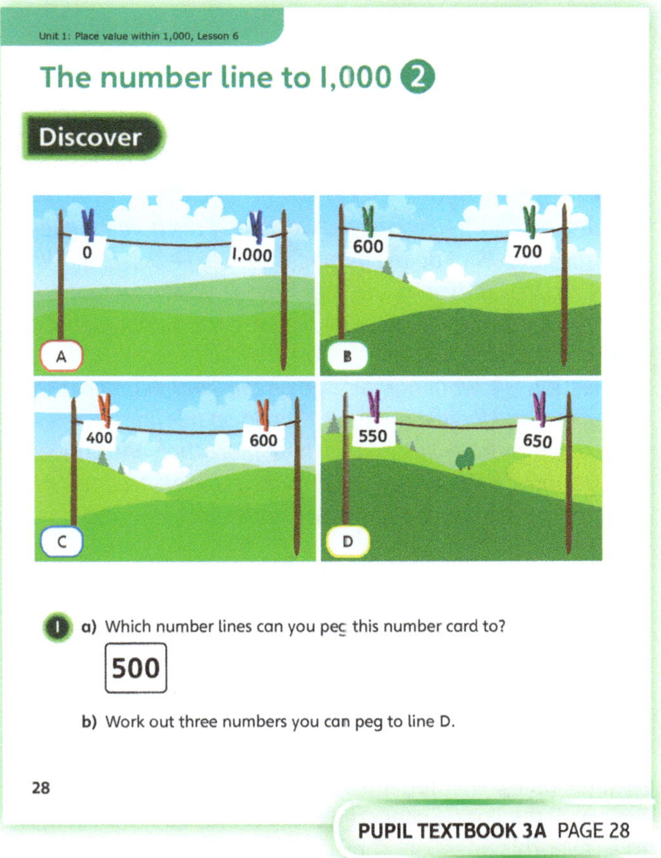

PUPIL TEXTBOOK 3A PAGE 28

Share

 Whole class teacher led

ASK

- Question ① a): *Can you see which lines will contain the number 500? How do you know?*
- Question ① b): *What number lies in the middle of this number line? Can you place the numbers you got on the number line? How do you know that they lie there?*

IN FOCUS In question ① a) children need to realise that they must look at the start and end points of each number line to see whether the number 500 will go on the line. In question ① b) the markers for the 10s between 550 and 650 have been added to help children identify numbers that lie on the number line and say where they are positioned.

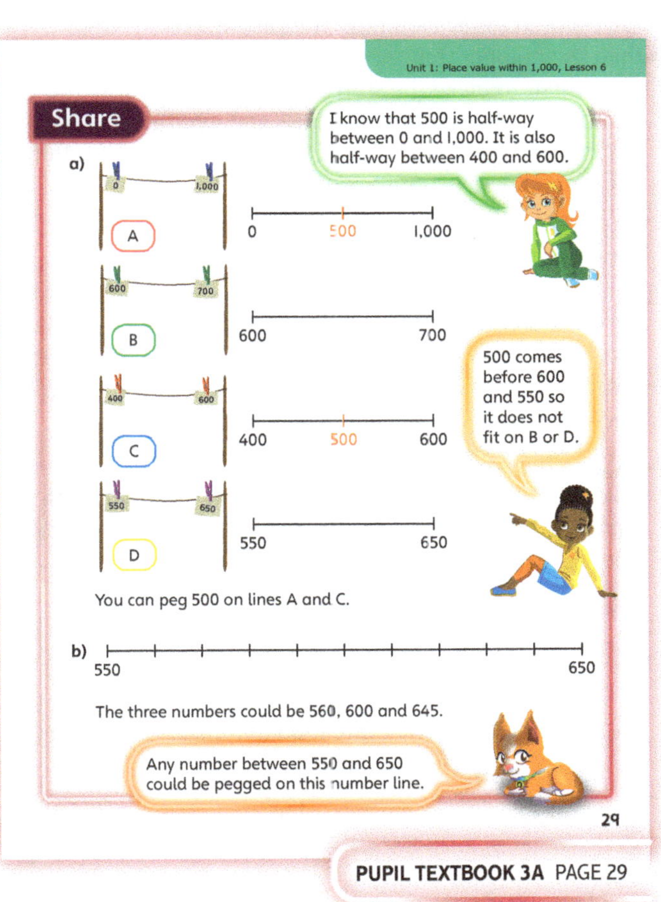

PUPIL TEXTBOOK 3A PAGE 29

Think together

WAYS OF WORKING Whole class teacher led (I do, We do, You do)

ASK

- Question **1** a): *How can you work out if the number goes on the line? What do the numbers have to lie between? What will you look at first on the line? How do you know that 400 does not go on the line? What digit will the numbers that can go on this line start with?*
- Question **2**: *What numbers go on this line? What are the smallest and largest numbers that go on this line? Where could the number line end? Is there more than one answer?*
- Question **3**: *What strategy can you use to work out whether or not there is a number that can appear on each line?*

IN FOCUS Question **2** is an open-ended question, where children work out the end point of the line by looking for the greatest number in the list. Some children will suggest that 560, the greatest number, should be the end point, and some may say the next multiple of 100 (so, 600). Discuss how there could be multiple end points for this number line, as it just needs to be as large as the largest number.
In question **3** children are asked to find a number that lies on all the lines. Encourage them to work out a strategy to answer this question. They could start by drawing the longest number line and marking all the other lines on this line. This will help them to see if there is a number where all the lines overlap.

STRENGTHEN To strengthen understanding in question **1**, ask children to say aloud the start and end points and then the number that they are looking at. Use base 10 equipment to make the start and end points and all the numbers in the question. Then ask children to make another number in base 10 equipment that lies between the start and end points.

DEEPEN Ask children to draw their own number line between two points, for example, 300 and 400, and to mark on three numbers that lie on the line. Ask what happens to the positions of the numbers if the start number is made smaller, for example, 200. They should be able to say that the numbers will all lie on the right-hand half of the line. Ask what happens if the end point changes.

ASSESSMENT CHECKPOINT Use question **1** to assess whether children know which numbers lie between any two given numbers. Also check that they can work out approximately where the numbers lie.

ANSWERS

Question **1** a): 220, 250, 275, 222

Question **1** b): 762, 755

Question **2**: The end number of the number line could be 560 or greater.

Question **3**: 600 lies on every line.

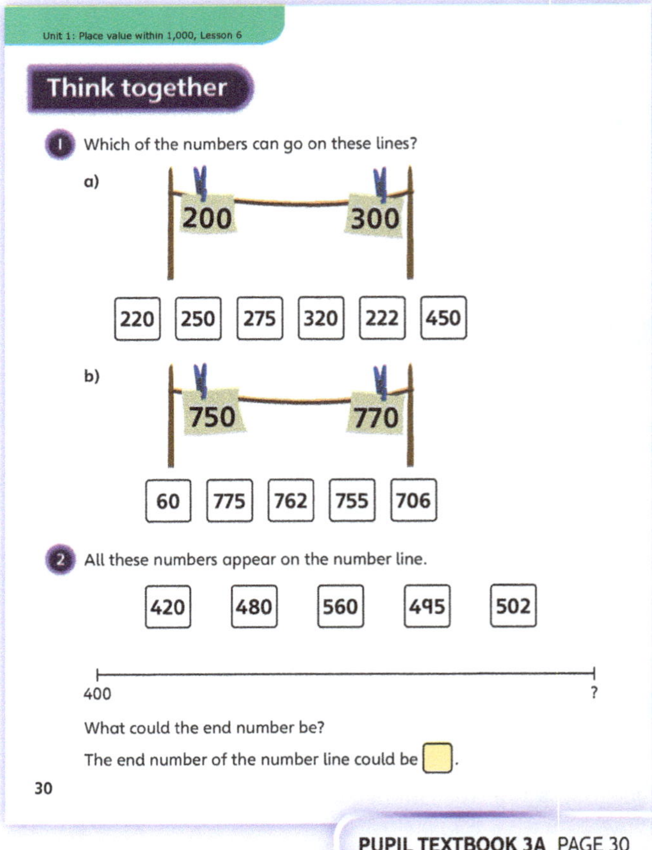

PUPIL TEXTBOOK 3A PAGE 30

PUPIL TEXTBOOK 3A PAGE 31

Practice

WAYS OF WORKING Independent thinking

IN FOCUS Questions **1** and **2** help children develop a sense of where numbers lie on a number line. Question **1** parts a) and b) give them the division of the number line and questions **1** c) and **2** then just give the start and end points. Question **1** includes a number (55) that starts with the same digit as the start point of the number line (500) to encourage them to look at the place value of the first digit as well as the digit itself.

STRENGTHEN In question **1** parts a) and b), support children by asking them to count aloud and see if they can work out what each number line goes up by. In the other questions, ask them to use base 10 equipment to make the numbers. This will help them see the size of the start and end numbers and this should help them work out numbers that lie in between. Make explicit reference to the starting digits and the number of digits that their number must contain.

DEEPEN In question **3**, ask children to mark the numbers on the number line for their suggested start and end numbers. Ask whether the numbers would lie in the same place if the end number was 1,000. Ask how the positions of the numbers would change.

THINK DIFFERENTLY Question **3** challenges thinking by asking children to work out where a number line could start and end given a set of numbers that lie on a particular number line. This makes them consider all the numbers on the line.

ASSESSMENT CHECKPOINT Use question **1** to assess whether children can identify which points from a list lie between two numbers. Use question **2** to assess whether children can list some numbers that lie between two points on a number line.

ANSWERS Answers for the **Practice** part of the lesson appear in the separate **Practice and Reflect answer guide**.

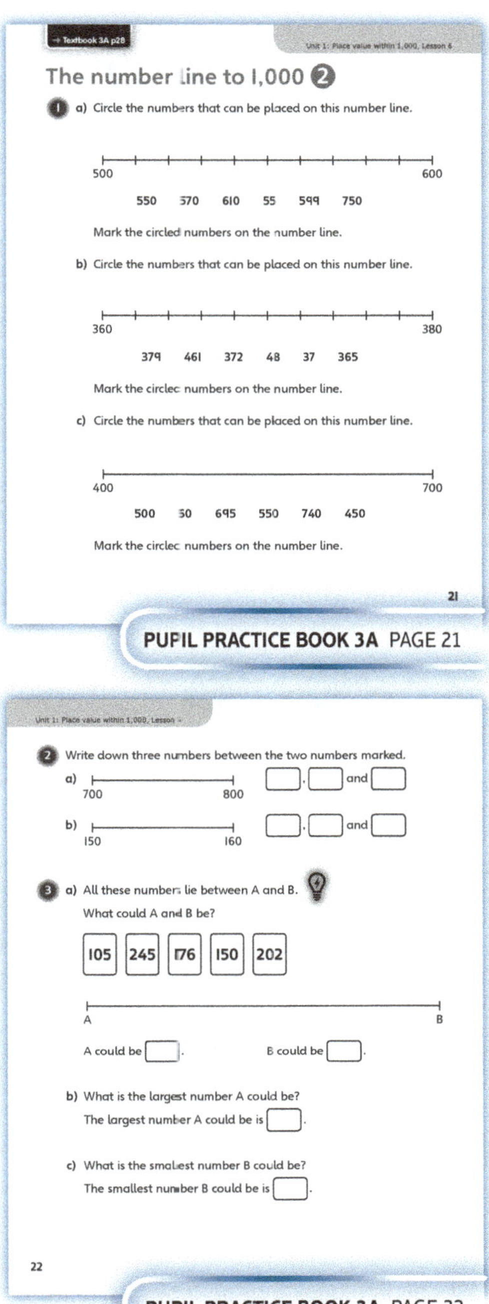

PUPIL PRACTICE BOOK 3A PAGE 21

PUPIL PRACTICE BOOK 3A PAGE 22

Reflect

WAYS OF WORKING Pair work

IN FOCUS Children have to work out the start and end points of a number line that has some numbers marked, justifying their answer. Children may discuss that it could be different numbers, but the spacing of the numbers means that the start and end points must be 200 and 400. This is higher level reasoning and some children may not see that 200 and 400 are the only options.

ASSESSMENT CHECKPOINT Check that children can identify start and end points based on the size of the numbers given.

ANSWERS Answers for the **Reflect** part of the lesson appear in the separate **Practice and Reflect answer guide**.

After the lesson ⏸

- Can children identify points between two numbers on a number line?
- Can they identify where a number line starts and/or ends based on numbers that are marked on the line?

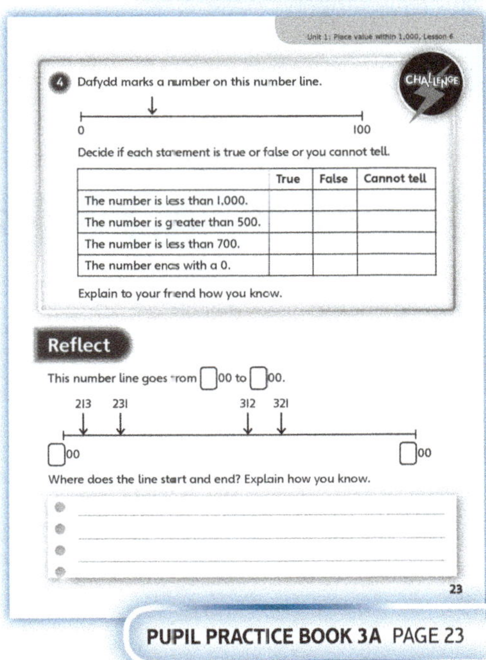

PUPIL PRACTICE BOOK 3A PAGE 23

Finding I, I0 and I00 more or less

Learning focus

In this lesson, children will find 1, 10, 100 more or less than a given number (including cases that require an exchange). They will also find the original number given the increase or decrease.

Small steps

→ Previous step: The number line to 1,000 (2)
→ **This step: Finding 1, 10 and 100 more or less**
→ Next step: Comparing numbers to 1,000 (1)

NATIONAL CURRICULUM LINKS

Year 3 Number – Number and Place Value

- Recognise the place value of each digit in a three-digit number (hundreds, tens, ones).
- Count from 0 in multiples of 4, 8, 50 and 100; find 10 or 100 more or less than a given number.
- Identify, represent and estimate numbers using different representations.

ASSESSING MASTERY

Children can find 1, 10, 100 more or less than a given number, including cases that involve an exchange. Children recognise which digit(s) will change. Children can also find the original number following an increase or decrease of 1, 10 and 100 by considering the inverse.

COMMON MISCONCEPTIONS

When answering a question such as 527 is 100 more than ___, children often think they have to find 100 more. Children need to see that 527 is the value after the increase, therefore they need to do the inverse. Ask:
- *Have you been given the original number or the end number? Should your answer be higher or lower than the given number?*

STRENGTHENING UNDERSTANDING

Some children may find it helpful to use base 10 equipment or place value counters, particularly with questions that involve exchange. Model the question using base 10 equipment or place value counters, then using a place value grid. When counting objects, encourage children to count the 100s, then the 10s and then the 1s. Some children may use a number line to help track the counting.

GOING DEEPER

Ask children questions about which digits can change if they find 1, 10, or 100 more. Ask whether it is possible for all the digits to change when they find 1 less or 1 more. Repeat for 10 and 100 more or less. Encourage children to experiment with different numbers.

KEY LANGUAGE

In lesson: more, less, **exchange**

Other language to be used by the teacher: inverse

STRUCTURES AND REPRESENTATIONS

Base 10 equipment, place value grid

RESOURCES

Mandatory: base 10 equipment, place value grids

Optional: number lines, place value counters

 In the eTextbook of this lesson, you will find interactive links to a selection of teaching tools.

Before you teach ⏸

- Can children represent a 3-digit number using base 10 equipment and on a place value grid with counters?

Discover

 Pair work

ASK

- Question **1** a): *What can you use to help you work out how many points Amal has? Which digit will change?*
- Question **1** b): *How can you work out 10 less than 204? How can you do this if there are no 10s?*

IN FOCUS In question **1** a), children can use their knowledge of place value to work out how many points the player has. They then need to work out which digit changes when he receives 100 more points. Discuss how the notes can be used like place value counters, as they are in 100s, 10s and 1s.

PRACTICAL TIPS The context for this lesson is a board game where players have notes with points on. You could give children notes from a game, such as Monopoly.

ANSWERS

Question **1** a): Amal has 253 points.
Amal adds 100 points, so he now has 353.

Question **1** b): Holly now has 194 points.

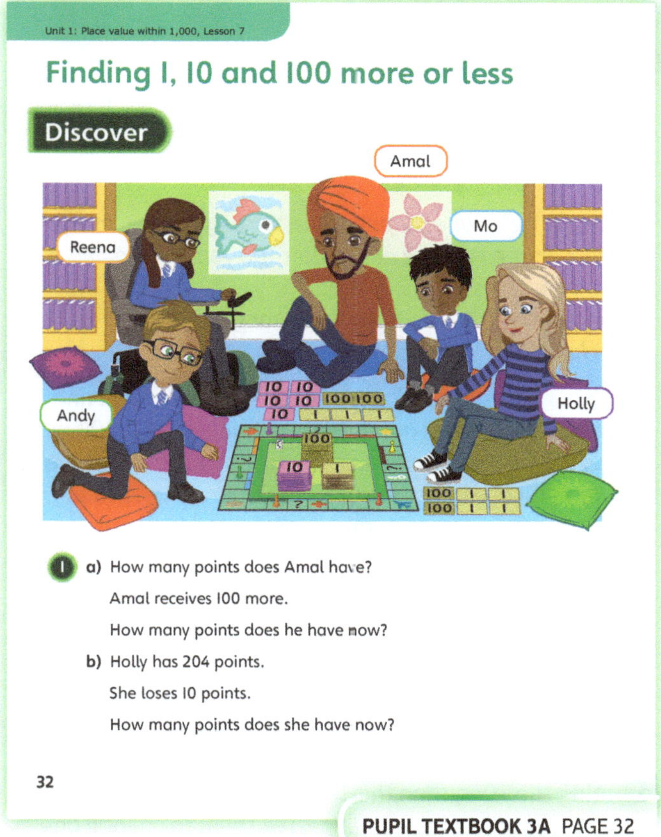

PUPIL TEXTBOOK 3A PAGE 32

Share

 Whole class teacher led

ASK

- Question **1** a): *What did Dexter use to represent the notes?*
- Question **1** b): *How many 10s make 100? How do you know that? Can you show me using base 10 equipment? How can you use this to answer the question? Which digits changed?*

IN FOCUS Question **1** b) involves finding 10 less where an exchange is necessary. Ensure that children understand that losing 10 points means that the player has 10 less. Children need to realise that they have no 10 point notes and so that they will need to exchange one of the hundreds for 10 tens. They did this in Year 2, but may need a reminder about how to approach this. You could use base 10 equipment to show that 10 tens make 100. Discuss with children which digits change and why.

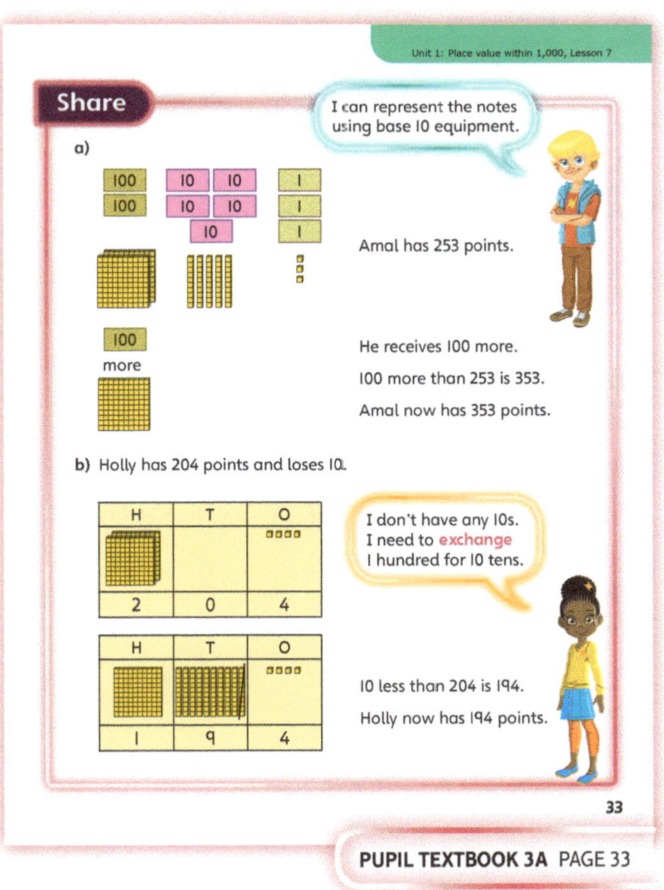

PUPIL TEXTBOOK 3A PAGE 33

Think together

WAYS OF WORKING Whole class teacher led (I do, We do, You do)

ASK

- Question **1** : *What happens to each player? How many points does each have at the start? Do you have to find 1, 10, 100 more or less?*
- Question **3** : *Do you need to exchange anything? How do you know?*

IN FOCUS Question **1** continues the game context and children have to work out if they are finding 1, 10, 100 more or less than the starting value. Questions **2** and **3** give children practice in finding 1, 10, 100 more or less than numbers, moving from numbers presented in concrete form to abstract numbers. In question **3** children need to realise that they do not have enough 10s and so they need to exchange. Question **4** addresses some common misconceptions that children make.

STRENGTHEN Provide children with base 10 equipment or place value counters to find 1, 10 100 more or less. Encourage them to be systematic in their counting, counting the 100s, then the 10s and then the 1s. In questions that involve exchange, tell them to change one of their 100 blocks for ten 10 blocks, or one 10 block for ten singles before trying to subtract 10 or 1.

DEEPEN After each question ask children to work out which digits changed. Give children some similar problems and ask them to predict which digits will change.

ASSESSMENT CHECKPOINT Use questions **1**, **2** and **3** to assess whether children can find 1, 10, and 100 more or less than any number, both concrete representations and abstract numbers. Check that children can identify when they need to carry out an exchange first.

ANSWERS

Question **1** : Reena 574, Andy 206, and Mo 783.

Question **2** a): 10 more than 350 is 360.

Question **2** b): 100 less than 648 is 548.

Question **2** c): 1 more than 248 is 249.

Question **3** : 10 less than 407 is 397.

Question **4** : Kate needs to exchange 10 tens for 100.

Ebo has found 100 more than 457 rather than 100 less.

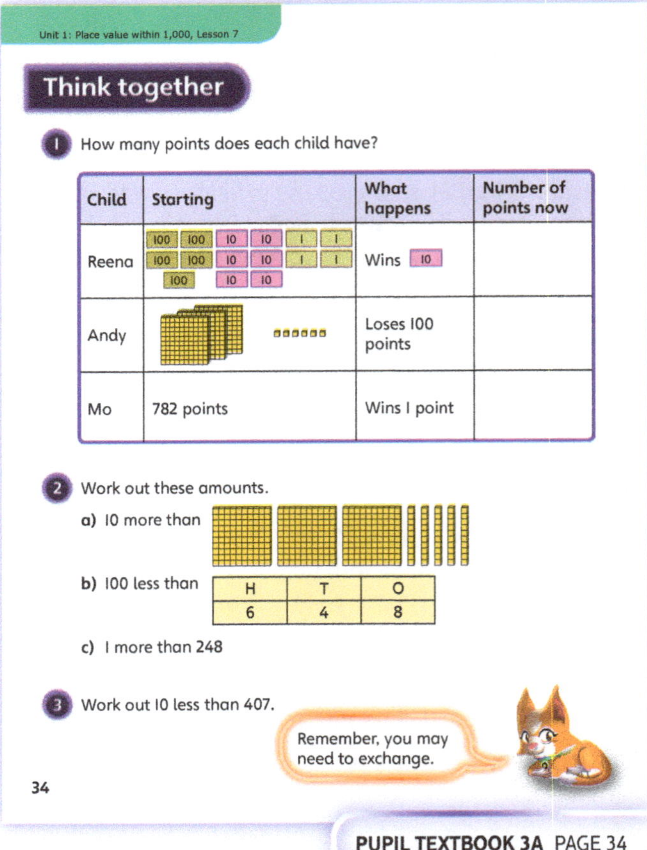

PUPIL TEXTBOOK 3A PAGE 34

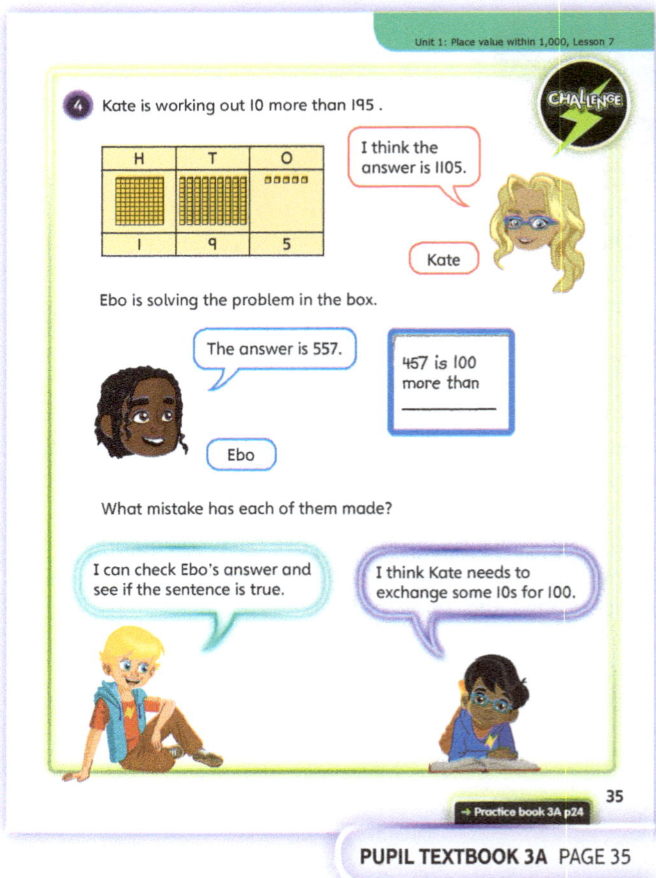

PUPIL TEXTBOOK 3A PAGE 35

Practice

WAYS OF WORKING Independent thinking

IN FOCUS Question ③ uses place value grids and then abstract numbers, encouraging children to use place value grids for the questions where they are not given. These questions should also develop understanding of which digits change. The last three parts of question ③ extend into finding the original number given the result of an increase or decrease. This develops the concept of needing to do the inverse or opposite operation to go back to the original number. Question ⑥ provides a deeper understanding where there are two steps involved. In question ⑥ b), children need to realise that they must first work out 100 less than 238.

STRENGTHEN In question ①, some children may find it helpful to represent the apples using base 10 equipment or place value counters. Using concrete objects helps them to physically understand what it means to find 10 or 100 more. Using base 10 equipment or place value counters will particularly help with questions that involve exchange.

DEEPEN When children have completed question ④, ask how many pieces of information in the table they need to find all the other numbers. Ask them to create their own tables, including cases where they need to exchange.

THINK DIFFERENTLY Question ④ gives further practice in finding the original number given the result of an increase or decrease. Children then need to use the original number to complete the table. In question ④ b), ask children to find the original number using both the numbers in the table and to check that they get the same number.

ASSESSMENT CHECKPOINT Use question ③ to assess whether children can find 1, 10 and 100 more or less than any number, including cases that require exchange. Children should be able to identify when they need to exchange. Question ③ also checks whether children can find the original number given the increase or decrease.

ANSWERS Answers for the **Practice** part of the lesson appear in the separate **Practice and Reflect answer guide**.

Reflect

WAYS OF WORKING Pair work

IN FOCUS The question involves children both in finding 1, 10 and 100 more or less and in working out the original number given the increase or decrease by using the inverse.

ASSESSMENT CHECKPOINT Check that children can find 1, 10, and 100 more and less than the number they have generated. Also, check they can work out the original number given an increase or decrease.

ANSWERS Answers for the **Reflect** part of the lesson appear in the separate **Practice and Reflect answer guide**.

After the lesson 

- Can children find 1, 10, 100 more or less than a given number?
- Do children know which digit to look at and which ones will change?
- Can children find 1, 10, 100 more or less than numbers that involve exchange?
- Can children find the original number given an increase or decrease?

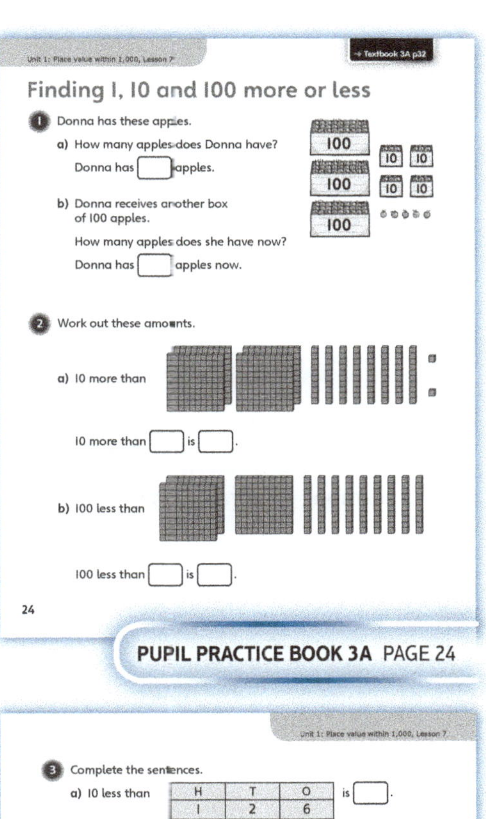

PUPIL PRACTICE BOOK 3A PAGE 24

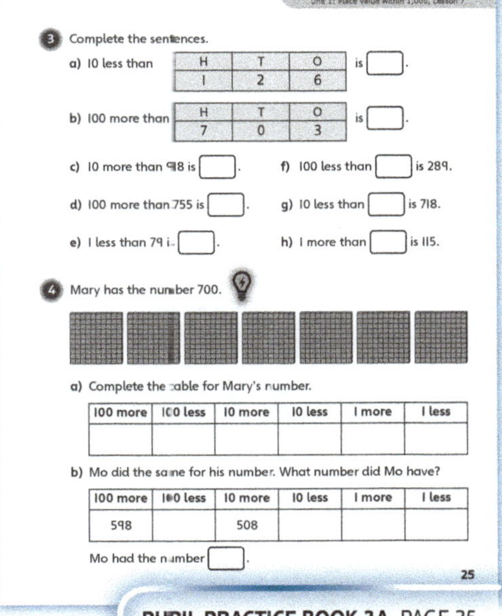

PUPIL PRACTICE BOOK 3A PAGE 25

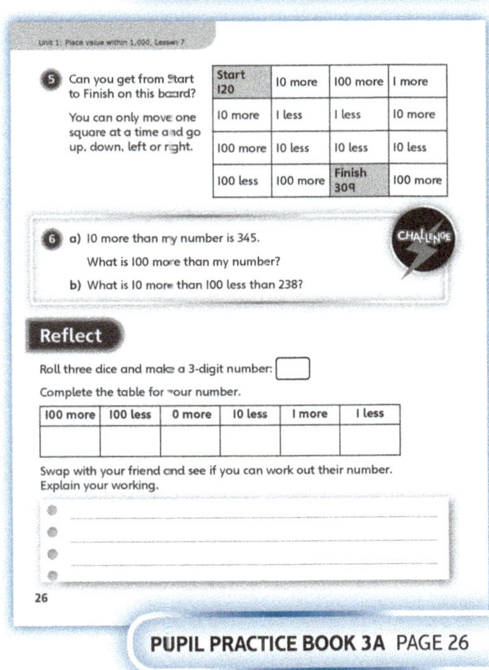

PUPIL PRACTICE BOOK 3A PAGE 26

Comparing numbers to 1,000 ❶

Learning focus

In this lesson, children will compare two groups of objects using <, > and = signs.

Small steps

→ Previous step: Finding 1, 10, 100 more or less
→ **This step: Comparing numbers to 1,000 (1)**
→ Next step: Comparing numbers to 1,000 (2)

NATIONAL CURRICULUM LINKS

Year 3 Number – Number and Place Value

- Compare and order numbers up to 1,000.
- Identify, represent and estimate numbers using different representations.
- Read and write numbers up to 1,000 in numerals and in words.

ASSESSING MASTERY

Children can compare two groups of objects, saying which is the smaller and which is the greater. Children form statements using <, > and = signs to compare two groups of objects. Children realise that they should start comparing the greatest place value first.

COMMON MISCONCEPTIONS

Children often get the signs the wrong way around. Encourage children to find a way of remembering. Ask:
- *Which end of the sign looks bigger?*

Some children think that the number of pieces of equipment determines which number is smaller or greater, saying, for example, that 3 × 100 blocks is less than 2 × 100, 8 × 10 and 7 × 1 blocks as there are only 3 blocks. Ask:
- *Can you exchange this 100 block for 10 blocks? Can you exchange one of your 10 blocks for singles? What happens when you compare the 100s, 10s, 1s?*

STRENGTHENING UNDERSTANDING

Children should make the numbers using base 10 equipment or place value counters. Encourage children to start by just comparing the 100s using the equipment. Line up the 100s in the two numbers. Ask whether they can tell which is the greater by comparing the 100s. If the 100s are the same then repeat with the 10s and, if necessary, the 1s.

GOING DEEPER

Encourage children to start comparing two amounts without supporting equipment. Ask them to work out the maximum and minimum number of comparisons they need to make to compare numbers.

KEY LANGUAGE

In lesson: more, less than (<), greater than (>), equal to (=)

STRUCTURES AND REPRESENTATIONS

Base 10 equipment, place value grid

RESOURCES

Mandatory: base 10 equipment

Optional: place value cards and counters

 In the eTextbook of this lesson, you will find interactive links to a selection of teaching tools.

Before you teach 

- Do children know what <, > and = mean?
- Can children use these signs in a statement to compare two 2-digit numbers?
- Can children explain how to compare two 2-digit numbers?

Discover

 Pair work

ASK

- Question **1** a): *How many of each shape of lolly do you have? How can you represent this? What can you compare to find the greater amount?*
- Question **1** b): *How many triangular lollies are there? Can you tell straight away which you have more of? How can you tell? Do you compare the 1s, 10s or 100s first? Why? What does each of the signs represent?*

IN FOCUS In Question **1** a), children compare two types of lollies. Children are likely to make the objects using base 10 or counters. Children should be able to compare the two numbers by just comparing the 100s. Question **1** b) requires children to compare the 100s first (which are equal), then the 10s (which are also equal) and then the 1s. Children start to realise that they compare the highest place value first and if they are the same move on to the next highest.

PRACTICAL TIPS Have base 10 equipment available for children to compare.

ANSWERS

Question **1** a): There are more square lollies as 215 > 183.

Question **1** b): 215 > 214, 214 < 215

PUPIL TEXTBOOK 3A PAGE 36

Share

 Whole class teacher led

ASK

- Question **1** a): *What did you compare first? How many comparisons did you need to make? How can you tell from the diagram you have more square lollies?*
- Question **1** b): *Can you remember what the signs < and > mean? How many comparison sentences can you form? Can you always make two comparison statements?*

IN FOCUS In question **1** b), the 100s and 10s are the same, so children need to make more than one comparison. Children need to understand that they should compare the highest place value first. This question uses the signs <, > and = that they met in Year 2. They may need reminding what each one means. The question requires them to form two comparison statements, one involving less than and one involving greater than.

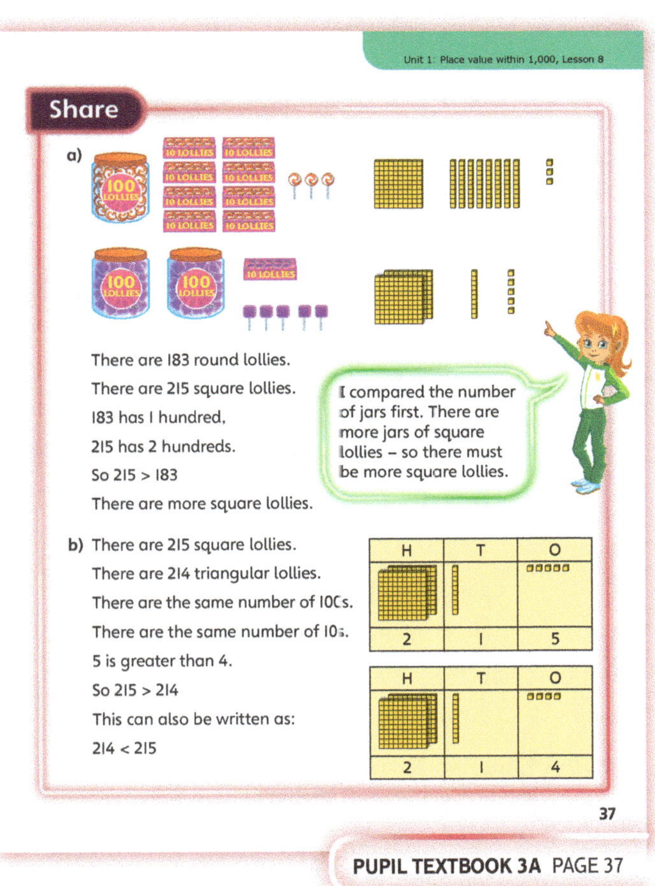

PUPIL TEXTBOOK 3A PAGE 37

Think together

WAYS OF WORKING Whole class teacher led (I do, We do, You do)

ASK

- Question **1** : *How many books does each teacher have? What place value should you compare first? Do you need to make any more comparisons?*
- Question **2** : *What are the numbers? What does each sign mean? Which values should you compare first?*
- Question **3** : *What is the same about these numbers? What is different?*
- Question **4** : *Is Dexter right? How can you work out who has more sweets? What do 10 packs of 10 sweets equal?*

IN FOCUS Question **2** moves on from the context questions used in **Discover** and question **1**, comparing numbers represented in base 10. Children need to understand that the approach is still to compare the 100s, then the 10s if the 100s are equal, and finally the 1s. It includes a comparison where the numbers have a different number of digits, to emphasise the importance of comparing digits with the same place value. Children need to realise that a 2-digit number has 0 hundreds.

STRENGTHEN Start by just comparing the 100s using base 10 equipment. Line up the 100s from the first number with the 100s from the second number. Ask whether children can tell which number is greater by just comparing the 100s. If necessary, repeat with the 10s and the 1s.

DEEPEN Develop the idea of comparing two amounts without supporting equipment. For example, Alex has 355 apples and Ebo has 378 apples. Children should understand that they still compare from the greatest place value. Ask children to work out the maximum and minimum number of comparisons they need to make to compare numbers. When children write their comparison using < or >, ask them to write another comparison sentence using the same numbers.

ASSESSMENT CHECKPOINT Use questions **2** and **3** to assess whether children can compare two numbers presented using mathematical equipment. Check also that children know how to use < and > signs correctly in mathematical comparison statements.

ANSWERS

Question **1** : Miss Hall has 524 books. Mr Jones has 350 books.

500 is greater than 300, so 524 > 350

Miss Hall has more books.

Question **2** a): 265 < 290

Question **2** b): 350 > 84

Question **3** a): 300 > 282

Question **3** b): 126 = 126

Question **4** : They have the same amount (265) because ten packs of 10 is equal to one pack of 100.

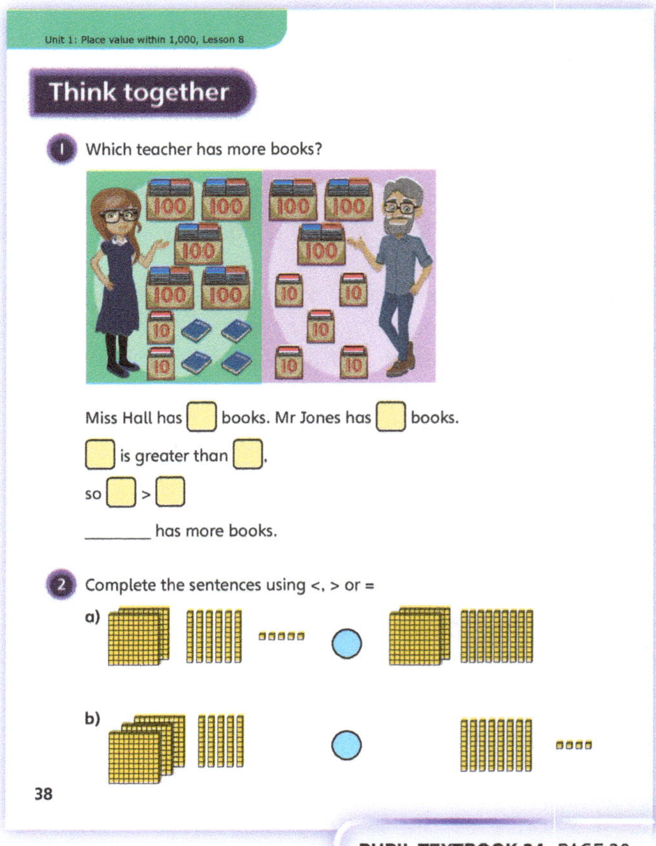

PUPIL TEXTBOOK 3A PAGE 38

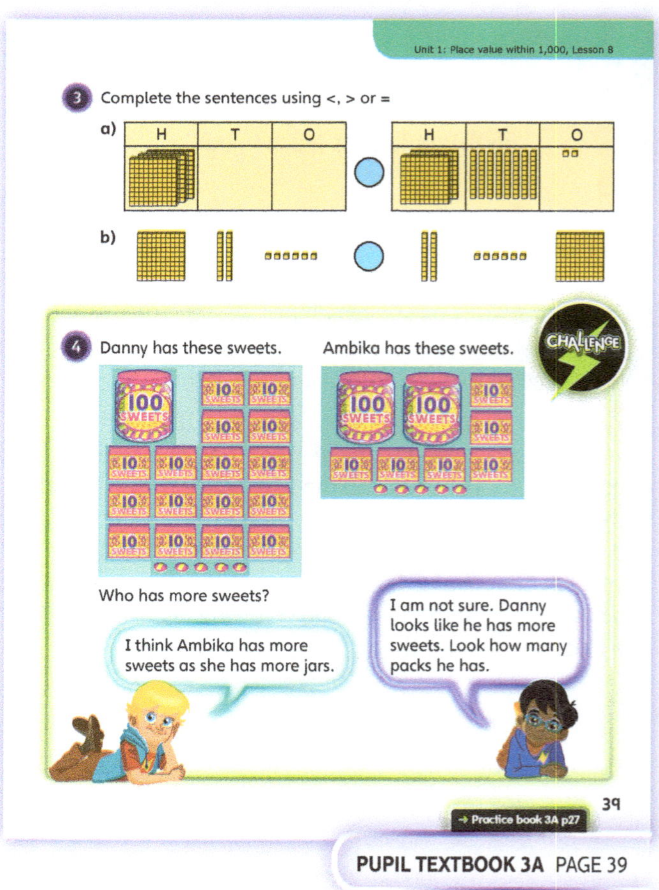

PUPIL TEXTBOOK 3A PAGE 39

Practice

WAYS OF WORKING Independent thinking

IN FOCUS Question ③ addresses two common misconceptions. Firstly, children may think that 3 × 100 base 10 is less than 2 × 100, 4 × 10 and 9 × 1 because there are more pieces required to make the second number. Secondly, children may think that the order of the base 10 pieces matters. Children need to understand that if the pieces are the same, then the number will be the same number, despite the pieces being in a different order.

STRENGTHEN Practise comparing two numbers using base 10 equipment and place value counters. Encourage children to put the equipment into a place value grid to emphasise the importance of comparing numbers with the same place value.

DEEPEN When children have completed question ④, ask how many different answers are possible. Ask how they can be confident that they have found them all. When they have completed question ⑤, challenge them to find a number that is greater than A but less than B. Ask how many numbers they can find and how they know that these are all the possible numbers.

ASSESSMENT CHECKPOINT Use questions ② and ③ to assess whether children can compare two groups of objects by first comparing the number of 100s, then the 10s (if necessary) and then the 1s.

ANSWERS Answers for the **Practice** part of the lesson appear in the separate **Practice and Reflect answer guide**.

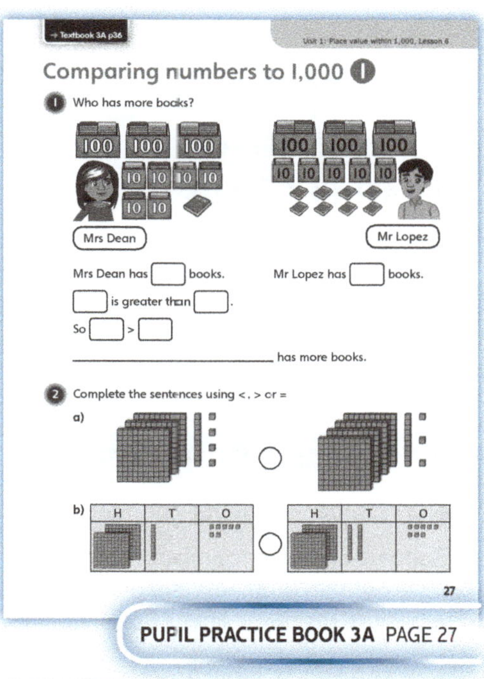

PUPIL PRACTICE BOOK 3A PAGE 27

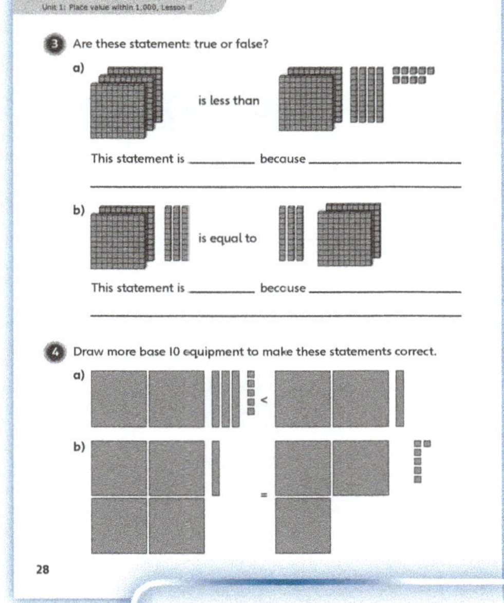

PUPIL PRACTICE BOOK 3A PAGE 28

Reflect

WAYS OF WORKING Pair work

IN FOCUS The question focuses on the first step of the comparison of two 3-digit numbers. Children need to reason that they should compare the 100s as comparing the 10s and 1s will not help until they have compared the 100s. Encourage children to demonstrate their answer by comparing two numbers. Some children may complete the full method.

ASSESSMENT CHECKPOINT Check that children understand that they compare the 100s first.

ANSWERS Answers for the **Reflect** part of the lesson appear in the separate **Practice and Reflect answer guide**.

After the lesson ⏸

- Do children know how to compare the numbers represented by two groups of objects?
- Do children compare the 100s first, then the 10s and then the 1s?
- Do children know that a 2-digit number has 0 hundreds?

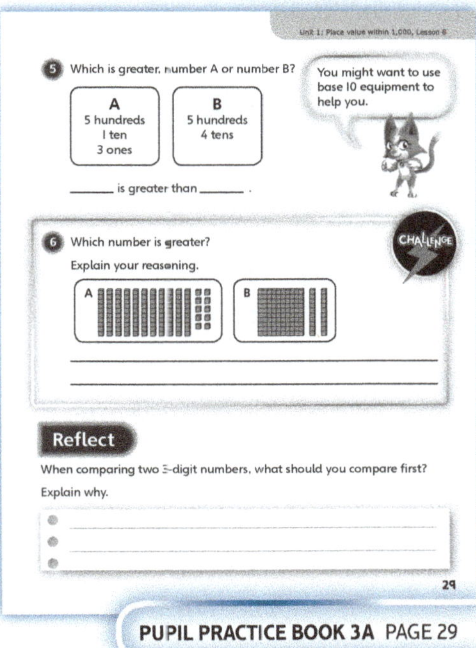

PUPIL PRACTICE BOOK 3A PAGE 29

Comparing numbers to 1,000 ❷

Learning focus

In this lesson, children will compare two 3-digit numbers. Children will also be able to work out missing digits to make an inequality statement correct.

Small steps

→ Previous step: Comparing numbers to 1,000 (1)
→ **This step: Comparing numbers to 1,000 (2)**
→ Next step: Ordering numbers to 1,000

NATIONAL CURRICULUM LINKS

Year 3 Number – Number and Place Value

- Compare and order numbers up to 1,000.
- Solve number problems and practical problems involving these ideas.
- Recognise the place value of each digit in a three-digit number (hundreds, tens, ones).

ASSESSING MASTERY

Children can compare two 3-digit numbers using <, > and = signs. Children should compare the numbers by first considering the 100s and then the 10s (if needed) and then the 1s. Children can also work out the missing digits in numbers.

COMMON MISCONCEPTIONS

Children sometimes do not think about the place value of a digit. For example, when comparing 586 and 84 they think that 84 is greater than 586 as the first digit of 84 is 8 and the first digit of 586 is 5. Encourage children to think about the place value of each digit. Ask:

- *What does the 8 represent in 84? What does the 5 represent in 586? How many 100s in 84? Which number has more 100s?*

STRENGTHENING UNDERSTANDING

Ask children to make the numbers using base 10 equipment and put them in a place value grid. Then get them to compare the 100s by lining them up. If they have the same amount then ask them to compare the 10s and so on. Each time, link the base 10 equipment to the numbers in the place value grid.

GOING DEEPER

Ask children to explain the greatest and least number of comparisons that need to be made to compare 3-digit numbers. Challenge them to write down two numbers where they need to make one, two and three comparisons.

KEY LANGUAGE

In lesson: more than, less than, greater than, smaller than, is equal to, <, >, =

STRUCTURES AND REPRESENTATIONS

Place value grid, number line

RESOURCES

Mandatory: place value grids

Optional: base 10 equipment

 In the eTextbook of this lesson, you will find interactive links to a selection of teaching tools.

Before you teach 

- Do children know what the signs <, > and = mean?
- Can children use base 10 equipment to compare two 3-digit amounts?
- Can children explain how to compare two 2-digit numbers?

Discover

 Pair work

ASK

- Question **1** a): *What are the numbers? Do you need to make the numbers out of base 10 equipment or counters to compare? Is there a way of working out which is smaller by just comparing the digits? Which digits should you compare first? What happens if they are the same?*
- Question **1** b): *Which digits are missing? What can you tell me about the second digit? Why? What can you tell me about the third digit? Is there more than one answer?*

IN FOCUS In Question **1** a), children compare two 3-digit numbers. This time the numbers are abstract (i.e. there are not 542 objects). Although some children may make the numbers using base 10 equipment, they should be looking to form a method that does not always require them to make the number. Eventually children should explore the numbers to realise that they can compare the 100s first, and then the 10s and 1s if necessary. They should compare the digits from left to right.

PRACTICAL TIPS Make the number cards for children to hold up at the front of the classroom. Have place value grids for children to put the numbers in. Some children may need base 10 equipment for support.

ANSWERS

Question **1** a): 542, 240

Question **1** b): 396, 397, 398 or 399

Share

 Whole class teacher led

ASK

- Question **1** a): *What did you compare first? How many comparisons did you need to make? Why do you compare from left to right? How can you use a number line to compare the numbers?*
- Question **1** b): *Explain why the 10s digit must be 9. What numbers could the 1s digit be?*

IN FOCUS For both questions, listen to children discuss their method. Highlight the method of comparing numbers without using base 10 equipment. Ensure children understand that they can compare the digits from left to right. In question **1** a), two comparisons are needed to compare 542 and 589. A number line is used to compare 240 and 395, which is an alternative method that children may not have thought of. Question **1** b) requires children to use three comparisons to find the digits.

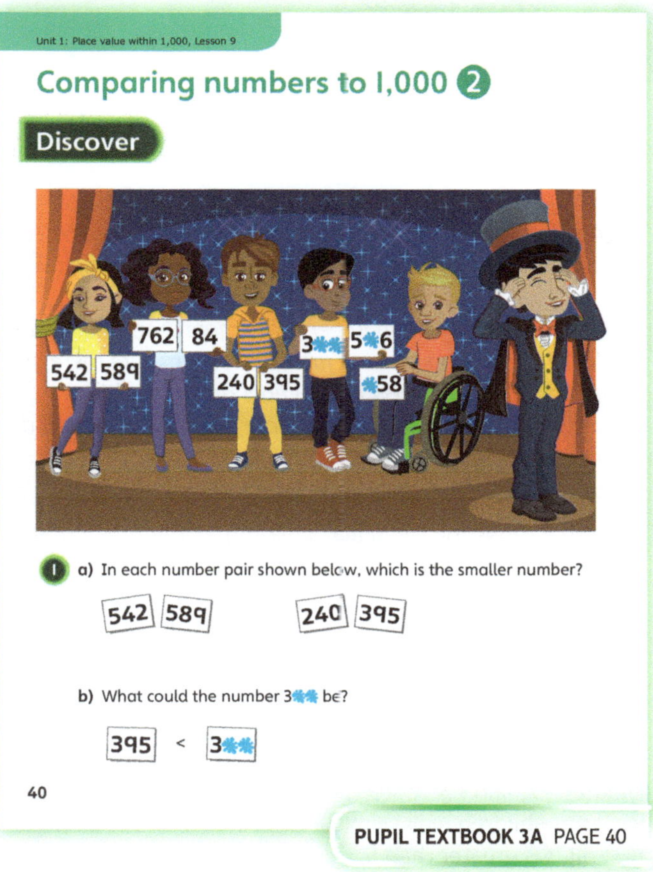

PUPIL TEXTBOOK 3A PAGE 40

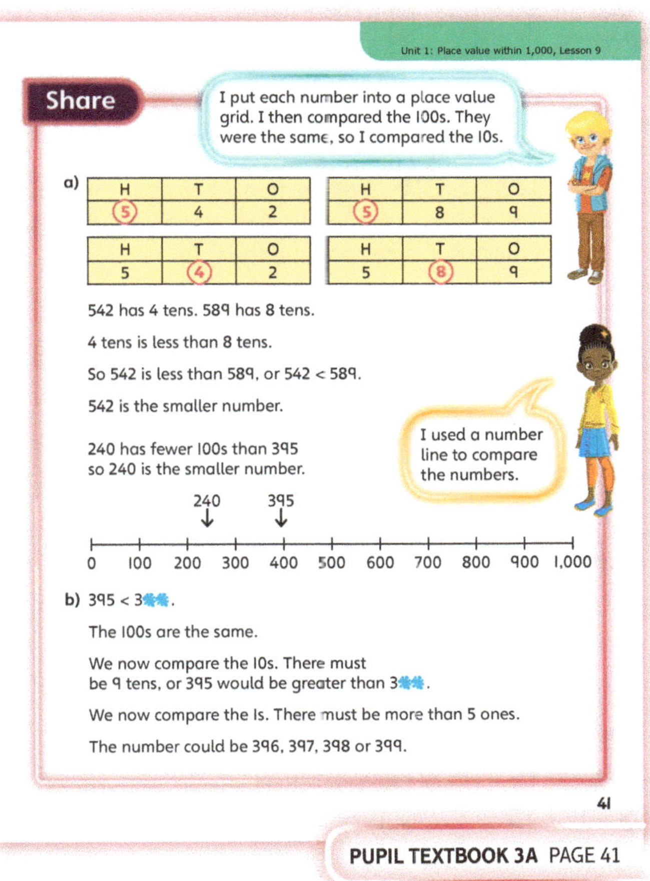

PUPIL TEXTBOOK 3A PAGE 41

Think together

 Whole class teacher led (I do, We do, You do)

ASK

- Question **1** a): *What numbers have fewer than 4 hundreds? Are there any numbers with 4 hundreds and fewer than 9 tens?*
- Question **1** b): *What numbers have more than 5 hundreds? Which have 5 hundreds and more than 8 tens? Are there any numbers that have the same number of 100s and 10s?*
- Question **2** : *How can you compare the numbers? What do you look at first? Then what do you look at?*
- Question **4** : *How many answers can you find?*

IN FOCUS Question **1** a) asks children to identify numbers smaller than 490. This only requires them to compare 100s. Question **1** b) asks children to identify numbers greater than 580, which requires them to looks at 100s, 10s and 1s. They first look for numbers that have 100s the same or greater and then look for ones that have 100s the same and 10s the same or greater. Finally, they realise they have to look for 100s and 10s the same and 1s greater. Ensure that children explain every step in their method. 84 could cause issues as children may think 84 is greater than 580 as the first digit is greater than 5.

STRENGTHEN Throughout, if children are struggling to compare numbers ask them to make the numbers using base 10 equipment and put them in a place value grid. Then use the base 10 equipment to compare the 100s, 10s and 1s, as necessary, by lining them up.

DEEPEN In question **4**, ask children to work out how many different solutions there are for each question. Ask them to explain what the least and greatest digits are that they can use in each number.

ASSESSMENT CHECKPOINT Use question **2** to assess whether children can compare two numbers by comparing the 100s first, then the 10s if necessary and then the 1s.

ANSWERS

Question **1** a): 395, 84 and 240

Question **1** b): 762 and 589

Question **2** a): 948 > 820

Question **2** b): 385 > 368

Question **2** c): 600 < 950

Question **3** a): 749 < 826

Question **3** b): 286 > 243

Question **3** c): 580 > 68

Question **3** d): 300 < 307

Question **4** a): 3, 2, 1 or 0

Question **4** b): 1, 2, 3 or 4

Question **4** c): 1, 2, or 3

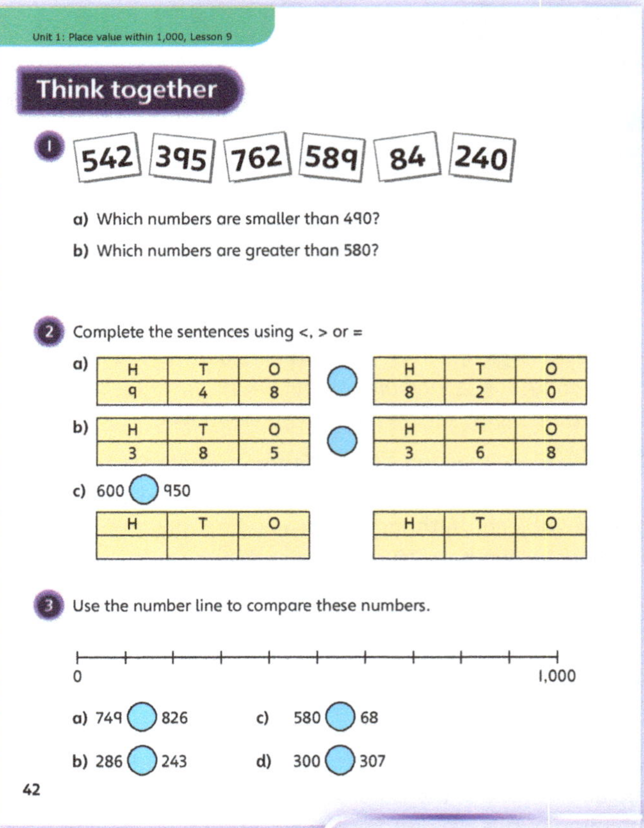

PUPIL TEXTBOOK 3A PAGE 42

PUPIL TEXTBOOK 3A PAGE 43

Practice

WAYS OF WORKING Independent thinking

IN FOCUS Questions **1**, **2** and **3** ask children to work out greater and smaller numbers. Children need to use the <, > and = signs correctly. There is variation in how the numbers are presented to reinforce earlier work. In question **4**, where children have to work out missing digits to make the inequality statements correct, they should realise that there is more than one answer. Question **5** provides description of numbers from three children and they have to reason their answer based on the information they have. Children need to work out if they have enough information to be able to make the comparison.

STRENGTHEN In question **4**, ask children to make the parts of the numbers that they know using base 10 equipment and put them in a place value grid. Then ask them what they could put in the missing columns to make the statement true. Then ask them to write the place value grid comparison that they have made using base 10 equipment.

DEEPEN In question **5**, ask children how much more information they would need to be able to compare Reena and Zac's numbers.

ASSESSMENT CHECKPOINT Use question **2** to assess whether children can compare any two 3-digit numbers to determine which is greater and which is smaller. Children should accurately use the > and < signs to compare numbers.

ANSWERS Answers for the **Practice** part of the lesson appear in the separate **Practice and Reflect answer guide**.

Reflect

WAYS OF WORKING Pair work

IN FOCUS Children are given two numbers to compare. They need to explain clearly the steps involved in comparing two 3-digit numbers.

ASSESSMENT CHECKPOINT Check that children know that they need to compare the 100s first and then, if these are the same, the 10s and finally, if necessary, the 1s. Children should use this method to compare any two 3-digit numbers.

ANSWERS Answers for the **Reflect** part of the lesson appear in the separate **Practice and Reflect answer guide**.

After the lesson

- Can children compare two 3-digit numbers by comparing the greatest place value first?
- Can children find missing numbers to ensure a mathematical statement is correct?
- Do children use the < and > signs accurately?

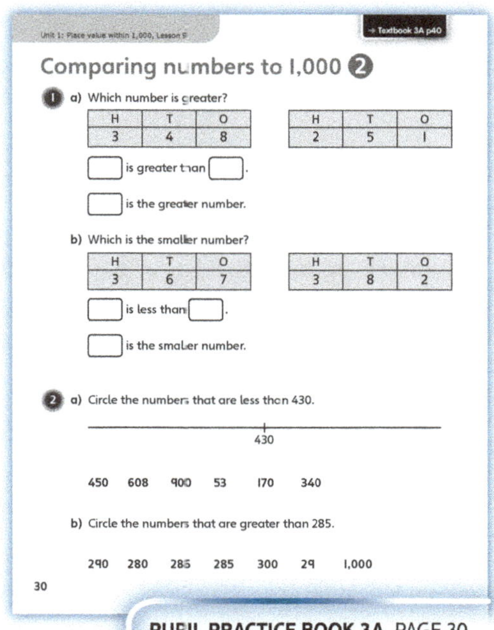

PUPIL PRACTICE BOOK 3A PAGE 30

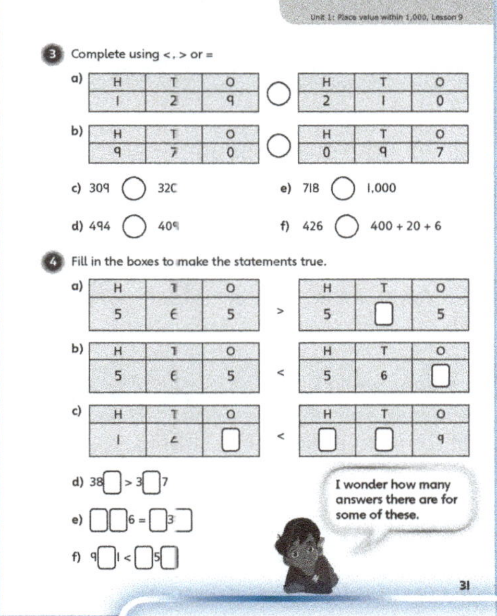

PUPIL PRACTICE BOOK 3A PAGE 31

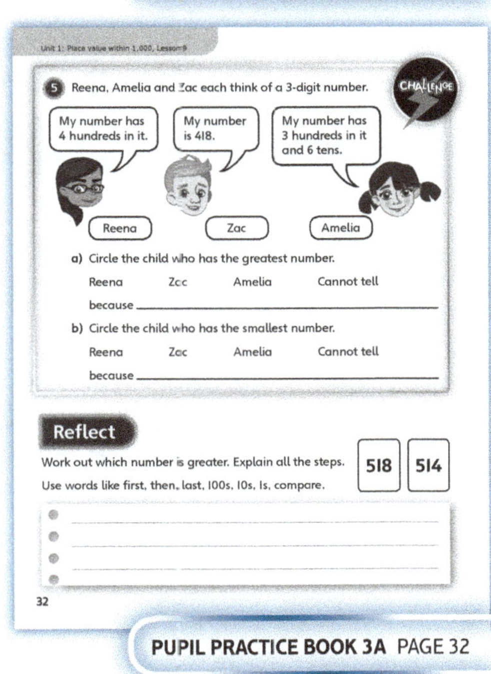

PUPIL PRACTICE BOOK 3A PAGE 32

Ordering numbers to 1,000

Learning focus

In this lesson, children will order three or more numbers up to three digits in length. They will also work out missing digits in lists of ordered numbers.

Small steps

→ Previous step: Comparing numbers to 1,000 (2)
→ **This step: Ordering numbers to 1,000**
→ Next step: Counting in 50s

NATIONAL CURRICULUM LINKS

Year 3 Number – Number and Place Value

- Compare and order numbers up to 1,000.
- Recognise the place value of each digit in a three-digit number (100s, 10s, 1s).
- Read and write numbers up to 1,000 in numerals and in words.

ASSESSING MASTERY

Children can order three or more 3-digit numbers. Children can also work out the missing digits in numbers listed in order.

COMMON MISCONCEPTIONS

Children sometimes do not think about the place value of a digit. For example, when ordering 850, 98 and 700 they think that 98 is the greatest number as this is the one with the greatest first digit. Ask:

- *What does the 8 represent in 850? What does the 9 represent in 95? What about the 7 in 700? Which number has more 100s?*

Children may think that when finding missing digits they only have to look at adjacent numbers. This is not always the case and numbers later in the list could have an impact on the missing digits. Ask:

- *Compare the first two numbers. What are the possible digits for the second number? Now compare the second number with the third number. What are the possible digits for the second number? Which digits are in both lists?*

STRENGTHENING UNDERSTANDING

Children can make the numbers using base 10 equipment to help them understand the size or using counters on a place value grid. This will help them to see which number has the greatest number of 100s, then 10s and then 1s.

GOING DEEPER

Ask children to explain the greatest and least number of comparisons that need to be made to order three 3-digit numbers.

KEY LANGUAGE

In lesson: greater than (>), taller, tallest, longest, shortest, greatest, smallest, most, least, fewest

Other language to be used by the teacher: more, less than (<), equal to (=)

STRUCTURES AND REPRESENTATIONS

Place value grid, number line

RESOURCES

Mandatory: place value grids

Optional: base 10 equipment, place value counters

 In the eTextbook of this lesson, you will find interactive links to a selection of teaching tools.

Before you teach 

- Do children know what the signs <, > and = mean?
- Can children compare two 3-digit numbers?
- Can children order a set of 2-digit numbers?

Discover

ASK

- Question **1** a): *How can you compare the two numbers? How many 100s in each number? Do you need to compare more than the 100s?*
- Question **1** b): *How can you put these buildings in height order? Which building is the shortest? How do you know?*

IN FOCUS Question **1** a) reinforces the work from the previous lesson. The question highlights a key misconception, with some children thinking that Big Ben is taller as the 9 is greater than 3. Remind children that the place value of the digit is important. Question **1** b) prompts children to think of different strategies for ordering the buildings. Some children may want to compare two buildings at a time, some may use a number line, and others may consider all the buildings all at once.

PRACTICAL TIPS Make cards of the buildings and their heights and peg these on to a line. Encourage children to move their positions on the line until they are in the correct order.

ANSWERS

Question **1** a): The Empire State Building is taller than Big Ben.

Question **1** b): Big Ben, Eiffel Tower, Empire State Building, Shanghai Tower, Burj Khalifa

PUPIL TEXTBOOK 3A PAGE 44

Share

ASK

- Question **1** b): *How does the table help you compare the numbers? Which column should you look at first? Why? What happens if the 100s are the same? What column do you look at then? Why?*

IN FOCUS Question **1** a) highlights the fact that there are 0 hundreds in 96 metres. This will help children understand that 381 is greater than 96 and so the Empire State Building is taller than Big Ben. Discuss if the number line is a helpful way to compare numbers. In question **1** b), the table shows the building heights in a place value grid. Discuss Astrid's method for ordering numbers. Talk children through the example. You may find it easier to write the numbers on the board and cross out numbers as they are used.

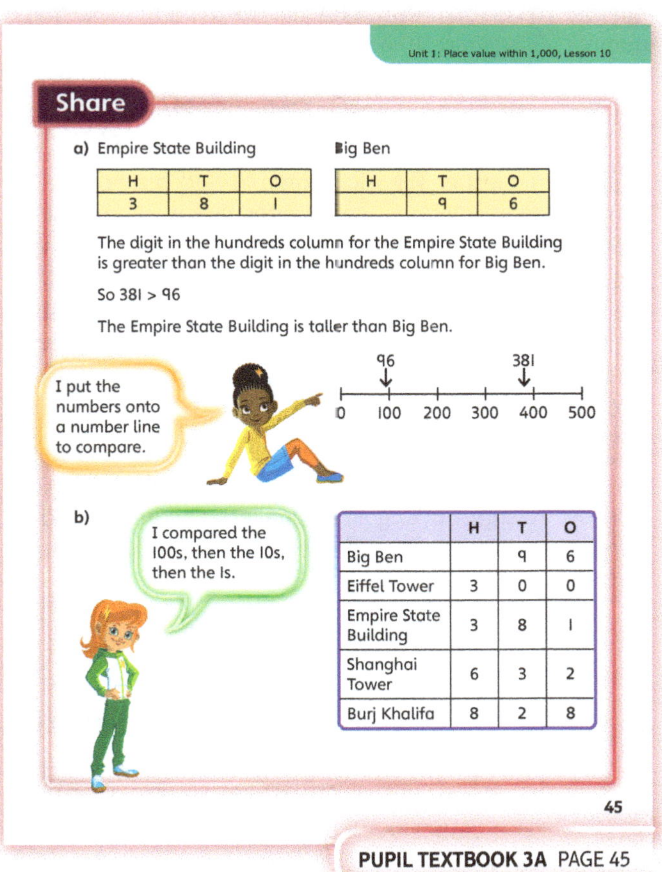

PUPIL TEXTBOOK 3A PAGE 45

Think together

WAYS OF WORKING Whole class teacher led (I do, We do, You do)

ASK

- Question **1**: *How is this similar to the buildings question? What is different? What numbers go into the table? How can you order these numbers?*
- Question **2**: *How can you organise your work? How do you know which box has the smallest number of counters in?*
- Question **3**: *Can you place the numbers on the number line? Are the numbers already in order? How do you know? How can you use the number line to write the numbers in order? How many different possible orders are there?*
- Question **4**: *Can you put a number in the first box without working out the second number? Did you write down any digits and then have to change them?*

IN FOCUS In question **2**, children need to think about making their own place value grid to compare the number of counters in each box. Question **3** uses a number line to help children compare numbers. The question helps children to realise that the same list of numbers can be written in order either from smallest to greatest, or greatest to smallest. Question **4** requires children to find missing digits in a list of numbers in order. Children should realise that they should start from left to right and they need to compare the numbers as a whole and not just in pairs.

STRENGTHEN Remind children that ordering numbers can be seen as simply making a series of comparisons of two numbers, as in the previous lesson. Ask them to make the numbers using base 10 equipment and put them in a place value grid. Start them comparing two numbers at a time, and then progress to more than two numbers and looking for the one with the smallest or greatest number of 100s.

DEEPEN In question **4** ask children to work out how many different solutions there are for each question. Ask them to explain how they know what are the smallest and greatest digits that can be used in each number. Ask how the smallest and greatest digits change depending on other digits used. For example, if they put an 8 in the second box, what numbers can go in the first box? What about if they put a 9 in the second box?

ASSESSMENT CHECKPOINT Use questions **1** and **2** to assess whether children can order three or more numbers from smallest to largest or largest to smallest. Children should also be able to recognise whether a list is already in order.

ANSWERS

Question **1**: ferry, cruise ship, container ship

Question **2**: B, D, A, C

Question **3**: Yes, the numbers are in order from largest to smallest.

Put them from smallest to largest on the number line.

Question **4**: There are many solutions, for example 4, 8, 2, 9. Possible values for the first digit depend on the value chosen for the second digit. The second digit must be 8 or 9. The third digit must be 2. The fourth digit can be any digit.

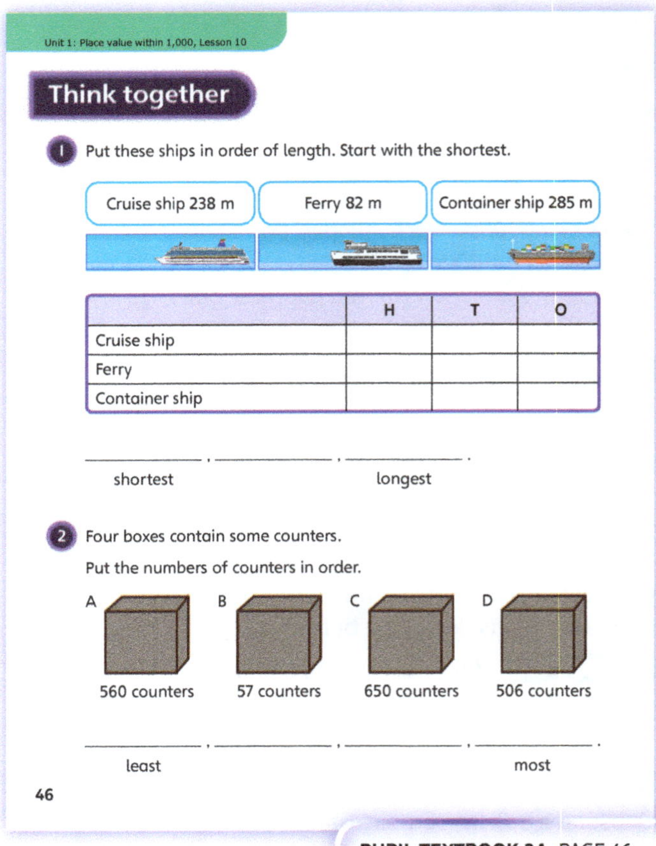

PUPIL TEXTBOOK 3A PAGE 46

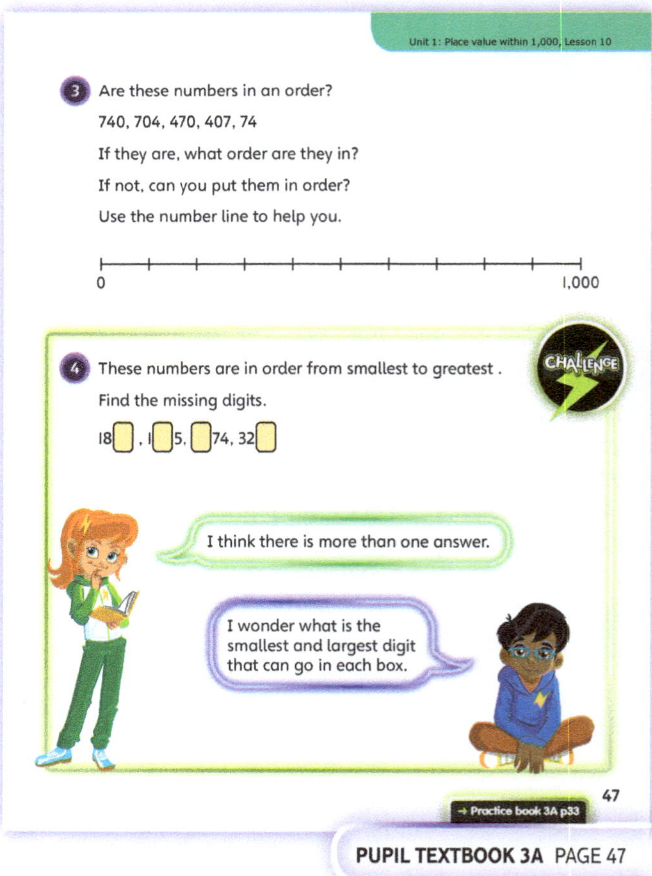

PUPIL TEXTBOOK 3A PAGE 47

Practice

WAYS OF WORKING Independent thinking

IN FOCUS Question ❶ mirrors the **Discover** task and asks children to compare heights. The table should be used as demonstrated in **Share**, with children looking for the tower with the smallest number of 100s. In question ❹, children are not given the prompt of a place value grid. Encourage children to use a place value grid if it helps them, although some may be able to work out the order by just looking at the digits. Parts c) and d) include numbers that have different number of digits, so children need to understand the importance of place value. Question ❺ asks children to reason more deeply by working out missing digits. Children need to understand that there are lots of different possible answers and their answers may depend on their choices for other digits.

STRENGTHEN In questions ❶, ❷ and ❸, children could make the numbers using base 10 equipment to help them understand the size. They may then find it easier to see when they need to compare just the 100s, and when they need to also compare 10s and 1s. In question ❹, encourage children to draw a place value grid like in the earlier questions to help them order the numbers, or provide them with blank grids to fill in.

DEEPEN In question ❺, ask how many different answers they can find. Encourage them to work systematically, to be sure that they have found all the possible solutions.

ASSESSMENT CHECKPOINT Use question ❹ to assess whether children can order three or more numbers by first comparing the 100s, then the 10s and then the 1s if needed. Use question ❺ to assess whether children can find missing digits to make a set of numbers in order.

ANSWERS Answers for the **Practice** part of the lesson appear in the separate **Practice and Reflect answer guide**.

Reflect

WAYS OF WORKING Pair work

IN FOCUS Children use their understanding from this lesson to order a set of numbers. They can use whatever method they feel most comfortable with, but they need to explain in words or show a clear method for what they did. For example, they may draw a number line or create a place value grid.

ASSESSMENT CHECKPOINT Check that children have a method for ordering numbers. Check that children know that there are two different orders that they can put the numbers in.

ANSWERS Answers for the **Reflect** part of the lesson appear in the separate **Practice and Reflect answer guide**.

After the lesson 

- Can children order a set of three or more numbers from smallest to largest, or largest to smallest?
- Do children know how to accurately use the <, > and = signs in mathematical statements?

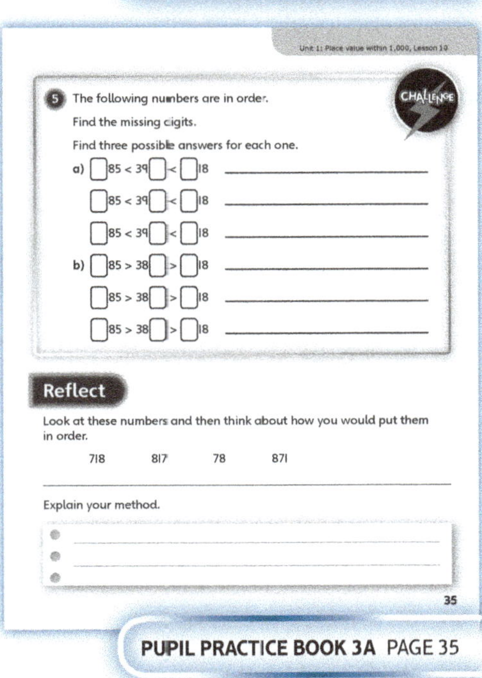

PUPIL PRACTICE BOOK 3A PAGE 33

PUPIL PRACTICE BOOK 3A PAGE 34

PUPIL PRACTICE BOOK 3A PAGE 35

Counting in 50s

Learning focus

In this lesson, children will count forwards and backwards in 50s from 0 to 1,000 and count from any multiple of 50. They will work out how many 50s in a number.

Small steps

→ Previous step: Ordering numbers to 1,000
→ **This step: Counting in 50s**
→ Next step: Adding and subtracting 100s

NATIONAL CURRICULUM LINKS

Year 3 Number – Number and Place Value

- Count from 0 in multiples of 4, 8, 50 and 100; find 10 or 100 more or less than a given number.
- Solve number problems and practical problems involving these ideas.

ASSESSING MASTERY

Children can count forwards and backwards in 50s from 0 to 1,000. Children should be able to start at any multiple of 50. They can work out how many 50s in a number by counting up to that number. Children can start to pattern spot and be able to identify numbers that are in the pattern.

COMMON MISCONCEPTIONS

Children may miscount because they are trying to keep track of too many things. For example, when counting on seven 50s from 300 they may struggle to keep track of how many 50s they have counted and where they are. Encourage them to use a number line to keep track. Ask:

- *What number are you starting from? Count on one 50. What number are you at now? How many 50s have you counted? How can you keep track of how many you have counted?*

STRENGTHENING UNDERSTANDING

To strengthen understanding, give children base 10 equipment to help them. Start with 5 base 10 blocks of ten and count in 10s with them. Explain that this is 50. Get them to take another 50 and count the total, which is 100. Ask them to count aloud from 0 and then show them how they could have counted in 50s: 0, 50, 100. Ask them what they can swap 10 tens for. Continue doing this, counting on, and children will start to notice the pattern.

GOING DEEPER

Give children some problems involving money. Ask them to work out how many 50ps are in £6. Ask how much they have in pounds and pence if they have 15 50p pieces. Tell them that banks keep 50p pieces in bags of 20. Ask how much this is in pounds.

KEY LANGUAGE

In lesson: fifty, 50s

Other language to be used by the teacher: count forwards, count backwards

STRUCTURES AND REPRESENTATIONS

Number line

RESOURCES

Mandatory: number line

Optional: base 10 equipment, place value counters, 50p coins

 In the eTextbook of this lesson, you will find interactive links to a selection of teaching tools.

Before you teach 

- Can children count forwards and backwards in 10s?
- Do children know that 10 tens make 100?
- Do children know the different coins and amounts and know how many pence make £1?

Discover

 Pair work

- Question **1** a): *Do you recognise this flag? How many stars are on the flag? How did you count the number of stars? How can you find the number of stars on 4 flags?*
- Question **1** b): *How could you work out the number of flags? Is there a quicker way?*

 In question **1** a) children firstly identify that there are 50 stars on each flag. Some may already know this, but encourage them to check. Encourage children to look for counting strategies other than just counting in 1s. Once children have worked out there are 50 stars on a flag, they need to look at methods for working out how many are on 4 flags. Children may count in 1s, 5s, 10s or 50s, but encourage them to find the most efficient method (i.e. counting in 50s). They could use a number line and base 10 equipment to help them. Encourage them to look for patterns. In question **1** b) children need to count in 50s until they get to 350. They need to develop a system for keeping count of how many 50s they have counted.

 Count aloud in 5s up to 50. Get children to stand at the front of the classroom holding number cards in 50s in order from 0 to 500. Count aloud in 50s.

Question **1** a): There are 50 stars on each flag.
There are 200 stars on 4 flags.

Question **1** b): Sylvie counted 350 stars on 7 flags.

Share

 Whole class teacher led

- Question **1** a): *How did Astrid count the stars? Did you count in the same way? What other ways could you have counted? What did you count in to work out how many stars are on 4 flags? Did you use a number line? Did you notice any patterns in the numbers?*
- Question **1** b): *Why did you count in 50s? Does the number line help? How can you keep track of how many 50s you have counted? Will you get the same answer if you count down from 350? Why?*

 Question **1** a) shows a different method for counting, rather than in 1s. Children counted in 5s in Year 1, so this is a useful opportunity to revisit that before counting in 50s. Question **1** b) looks at the opposite to what they did in part a), counting up in 50s to get to a specific number.

PUPIL TEXTBOOK 3A PAGE 48

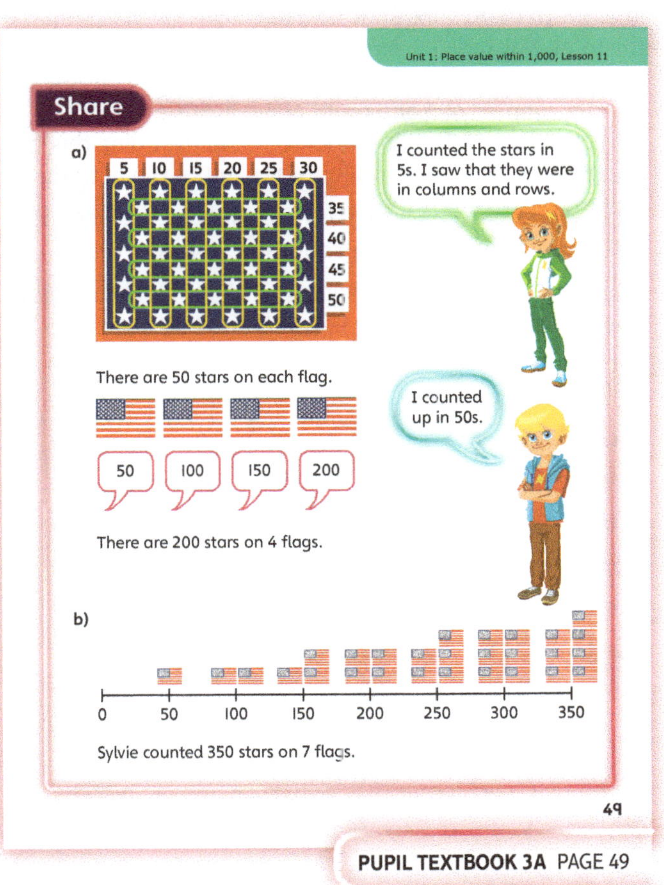

PUPIL TEXTBOOK 3A PAGE 49

Think together

WAYS OF WORKING Whole class teacher led (I do, We do, You do)

ASK

- Question ❶: *What does the table tell you? What do the numbers go up in on the bottom row? Why?*
- Question ❷: *How did you work out part a)? Can you use your table in question ❶ to help you? Do you have to start again from 0 for 16 flags?*
- Question ❹: *How do you know which numbers will appear in the count? Is there a way you can tell without counting up in 50s from 0 to 1,000? How?*

IN FOCUS Question ❶ looks at the sequence of multiples of 50. Children by this time should be learning the sequence, but they can link it to a number line to help them if they need. Children should start noticing patterns in the last two digits. Question ❷ looks to see if they can extend the table themselves and find the number of stars on 15 flags. Some children may need to write all the numbers down, some may be able to keep track in their heads. Discuss around whether they need to start at 0 each time for the 16 and 17 flags. Children should start to realise they just need to count on 50 more from the previous total. Question ❹ looks at the pattern of the last two digits and gets children to think about how to identify numbers in the count without actually counting up in 50s from 0 to 1,000.

STRENGTHEN To strengthen understanding, give children place value counters. Start with 5 ten counters and count in 10s with them. Take another 5 ten counters and count the total, which is 100. Ask them to count aloud from 0 to 100 in 10s, counting the separate counters. Then put the counters in piles of 5 and count aloud in 50s. Swap the 10 ten counters for 1 hundred counter. Continue doing this, counting on, and children will start to notice the pattern.

DEEPEN Ask children how many 50s there are between 350 and 800. Ask them to explain how they worked it out.

ASSESSMENT CHECKPOINT Use questions ❸ and ❹ to assess whether children can count forwards and backwards in 50s within 1,000. Check that children can recognise numbers that are multiples of 50.

ANSWERS

Question ❶: 150, 200, 250, 300, 350, 400, 450, 500

Question ❷ a): There are 750 stars on 15 flags.

Question ❷ b): There are 800 stars on the 16 flags.

Question ❷ c): There are 850 stars on 17 flags.

Question ❸: There are 1,000 stars on 20 flags.

Question ❹: 750, 650, 400, 50, 250

The last two digits are 00 or 50.

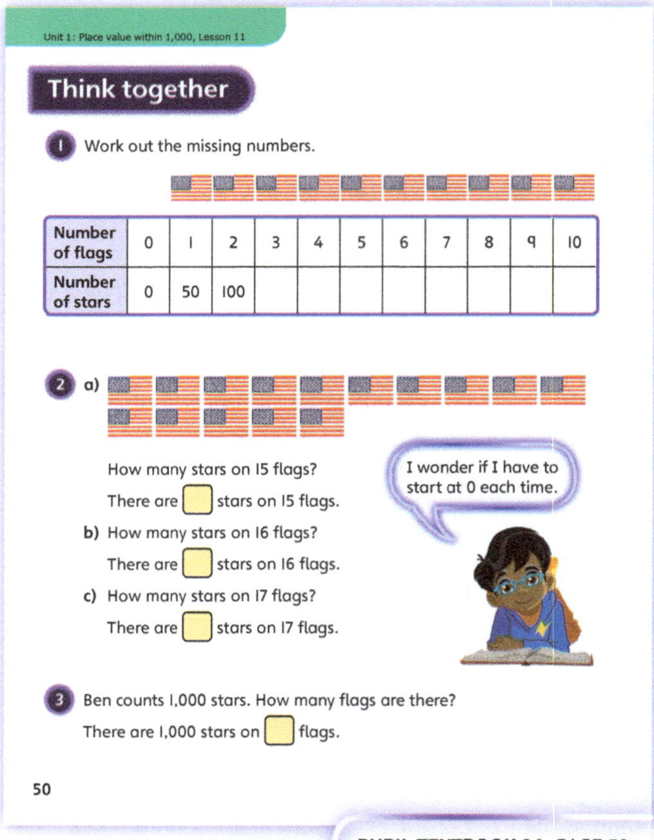

Number of flags	0	1	2	3	4	5	6	7	8	9	10
Number of stars	0	50	100								

PUPIL TEXTBOOK 3A PAGE 50

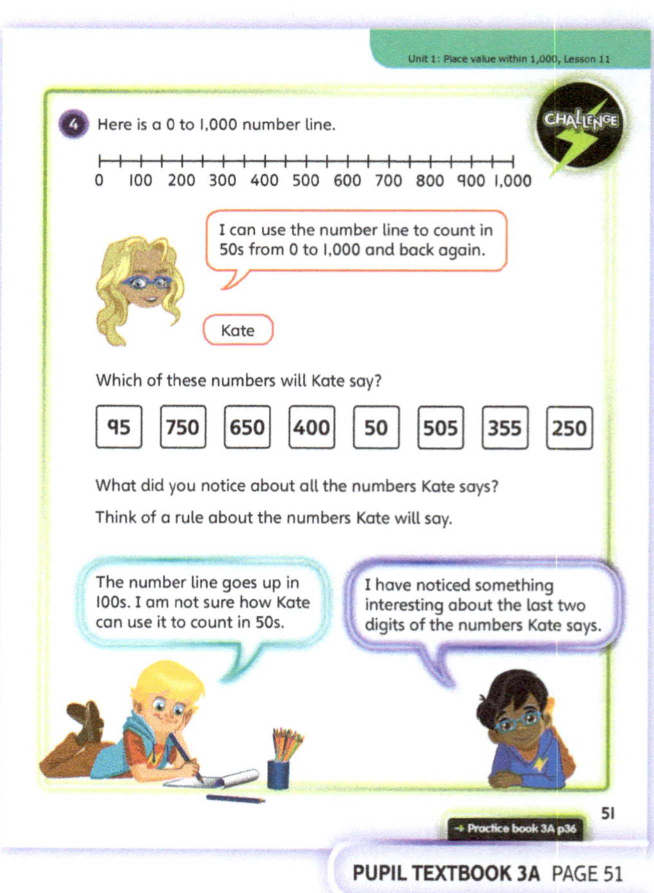

PUPIL TEXTBOOK 3A PAGE 51

Practice

WAYS OF WORKING Independent thinking

IN FOCUS Question ❶ reinforces counting in 50s. The table does not contain all the numbers from 1 to 10, so children should be careful when filling in the later values. Question ❺ links what children did in Year 2 on money to counting up in 50p pieces. They first need to convert £7 to pence and then count up in 50s to make 700. Children may need to be reminded that there are 100 pence in a pound.

STRENGTHEN In question ❺, children could use 50p coins to count up to £7. Provide a number line for them to put their 50p coins on. Ensure that they understand that two 50p coins make 100p, which is the same as £1.

DEEPEN Give children some more money problems. For example, Bella has twelve 50p coins and Jack has seven £1 coins. Who has more money? Ask them to write their answer as an inequality.

THINK DIFFERENTLY Question ❹ requires children to count in mixed amounts, mixing counting in 100s (lesson 1) and counting in 50s. Children could count in 50s by considering the 100s as two 50s or they could count the 100s first and then start counting on in 50s.

ASSESSMENT CHECKPOINT Use questions ❶, ❷ and ❸ to assess whether children can count forwards and backwards in 50s from any multiple of 50 between 0 and 1,000. Children should be able to apply their knowledge in a variety of contexts, such as money, number tracks and number lines.

ANSWERS Answers for the **Practice** part of the lesson appear in the separate **Practice and Reflect answer guide**.

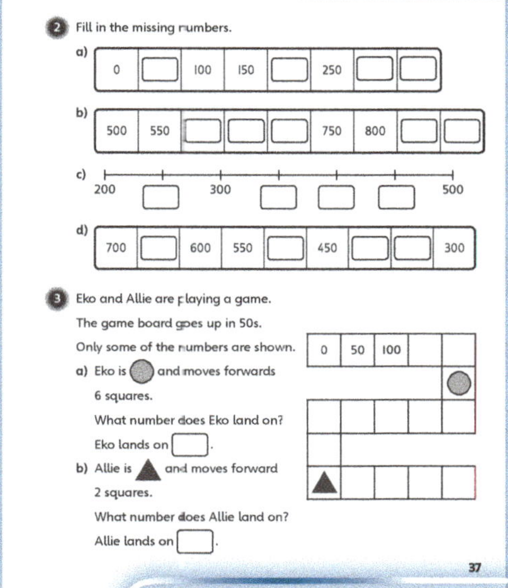

PUPIL PRACTICE BOOK 3A PAGE 36

PUPIL PRACTICE BOOK 3A PAGE 37

Reflect

WAYS OF WORKING Pair work

IN FOCUS Children try to spot the pattern of the numbers that will be in the count from 0 to 1,000 in 50s. They should count up from 0 to 1,000 and then look for any similarities between the numbers.

ASSESSMENT CHECKPOINT Check that children can count forwards in 50s and ensure that they can identify that every number ends in either 00 or 50.

ANSWERS Answers for the **Reflect** part of the lesson appear in the separate **Practice and Reflect answer guide**.

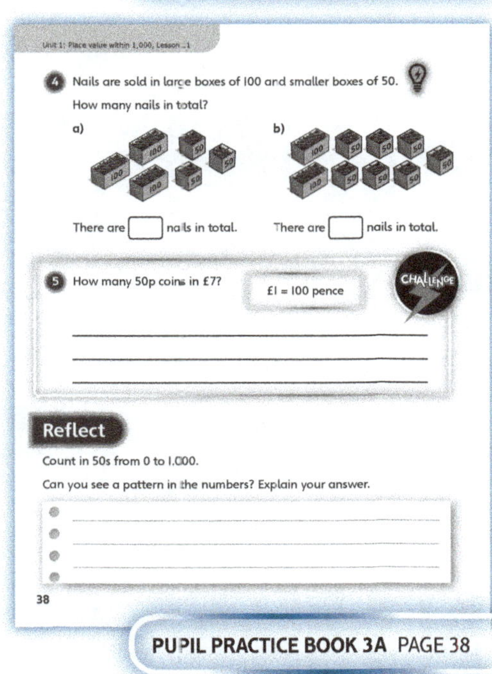

PUPIL PRACTICE BOOK 3A PAGE 38

After the lesson 

- Can children count in 50s forwards and backwards between 0 and 1,000?
- Can children count in 50s forwards and backwards from any multiple of 50?
- Do children know that all the multiples of 50 end in 00 or 50?

End of unit check

Don't forget the *Power Maths* unit assessment grid on p26.

WAYS OF WORKING Group work adult led

IN FOCUS In question **2**, children must accurately read from a number line, knowing when a number is half-way between two markers. Question **3** is designed to check that they know when to add 10 to the given number and when to do the inverse operation. Question **7** is a SATS-style problem and interweaves several topics within this unit. Children must know how to find 100 more and how to order numbers. Look for different approaches that children take to a two-step problem.

ANSWERS AND COMMENTARY Children who have mastered the concepts in this unit will be able to write down a 3-digit number given a representation in base 10 equipment. They will also realise that it does not matter which order the base 10 equipment appears in. Children will show that they can work out 1, 10 and 100 more or less than a given number and can work out the original number given an increase or decrease. Children will compare and order numbers, using the correct inequality signs where appropriate.

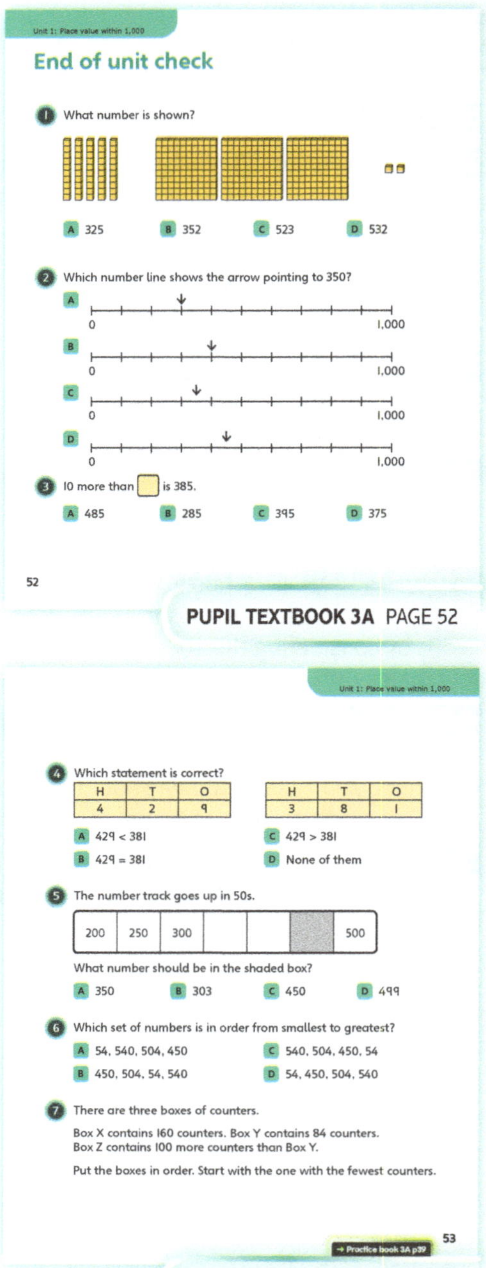

PUPIL TEXTBOOK 3A PAGE 52

PUPIL TEXTBOOK 3A PAGE 53

Q	A	WRONG ANSWERS AND MISCONCEPTIONS	STRENGTHENING UNDERSTANDING
1	B	A, C or D indicates that children do not know that the order of the digits is 100s, 10s or 1s or do not know the size of the base 10.	For question 1, emphasise and show using base 10 equipment that 100s come first, then 10s and then 1s. Show this on a place value table and next to this a part-whole model to secure understanding.
2	C	A or B could indicate a lack of understanding of numbers half-way between two marks on number line.	
3	D	A suggests that children have found 100 more. Choosing C indicates a lack of understanding of how to find the original number.	When finding 1, 10 and 100 more, ask children to focus on the order of the language. For example, '285 is 10 more' is different to 'what is 10 more than 285?' Get them to make the number with base 10 equipment or place value counters if necessary and ask: *Is this 10 more or do we need to work out 10 more?*
4	C	A, B or D indicates confusion with inequality signs.	
5	C	B suggests that they counted on from 300 in 1s. D suggests that they counted 50 on from 500.	
6	D	A or C indicates a lack of understanding of place value.	
7	Y, X, Z	Some children may think box Y contains the most counters as it starts with 8 and they don't understand that there are 0 hundreds.	

My journal

WAYS OF WORKING Independent thinking

ANSWERS AND COMMENTARY

In question ❶ , children may:
- say the number (for example, 415 or 4 hundred and 15)
- describe how the number is made (4 hundreds, 1 ten and 5 ones or 400 + 10 + 5)
- make a comparison (for example, 415 is 100 more then 315)
- show the number on a number line
- show the number using place value counters
- show the number using a part-whole model.

Encourage children to do as many different variations as they can.

In question ❷ , children should be able to make:
- 502, 511, 520, 601, 610 using seven counters
- 503, 512, 521, 530, 602, 620 using eight counters.

Some children may include 700. Explain that this is not less than 700. If children are struggling, ask them how many 100s the number must have if it lies between 500 and 700.

To extend ask children to put the numbers in order or represent them on a number line. Ask what would happen if they had nine counters. Encourage children to explain what strategy they are using.

Power check

WAYS OF WORKING Independent thinking

ASK

- *How many ways do you think you can represent 859?*
- *How confident are you of partitioning a number into 100s, 10s and 1s?*
- *What is 1, 10 and 100 more than a number?*
- *How confident do you feel comparing and ordering two or more numbers?*

Power play

WAYS OF WORKING Pair work

IN FOCUS Use this Power play to assess whether children understand the key concepts in this lesson. The criteria in the table make children think about the place value of the numbers. For example, a number greater than 200 must have 2 or more counters in the hundreds column, an odd number must have an odd number of counters in the ones column and so on.

ANSWERS AND COMMENTARY Largest number: 600; smallest number: 6; odd number greater than 200: 213, 231, 303, 321, 411 or 501; even number less than 200: 6, 24, 42, 60, 114, 132 or 150; same number of 100s and 1s: 60, 141, 222 or 303; 10 more is 241: 231.

For questions where there is more than one answer, encourage children to find as many answers as they can. Suggest that they make up their own questions for their friend. Ask what they notice that the digits add up to in each number. Ask why this is the case.

After the unit ⏸

- Can children represent 3-digit numbers in multiple ways?
- Do children display flexibility with 3-digit numbers and are they able to make numbers to fit certain criteria, and to compare and order numbers?

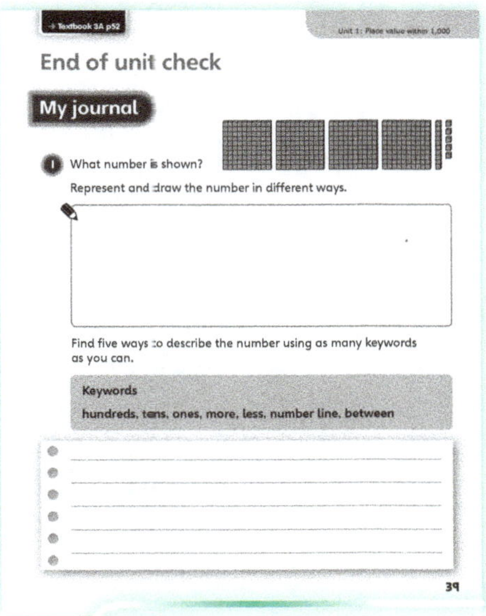

PUPIL PRACTICE BOOK 3A PAGE 39

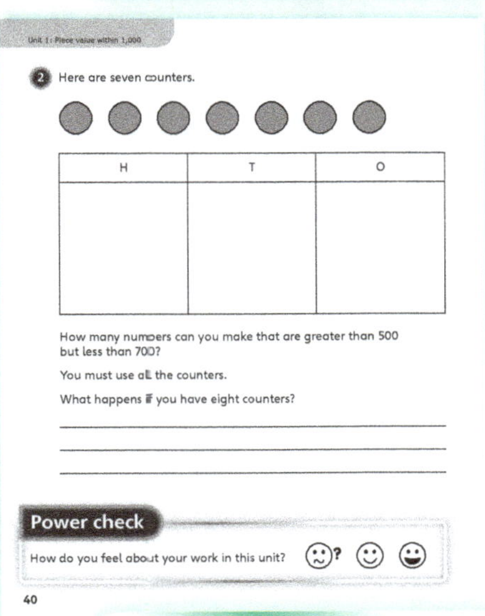

PUPIL PRACTICE BOOK 3A PAGE 40

PUPIL PRACTICE BOOK 3A PAGE 41

Strengthen and **Deepen** activities for this unit can be found in the *Power Maths* online subscription.

Unit 2
Addition and subtraction ❶

Don't forget to watch the Unit 2 video!

WHY THIS UNIT IS IMPORTANT

This unit is important because it builds key concepts in addition and subtraction on the strong foundation of place value from Unit 1. Children explore additions and subtractions gradually, beginning with adding and subtracting 1s, until by the end of the unit they are adding and subtracting 2-digit numbers. This unit prepares children to understand these calculations in formal methods, though the focus is on making decisions regarding the parts and wholes of numbers and on justifying the accuracy of mental methods where appropriate.

WHERE THIS UNIT FITS

→ Unit 1: Place value within 1,000
→ **Unit 2: Addition and subtraction (1)**
→ Unit 3: Addition and subtraction (2)

This unit builds upon the previous work children have done on place value within 1,000. It also builds upon children's ability to solve addition and subtraction calculations, by introducing exchange and formal written methods. This unit also develops children's reasoning and justifying skills which they are developing throughout the year.

Before they start this unit, it is expected that children:
- understand how to represent numbers to 1,000 on a number line and using place value equipment
- are able to partition numbers flexibly
- know how parts and wholes are related in additions and subtractions.

ASSESSING MASTERY

Children who have mastered this unit will be able to use mental methods, diagrams and place value grids to add to and subtract from 3-digit numbers. Children will be able to explain where an exchange was necessary, and how this relates to bridging a ten or hundred.

COMMON MISCONCEPTIONS	STRENGTHENING UNDERSTANDING	GOING DEEPER
Children may resort to counting strategies rather than using known number bonds to add digits within an addition.	Use part-whole models to display partially completed number bonds, to support children to think of these and use them effectively in their calculations.	Explore problems with multiple solutions and challenge children to explain how they have found all possible solutions in terms of the number bonds they used.
Children may choose one method for addition and subtraction, rather than learning to apply different written, pictorial or mental methods appropriate to particular calculations.	Ask: *Look at the numbers involved. What do you notice? Are there any signs about a good way to begin? What equipment would you use and how would you use the equipment?* Children should discuss these questions before launching into any of the processes.	Encourage children to discuss efficiency and accuracy of certain methods over others, and to justify which methods suit different calculations using reasoned arguments.

WAYS OF WORKING

Use these pages to introduce the unit focus to children as a whole class. You can use the different characters to explore different ways of working, and to begin to discuss and develop their reasoning skills relating to addition and subtraction.

STRUCTURES AND REPRESENTATIONS

Place value equipment: Can be manipulated by children to model the differences between addition and subtraction.

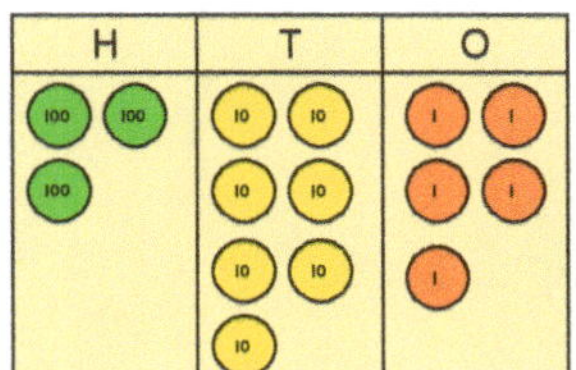
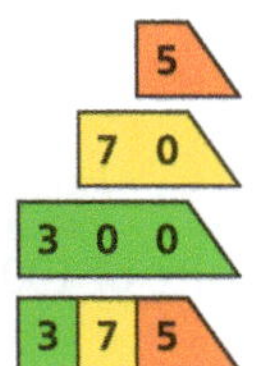

Number lines: This is a very useful model which in this unit enables children to understand how exchange is related to bridging 10s and 100s.

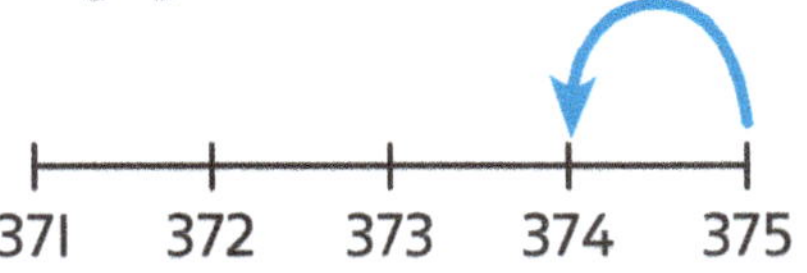

Part-whole models: This model is vital for children to be able to visualise how number bonds are related to the calculations involving 100s, 10s and 1s, and also for representing the flexible partitioning of numbers as it relates to exchange.

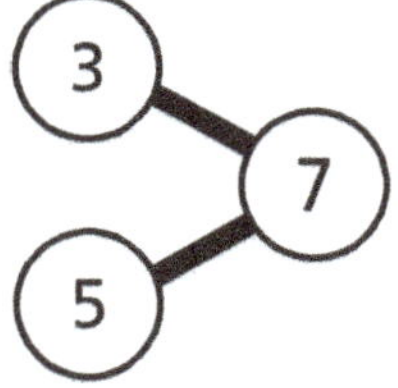

KEY LANGUAGE

There is some key language that children will need to know as part of the learning in this unit.

- add, addition
- subtract, subtraction, take away, difference
- exchange, pattern, variation, column method, mental method, part-whole model, number line
- total, altogether, calculations, regroup, partition, solutions
- place value, number bonds, fact family, related facts, number statements, method, order
- hundreds (100s), tens (10s), ones (1s), digits, zero (0)
- multiple of 10, multiples of 100, 3-digit number, 2-digit number, 10 ones, 10 tens
- left, greater than (>), less than (<), fewer, more, metres (m), miles, centimetres (cm), symbol

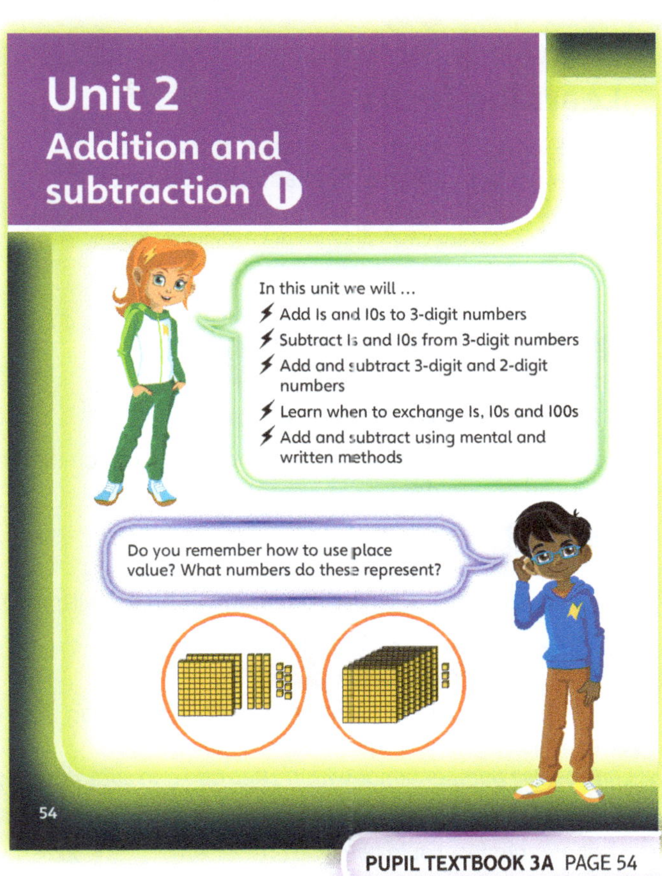

PUPIL TEXTBOOK 3A PAGE 54

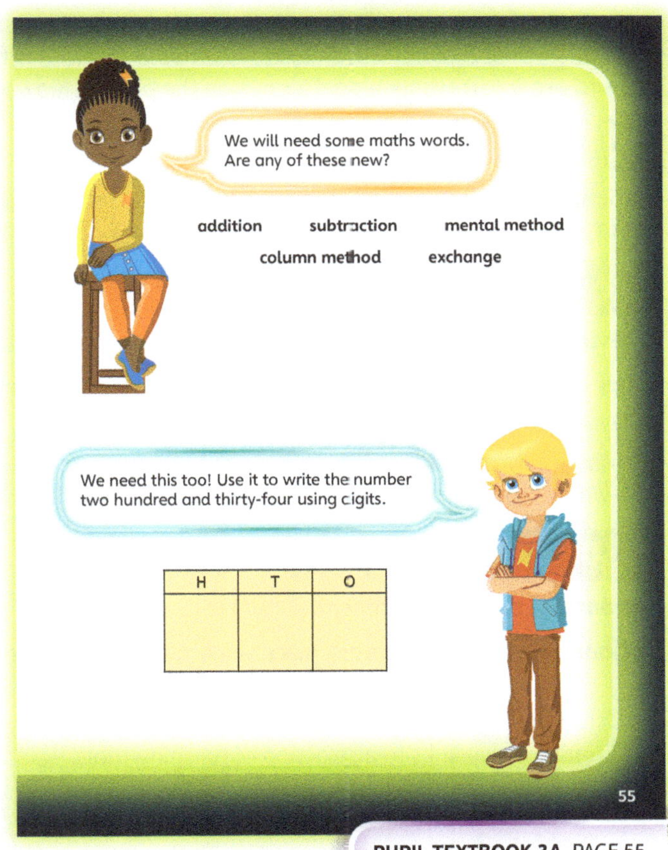

PUPIL TEXTBOOK 3A PAGE 55

Adding and subtracting 100s

Learning focus

In this lesson, children will use their knowledge of number bonds within 10 to add and subtract multiples of 100, up to 1,000.

Small steps

→ Previous step: Counting in 50s
→ **This step: Adding and subtracting 100s**
→ Next step: Adding and subtracting a 3-digit number and 1s

NATIONAL CURRICULUM LINKS

Year 3 Number – Addition and Subtraction

Add and subtract numbers mentally, including: a three-digit number and ones, a three-digit number and tens, a three-digit number and hundreds.

ASSESSING MASTERY

Children can use their knowledge of number bonds within 10 to add and subtract multiples of 100 up to 1,000.

COMMON MISCONCEPTIONS

Children may try to use a counting strategy rather than their knowledge of number bonds to derive additions and subtractions. Ask:

• *Can you use number bonds within 10 to add and subtract 100s?*

Children may struggle to unitise 100. Ask:
• *What number fact could help you answer 200 + ___ = 500?*

STRENGTHENING UNDERSTANDING

Use place value equipment such as base 10 equipment or place value counters to model the link between 2 + 3 = 5 and 200 + 300 = 500. Explore with children how this works with other part-whole relationships within 10. Children may find it helps to use a part-whole model alongside the place value counters.

GOING DEEPER

Challenge children to derive a family of 8 addition and subtraction facts for multiples of 100 related to 2 + 3 = 5, or other number bonds within 10. Then, challenge children to find 8 related facts for multiples of 10 as well.

KEY LANGUAGE

In lesson: number bond, addition, subtraction, left, hundreds (100s), total, altogether, fact family, part-whole model

Other language to be used by the teacher: related facts, take away

STRUCTURES AND REPRESENTATIONS

Part-whole model, number line, place value equipment

RESOURCES

Mandatory: base 10 equipment

Optional: place value counters, place value cards, place value grids

 In the eTextbook of this lesson, you will find interactive links to a selection of teaching tools.

Before you teach 

• Can children find related additions and subtractions to a number fact such as 3 + 4 = 7?
• Can children use equipment to represent 3 hundreds?
• Can children count up and back in 100s?

Discover

WAYS OF WORKING Pair work

ASK

- Question **1**: *Can you represent the number of bricks using place value equipment?*
- Question **1**: *What calculation matches the story?*
- Question **1**: *How can you decide whether to add or subtract?*

IN FOCUS Discuss the fact that each pallet holds 100 bricks in question **1**, and how this can help work out the number of bricks in an efficient way.

PRACTICAL TIPS Base 10 equipment or other place value equipment should be used to represent the number of 100s, alongside a part-whole model to represent the additive relationships.

ANSWERS

Question **1** a): 300 + 200 = 500. Amal has 500 bricks in total now.

Question **1** b): 400 – 200 = 200. There are 200 bricks left on the lorry.

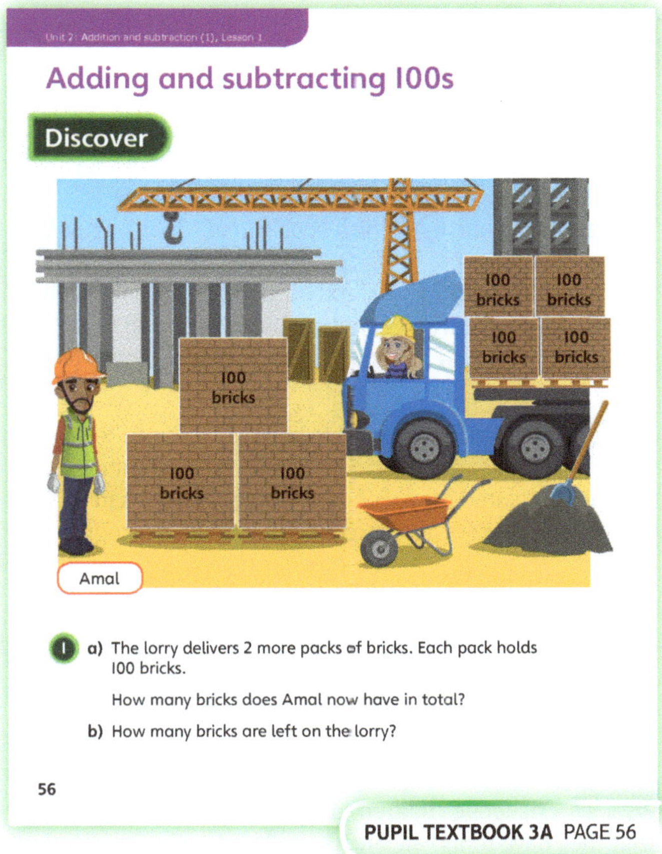

PUPIL TEXTBOOK 3A PAGE 56

Share

WAYS OF WORKING Whole class teacher led

ASK

- Question **1** a): *Can you show the parts and the whole in a part-whole model?*
- Question **1** a): *Which method is more efficient: counting or using bonds?*
- Question **1** a): *How is the number bond 3 + 2 = 5 related to the question?*
- Question **1** b): *Which part-whole model would suit 400 – 200 = 200? Can you think of a bond within 10 that matches?*

IN FOCUS In question **1** a), discuss the difference between finding the answer by counting in 100s, and thinking of the number bonds as 3 hundreds + 2 hundreds = 5 hundreds. It is important that children recognise that knowing and using known number bonds is a very efficient strategy. It will be a vital skill for children to use as they continue to develop addition and subtraction methods for more complicated additions.

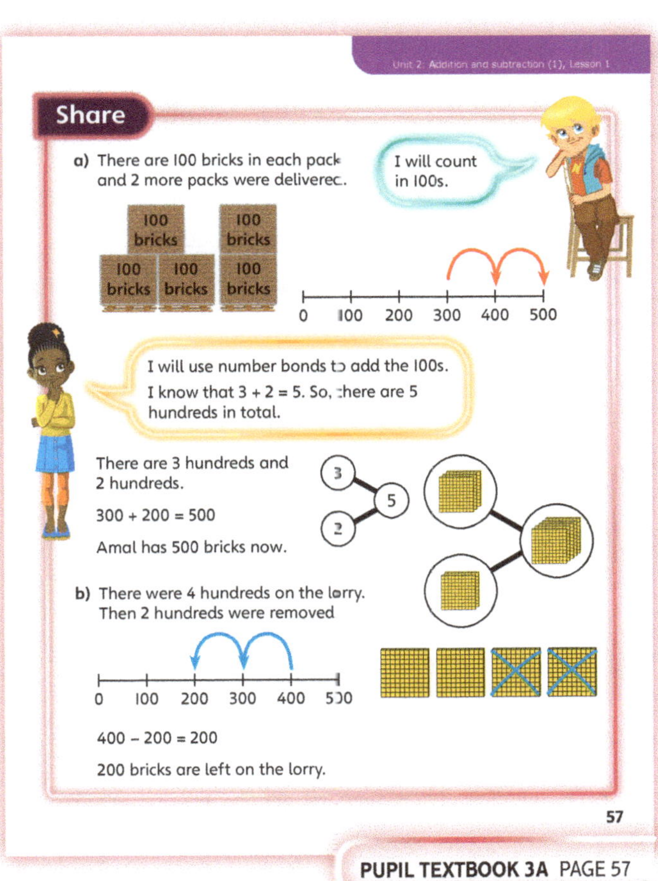

PUPIL TEXTBOOK 3A PAGE 57

Think together

WAYS OF WORKING Whole class teacher led (I do, We do, You do)

ASK

- Question **1** and **2**: *Which bond within 10 will help solve each calculation?*
- Question **1** and **2**: *Will you need to find a whole by addition, or a part, using subtraction? How do you know?*
- Question **2**: *How can you represent this calculation using equipment?*

IN FOCUS Question **4** really prompts children to recognise that knowing bonds is a good way to derive and find a number of related facts. For each of the questions in the **Think together** section encourage children to recognise the relationship between addition and subtraction within 10, and the related facts for multiples of 100s.

STRENGTHEN Use part-whole models and place value equipment to model the relationships between number bonds within 10 and related facts for multiples of 100. Model the language to unitise hundreds: 3 hundreds plus 4 hundreds is 7 hundreds, so 300 + 400 = 700.

DEEPEN Challenge children to convince you or a partner that they have found all 8 facts in a fact family. Ask: *How can you be sure that you have found all 8?*

ASSESSMENT CHECKPOINT Can children explain how the part-whole relationships for bonds within 10 are related to adding and subtracting 100s? Can children explain the place value of their additions and subtractions? For example, 3 hundreds plus 5 hundreds is 8 hundreds, because 3 + 5 = 8.

ANSWERS

Question **1**: 7 – 3 = 4. 700 – 300 = 400. There are 4 hundreds left. There are 400 hinges left.

Question **2** a): 600 – 400 = 200. She has 200 nails left.

Question **2** b): 400 + 400 = 800. He has 800 screws in total.

Question **3**: Astrid has added instead of subtracted. Instead of adding 6 boxes she should have subtracted 3 from 6: 6 – 3 = 3. She has also confused boxes with bolts. Rather than needing 100s of boxes, they need 100s of bolts. Each box contains 100 bolts, so they only need a few boxes. They need 3 more boxes, which means they need 300 more bolts.

Question **4**: 200 + 700 = 900, 700 + 200 = 900, 900 = 200 + 700, 900 = 700 + 200, 900 – 700 = 200, 900 – 200 = 700, 200 = 900 – 700, 700 = 900 – 200

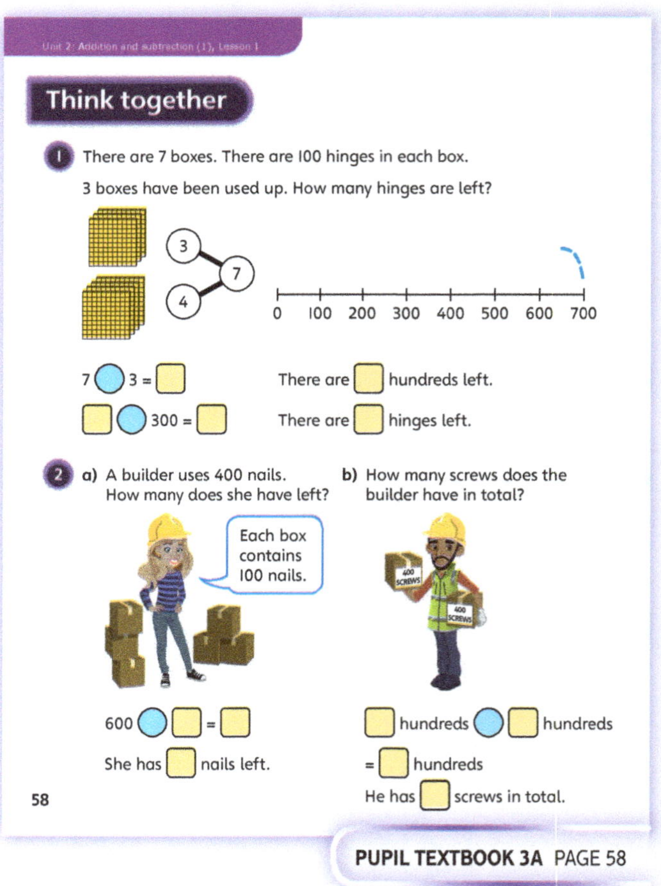

PUPIL TEXTBOOK 3A PAGE 58

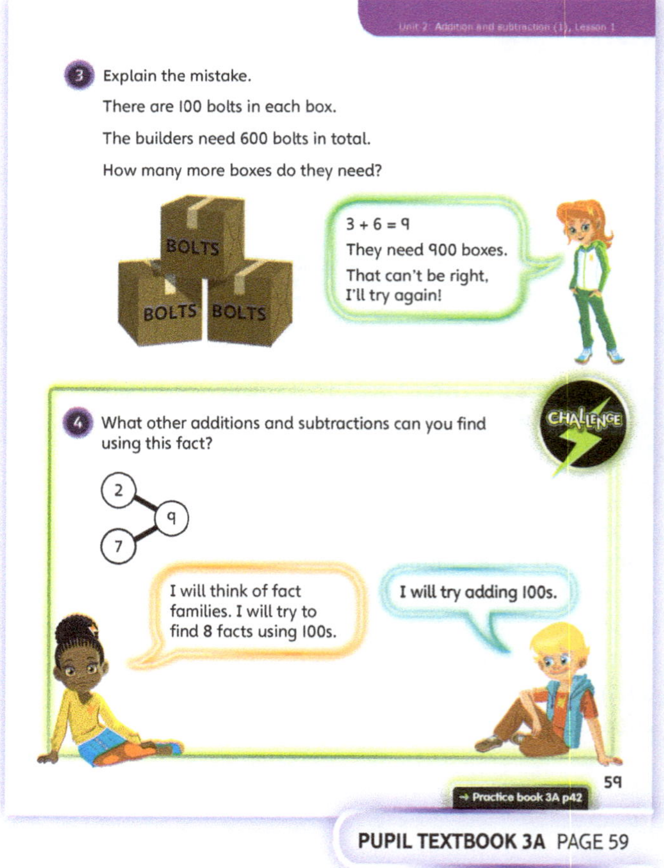

PUPIL TEXTBOOK 3A PAGE 59

Practice

WAYS OF WORKING Independent thinking

IN FOCUS Question ❶ requires children to recognise whether the context of each problem requires addition or subtraction to solve it, and for parts c) and d) children will need to choose the appropriate operations themselves.

The mathematical representations in questions ❶ and ❷ draw children's attention to the underlying structure of place value. Encourage them to notice the number of hundreds in each calculation.

Question ❹ prompts children to make decisions about which calculations are related to particular bonds within 10, and also requires them to use the concept of the lesson to solve missing number calculations.

STRENGTHEN Children could use place value equipment to model the calculations, and use them to represent the parts and wholes in a part-whole model. Encourage children not to rely on a method of counting all or counting on. They should use the place value equipment to represent the additive relationships, but use known bonds to support them. If children are not secure in bonds within 10, they will require significant development in this area. It may help to use partially completed part-whole models to represent the bonds that are relevant to the questions.

DEEPEN Question ❻ requires logical reasoning based on the symbols of the 'code'. If children manage to complete it and convince themselves or a partner that their answer satisfies all the parts of the problem, challenge them to invent their own puzzle. Other children could be challenged to write or draw a word problem for a particular calculation, such as 500 − 400 = ___.

THINK DIFFERENTLY Question ❺ addresses a potential uncertainty about the connection between three 3-digit numbers in a part-whole model. As Astrid hints, children should notice that 300 and 500 belong in the bottom circles, and their sum (800) in the top one. Encourage children to think about the relationship between number bonds within 10 and related facts for multiples of 100.

ASSESSMENT CHECKPOINT Can children explain how their answers are accurate, by justifying them using number bonds? Can they solve missing number problems such as 400 + ___ = 900?

ANSWERS Answers for the **Practice** part of the lesson appear in the separate **Practice and Reflect answer guide**.

Reflect

WAYS OF WORKING Pair work

IN FOCUS Children should work independently to create their fact families and then compare with their partner to see if their fact families agree. Can they convince one another they have found all eight facts?

ASSESSMENT CHECKPOINT Can children explain the relationship between the bond within 10 and the addition and subtraction of 100s?

ANSWERS Answers for the **Reflect** part of the lesson appear in the separate **Practice and Reflect answer guide**.

After the lesson ⏸

- Can children use number bonds efficiently to add and subtract 100s?
- Can children solve missing number problems for additions and subtractions?
- Are children able to represent the calculations using place value equipment?

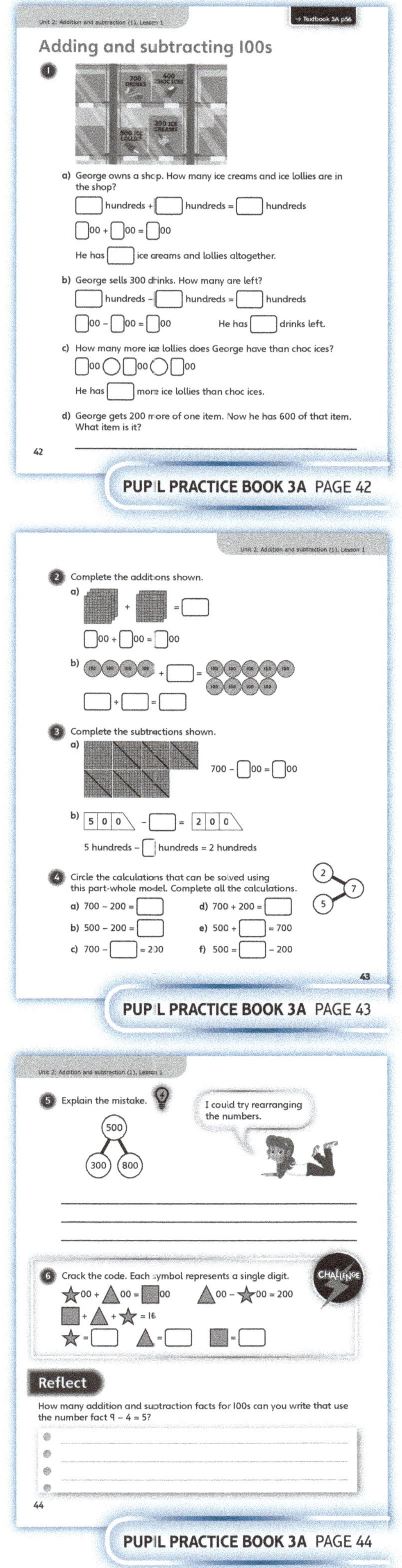

PUPIL PRACTICE BOOK 3A PAGE 42

PUPIL PRACTICE BOOK 3A PAGE 43

PUPIL PRACTICE BOOK 3A PAGE 44

Adding and subtracting a 3-digit number and 1s

Learning focus

In this lesson, children will add and subtract a single digit to and from a 3-digit number, using their understanding of place value. In this lesson, children will not be required to bridge 10s.

Small steps

→ Previous step: Adding and subtracting 100s
→ **This step: Adding and subtracting a 3-digit number and 1s**
→ Next step: Adding a 3-digit number and 1s

NATIONAL CURRICULUM LINKS

Year 3 Number – Addition and Subtraction

- Add and subtract numbers mentally, including: a three-digit number and ones, a three-digit number and tens, a three-digit number and hundreds.
- Solve problems, including missing number problems, using number facts, place value, and more complex addition and subtraction.

ASSESSING MASTERY

Children can add single-digit numbers to a 3-digit number by adding the 1s digits of both numbers. Children can subtract single-digit numbers from a 3-digit number by taking away single-digit numbers from the 1s of 3-digit numbers in calculations that do not require exchange.

COMMON MISCONCEPTIONS

Children may use a counting on strategy, where adding the 1s digits is a more efficient strategy. Ask:
- *How many 1s in the total? Can you work it out without counting on or back?*

Children may struggle if their understanding of place value above 100 is not secure. Ask:
- *What 3 numbers come next: 207, 208, 209 …?*

STRENGTHENING UNDERSTANDING

Children should represent the calculations using place value equipment to reinforce the concept of using place value to make their methods efficient.

GOING DEEPER

Open questions, such as: *I add a 3-digit number and a single-digit number. The total is 489. What could my numbers have been?* should prompt children to explore multiple solutions and develop systematic problem solving.

KEY LANGUAGE

In lesson: addition, subtraction, number bonds, hundreds (100s), tens (10s), ones (1s), solutions, total, altogether, number statements

STRUCTURES AND REPRESENTATIONS

Number line, place value equipment

RESOURCES

Mandatory: base 10 equipment, place value grids, place value cards, 0–9 digit cards

Optional: place value counters

 In the eTextbook of this lesson, you will find interactive links to a selection of teaching tools.

Before you teach 

- Do children understand what a '3-digit number' means?
- Do children know the difference between a digit and a number?
- Can children represent a 3-digit number using place value equipment?

Discover

 Pair work

ASK

• Question ❶ a): *How could you arrange the equipment to represent the place value clearly?*
• Question ❶ a): *How many 1s, 10s and 100s will you have in total?*

IN FOCUS In question ❶ a) children should discuss how the number being added is only a single-digit number. They may see that you can combine the 1s rather than have to add any 10s or 100s.

PRACTICAL TIPS Children should represent the number of people in the museum and also the number of people about to enter the museum using place value equipment.

ANSWERS

Question ❶ a): 245 + 4 = 249

Question ❶ b): 249 − 1 = 248

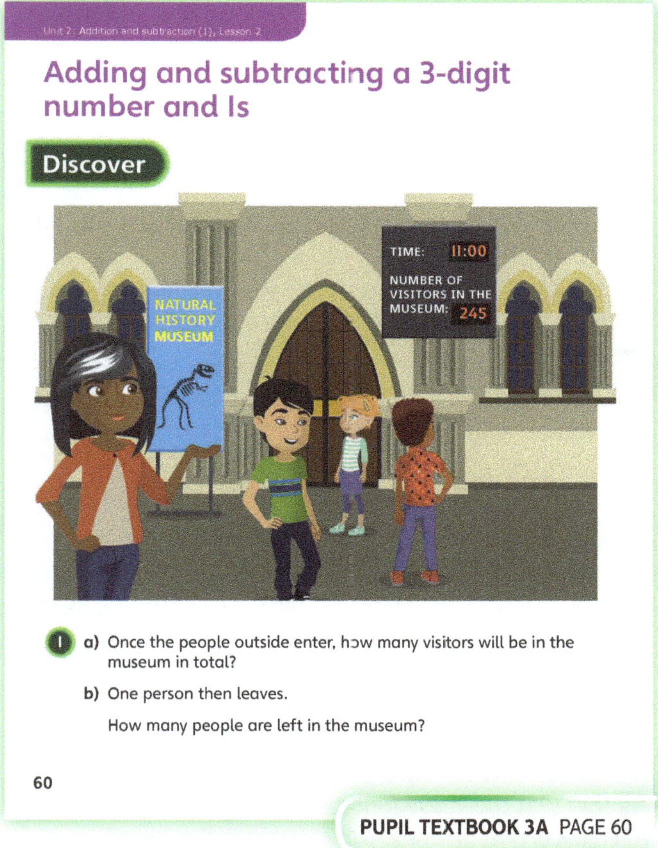

PUPIL TEXTBOOK 3A PAGE 60

Share

 Whole class teacher led

ASK

• Question ❶ a): *Can you see how the base 10 equipment has been organised to show the addition clearly?*
• Question ❶ a): *Why do the 10s digit and the 100s digit not change?*
• Question ❶ : *What mistakes might you make if you did not organise the equipment clearly?*
• Question ❶ : *Could you still do the calculations in the same way if you did not have equipment?*

IN FOCUS The important concept in this part of the lesson is to notice how the calculation can be solved by using knowledge of bonds within 10. The place value equipment represents the concept of adding the 1s digits, but the calculation can be solved without resorting to counting.

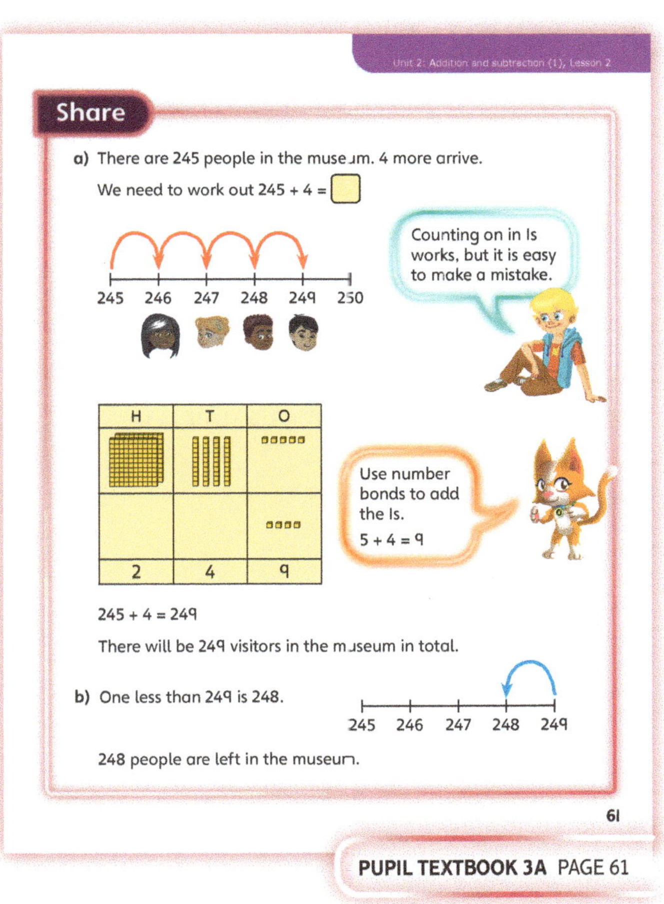

PUPIL TEXTBOOK 3A PAGE 61

Think together

WAYS OF WORKING Whole class teacher led (I do, We do, You do)

ASK

- Question **2** : *Can you tell whether to add or subtract? How do you know?*
- Question **3** : *Do you need to use equipment to find the answer by counting? How could you use bonds?*

IN FOCUS Question **3** requires children to recognise how to solve missing number problems, and when to use an inverse operation. Encourage children to model the calculations as additions and subtractions, and to solve them using their knowledge of number bonds.

STRENGTHEN Encourage children to represent the calculations using place value equipment, to gain an understanding of why they only need to add the 1s digits. However, also model for children how they can complete the calculations by using just their knowledge of bonds.

DEEPEN Question **4** prompts children to collect multiple solutions by understanding the place value of the calculations deeply. Challenge them to convince one another that they have found all possible solutions. Ask: *Can you think of a word problem to match this challenge?*

ASSESSMENT CHECKPOINT Can children explain and justify their calculations based on adding or subtracting 1s digits from 3-digit numbers?

ANSWERS

Question **1** : 9 – 7 = 2
Now there are 2 ones.
319 – 7 = 312
312 people are left in the museum.

Question **2** : 1 + 6 = 7
291 + 6 = 297
Now there are 297 people.

Question **3** : 2 pm: 204 + 3 = 207 people
3pm: 198 – 8 = 190 people
4 pm: 158 – 6 = 152 people
5 pm: 119 – 4 = 115 people

Question **4** a): 6 solutions: 430 + 5, 431 + 4, 432 + 3, 433 + 2, 434 + 1, 435 + 0

Question **4** b): 5 solutions: 439 – 4, 438 – 3, 437 – 2, 436 – 1, 435 – 0

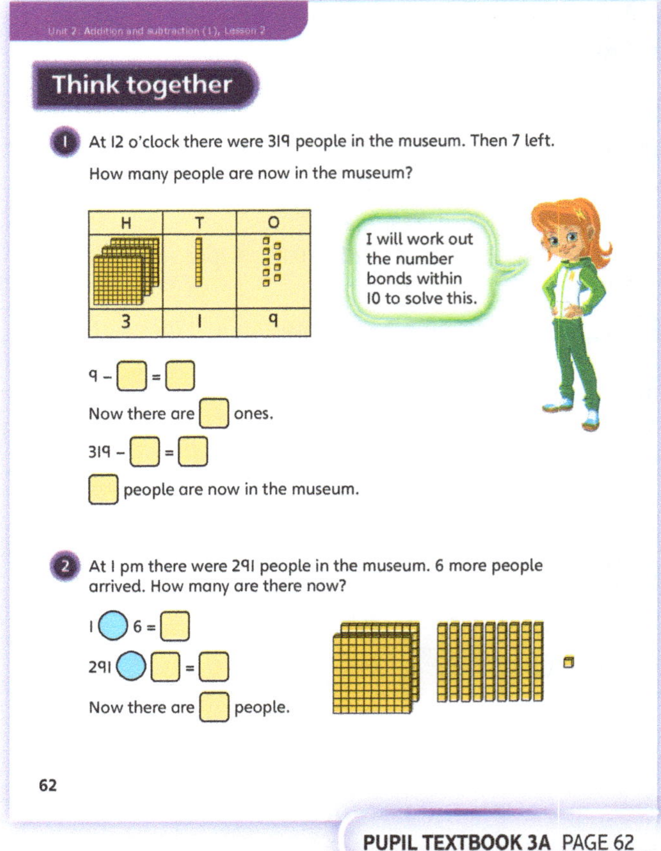

PUPIL TEXTBOOK 3A PAGE 62

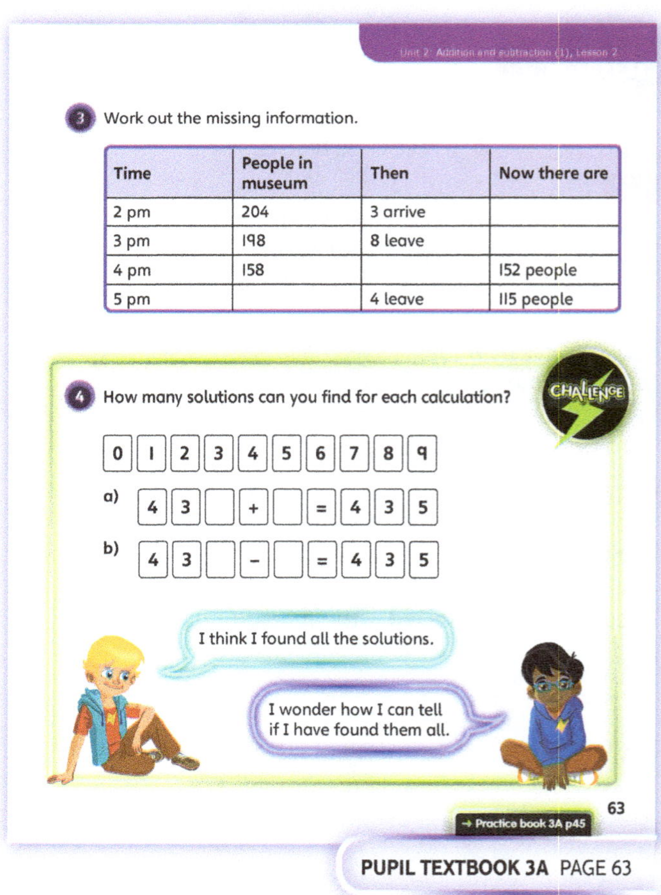

Time	People in museum	Then	Now there are
2 pm	204	3 arrive	
3 pm	198	8 leave	
4 pm	158		152 people
5 pm		4 leave	115 people

PUPIL TEXTBOOK 3A PAGE 63

Practice

WAYS OF WORKING Independent thinking

IN FOCUS Questions 4 and 5 will require children to have become confident enough to be able to solve abstract calculations including missing numbers and missing digits. They will need to understand the place value concept deeply to be able to justify their answers.

STRENGTHEN Encourage children to use a place value grid to organise their thinking, and to recognise how to combine or subtract the 1s depending on the context of the problem. Again, encourage them to use knowledge of number bonds to find the answers, but then check by counting the equipment.

DEEPEN Look at question 5 more deeply. Ask: *Can you explain the patterns in the answers, by describing the variations?*

Challenge children to create their own set of problems, where the answers are related through variation.

THINK DIFFERENTLY Question 4 contains the problem 200 + ___ = 200. This may cause some confusion, as children need to add a 0. Deepen their understanding further by asking about such examples as 218 – 8 = ___

ASSESSMENT CHECKPOINT Do children feel confident enough to justify their answers to abstract calculations by explaining the place value of their additions and subtractions?

ANSWERS Answers for the **Practice** part of the lesson appear in the separate **Practice and Reflect answer guide**.

Reflect

WAYS OF WORKING Independent thinking

IN FOCUS Notice whether children can organise their drawings to represent the place value accurately. Do they use base 10 equipment? Do they resort to counting on a number line or number track?

ASSESSMENT CHECKPOINT Do children's drawings represent the addition and subtraction of 1s digits?

ANSWERS Answers for the **Reflect** part of the lesson appear in the separate **Practice and Reflect answer guide**.

After the lesson

- Do children recognise how place value helps them solve these calculations efficiently?
- Can children justify their answers to themselves and others?
- Are children confident applying known number bonds to these calculations?

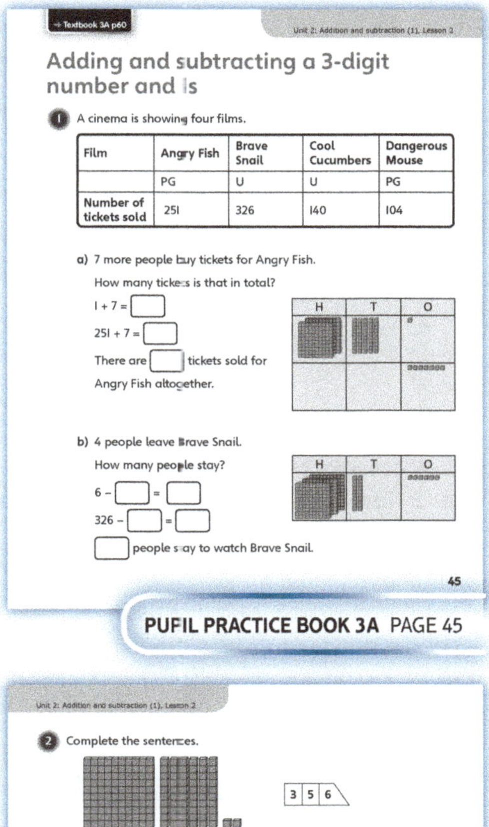

PUPIL PRACTICE BOOK 3A PAGE 45

PUPIL PRACTICE BOOK 3A PAGE 46

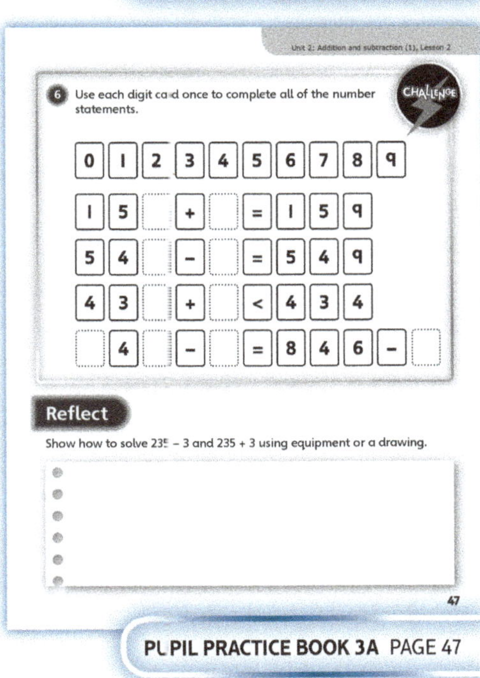

PUPIL PRACTICE BOOK 3A PAGE 47

Adding a 3-digit number and 1s

Learning focus

In this lesson, children will understand how to recognise additions where they will bridge a ten, and know how to use exchange of 10 ones for 1 ten.

Small steps

- → Previous step: Adding and subtracting a 3-digit number and 1s
- → **This step: Adding a 3-digit number and 1s**
- → Next step: Subtracting 1s from a 3-digit number

NATIONAL CURRICULUM LINKS

Year 3 Number – Addition and Subtraction

- Add and subtract numbers mentally, including: a three-digit number and ones, a three-digit number and tens, a three-digit number and hundreds.
- Solve problems, including missing number problems, using number facts, place value, and more complex addition and subtraction.

ASSESSING MASTERY

Children can add a single digit number to a 3-digit number and bridge the ten by exchanging 10 ones for 1 ten. Children can recognise when exchange of 10 ones for 1 ten is required.

COMMON MISCONCEPTIONS

Children may not understand that exchange is a different way of partitioning the number, and think that they are changing the number. Ask:

- *What is the same and what is different about 100 + 50 + 5 and 100 + 40 + 15?*

The concept of exchange may cause difficulty for some children, and they may assume you have to exchange for every addition. Ask:

- *Can you tell if the 10s digit will increase before working out the exact answer?*

STRENGTHENING UNDERSTANDING

Children will need to link this learning to knowledge of bonds within 20. It may help children to imagine or use a ten frame to support this thinking initially. Use of base 10 equipment to represent change from 1s to 10 in a concrete way will also support understanding.

GOING DEEPER

Encourage children to recognise whether the 10s digit will increase in an addition even before calculating the exact answer. Challenge children to recognise this as a checking strategy.

KEY LANGUAGE

In lesson: exchange, addition, subtraction, 10s digit, tens (10s), ones (1s), 10 ones, hundreds (100s), solutions, altogether, pattern

Other language to be used by the teacher: total, variation

STRUCTURES AND REPRESENTATIONS

Number line, part-whole model, place value equipment

RESOURCES

Mandatory: base 10 equipment, 0–9 digit cards

Optional: ten frame

 In the eTextbook of this lesson, you will find interactive links to a selection of teaching tools.

Before you teach 

- Do children know number bonds for 11?
- Can children tell if two numbers will add to 10 or greater?
- Do children know what to add to 135 to make 140?

Discover

WAYS OF WORKING Pair work

ASK

- Question **1** a): *What is the same and what is different about the calculations?*
- Question **1** a): *Can you use the method you learnt in the previous lesson?*
- Question **1** : *What equipment would you use to represent the calculations?*

IN FOCUS In question **1** a) the focus is to notice that they can apply the method of adding 1s from the previous lesson, though there is a complication that had not been covered previously, as now the 1s digits may total to ten or more.

PRACTICAL TIPS Children can replicate the context using digit cards. This will require them to think carefully about the place value of each digit.

ANSWERS

Question **1** a): 571 + 3 = 574 and 135 + 7 = 142

Question **1** b): 157 + 3 = 160 and 153 + 7 = 160

PUPIL TEXTBOOK 3A PAGE 64

Share

WAYS OF WORKING Whole class teacher led

ASK

- Question **1** a): *Can you explain why the number line uses a jump of 5 first? Why is there a jump of 2 next?*
- Question **1** a): *How does the method showing base 10 equipment link to the number line method?*
- Question **1** a): *How do you know the number of 1s to exchange?*

IN FOCUS In question **1** a) the focus is first to recognise that only one of the calculations requires exchange, or bridging of the next ten. Children should understand the decision about when to exchange and when not to exchange.

Ask children to look carefully at how the same concept is represented differently on the number line and with the base 10 equipment. Can children 'see' where the jump of 5 on the number line relates to the exchange of base 10 equipment?

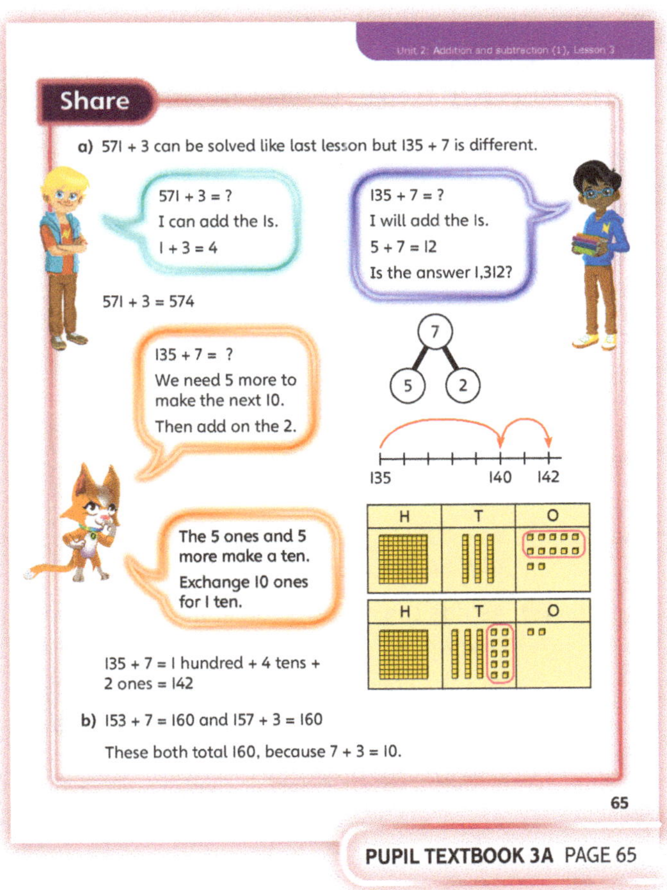

PUPIL TEXTBOOK 3A PAGE 65

Think together

WAYS OF WORKING Whole class teacher led (I do, We do, You do)

ASK

- Question ❶ : *How do you know whether the calculation requires an exchange of 10 ones?*
- Question ❶ : *Which method do you think is more effective: a number line or an exchange? Are they different methods, or are they just different ways of showing the same idea?*

IN FOCUS In question ❶ encourage children to recognise whether or not exchange is necessary, rather than always assuming they will have to do so. There are two methods shown: a number line and exchange of 10 ones in base 10 equipment. These represent the same concept in slightly different ways. It is important that children understand the link between the two methods. Encourage children to represent their thinking using place value equipment, and to justify the exchange of 10 ones, as this will build later into more formal column methods.

STRENGTHEN Support children by using knowledge of number bonds within 20 to justify whether or not the 10s digit will increase. Children should predict, by considering the addition of the 1s, then check by using place value equipment.

DEEPEN The conjecture in question ❹ is a good opportunity for children to engage in deep mathematical thinking. Challenge them to find examples and counter-examples for Astrid's conjecture, and ask if they could refine her wording to make it a more accurate statement.

ASSESSMENT CHECKPOINT Can children explain why the 10s digit increases by 1 for some calculations, justifying their reasoning based on number bonds within 20?

ANSWERS

Question ❶ : 6 ones + 5 ones = 11 ones
11 ones = 1 ten and 1 ones
316 + 5 = 3 hundreds + 2 tens + 1 ones = 321

Question ❷ a): The addition in 2 a) changes the 10s digit.
8 ones + 6 ones = 14 ones
14 ones = 1 ten + 4 ones
248 + 6 = 254

Question ❷ b): 2 ones + 6 ones = 8 ones
842 + 6 = 848

Question ❸ : 442 + 9, 443 + 8, 444 + 7, 445 + 6, 446 + 5, 447 + 4, 448 + 3, 449 + 2

Question ❹ : Astrid's conjecture is sometimes true. It is not true when the 10s digits total to less than ten.

The 10s digit will never increase by 2 if only a single digit number is added, as the largest total of the 1s digits possible is 18.

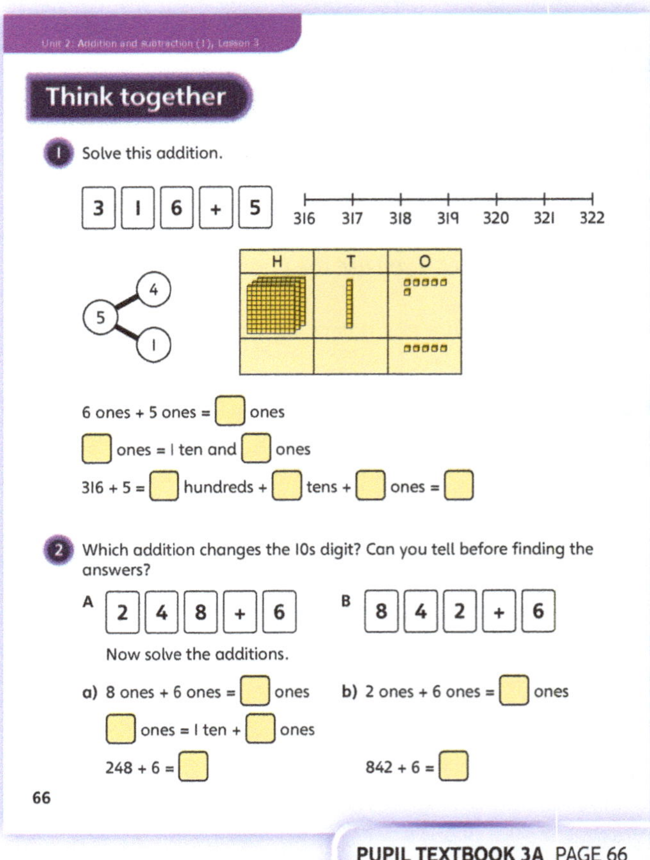

PUPIL TEXTBOOK 3A PAGE 66

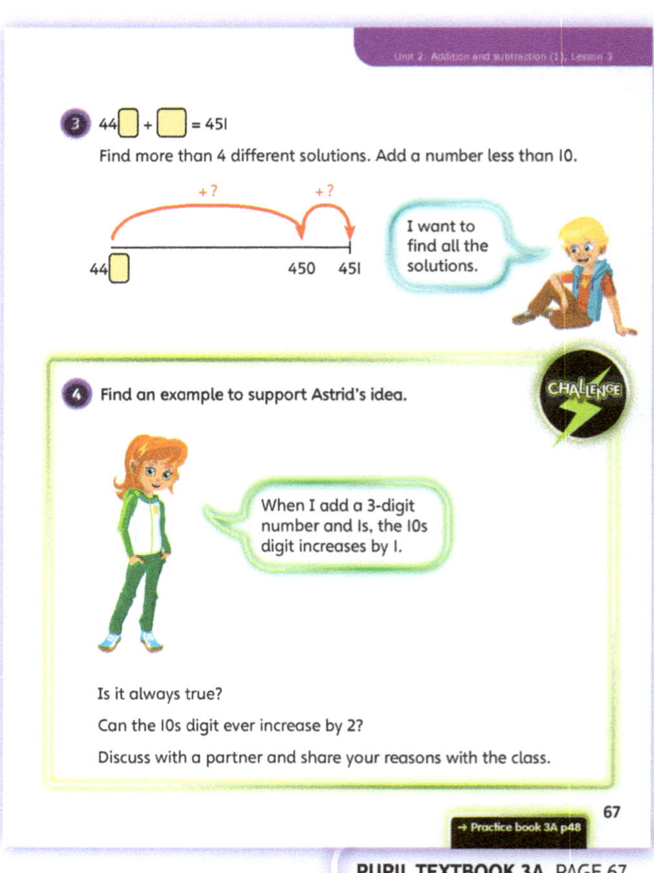

PUPIL TEXTBOOK 3A PAGE 67

Practice

WAYS OF WORKING Independent thinking

IN FOCUS In questions **1** and **2** the key aspect is that children are able to decide based on understanding of the addition of 1s, whether the 10s will increase or not. Through completing questions **3**, **4** and **5** children should make decisions about how to represent their thinking, but during the practice they should become fluent enough to be able to calculate mentally.

STRENGTHEN Display some partially completed bonds within 20 to support children in their thinking. It may also help children to display bonds to 10, so that they can reason whether their additions will cross the 10s barrier.

DEEPEN Challenge children to explain the patterns in question **3** deeply, in terms of the variation between the calculations.

THINK DIFFERENTLY Question **4** requires children to notice whether or not to exchange. It is important to challenge the misconception that an exchange is always necessary.

ASSESSMENT CHECKPOINT Can children add a single digit number to a 3-digit number accurately, deciding whether or not to exchange? Are children able to justify their reasoning?

ANSWERS Answers for the **Practice** part of the lesson appear in the separate **Practice and Reflect answer guide**.

Reflect

WAYS OF WORKING Pair work

IN FOCUS Children should discuss how to word their responses in pairs. Giving children the chance to rehearse their reasoning should improve their precision and accuracy.

ASSESSMENT CHECKPOINT Can children explain that only one calculation requires an exchange and justify why?

ANSWERS Answers for the **Reflect** part of the lesson appear in the separate **Practice and Reflect answer guide**.

After the lesson ⏸

- Can children solve calculations by exchanging 10 ones for 1 ten?
- Can children recognise when an exchange is not necessary?

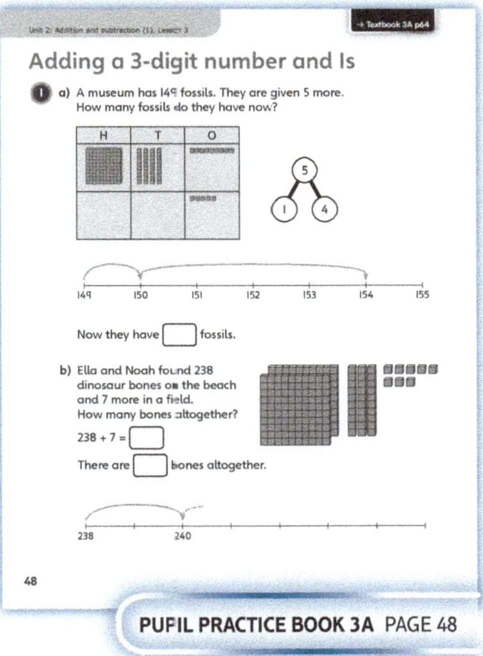

PUPIL PRACTICE BOOK 3A PAGE 48

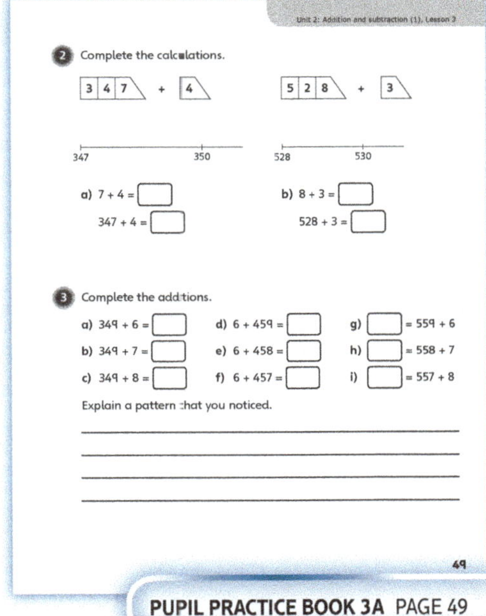

PUPIL PRACTICE BOOK 3A PAGE 49

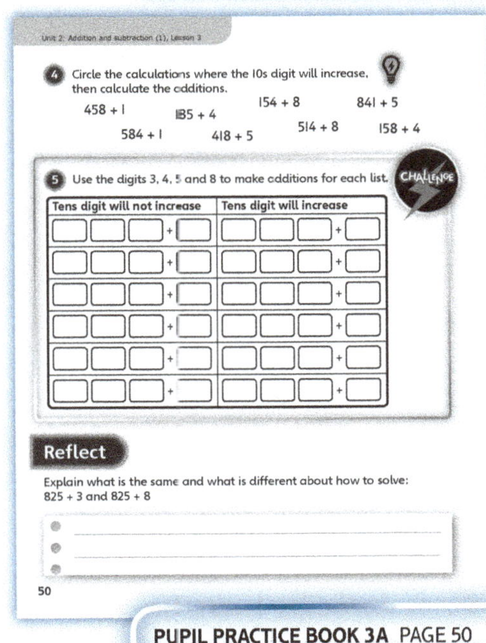

PUPIL PRACTICE BOOK 3A PAGE 50

Subtracting Is from a 3-digit number

Learning focus

In this lesson, children will subtract a single digit number where the subtraction bridges a 10. Children understand how to exchange 1 ten for 10 ones.

Small steps

→ Previous step: Adding a 3-digit number and 1s
→ **This step: Subtracting 1s from a 3-digit number**
→ Next step: Adding and subtracting a 3-digit number and 10s

NATIONAL CURRICULUM LINKS

Year 3 Number – Addition and Subtraction

Add and subtract numbers mentally, including: a three-digit number and ones, a three-digit number and tens, a three-digit number and hundreds.

ASSESSING MASTERY

Children can explain how to use exchange of 1 ten for 10 ones to subtract a single-digit number. Children can justify their reasoning using their knowledge of number bonds within 20.

COMMON MISCONCEPTIONS

One of the most common misconceptions children make is to subtract the digits in the wrong order. For example, when calculating 234 – 7, children very often do 7 ones take away 4 ones, as they think you have to subtract the smaller digit from the larger. This is one of the most common errors in learning to subtract. Ask:

• *What is the whole, and what part are you subtracting?*

STRENGTHENING UNDERSTANDING

It may help children to represent the subtraction as two jumps on a number line, then explore the link with the exchange of place value equipment.

GOING DEEPER

It may be that imagining a number line to subtract is a more efficient mental method than writing a subtraction involving exchange. Challenge children to discuss this and decide which methods are most efficient and accurate for different kinds of subtraction.

KEY LANGUAGE

In lesson: subtraction, number line, exchange, tens (10s), ones (1s), calculations, zero (0), greater than (>)

Other language to be used by the teacher: hundreds, less than (<)

STRUCTURES AND REPRESENTATIONS

Number line, part-whole model, place value equipment

RESOURCES

Mandatory: place value grids, place value cards, base 10 equipment

Optional: bead strings, 0–9 digit cards

 In the eTextbook of this lesson, you will find interactive links to a selection of teaching tools.

Before you teach 

• Can children answer questions such as 'Will the answer to 15 – 8 be greater than or less than 10?'

Discover

 Pair work

ASK

- Question **1** a): *What is the whole?*
- Question **1** a): *What is being taken away?*
- Question **1** a): *Is it easy to subtract the 1s?*

IN FOCUS The focus of this question is for children to recognise that the number of 1s to be subtracted is greater than the 1s digit in the 3-digit number. This means they will require a different strategy from the one learnt in lesson 2. They will need to bridge the 10.

PRACTICAL TIPS Represent the total number of parcels as base 10 equipment, and also locate it on a number line.

ANSWERS

Question **1** a): 151 − 6 = 145

Question **1** b): Number line showing jump of 1 back from 151 and another jump back from 150 to 145.

PUPIL TEXTBOOK 3A PAGE 68

Share

 Whole class teacher led

ASK

- Question **1** a): *Why is Ash's suggestion wrong?*
- Question **1** a): *Can you see two different ways of partitioning 151?*
- Question **1** : *Can you use your knowledge of number bonds to work out how many 1s are left after subtracting 6?*

IN FOCUS Ash's comment voices the misconception about just subtracting the smaller digit from the larger.

The focus of the question is to understand that the exchange of 1 ten for 10 ones is necessary in order to be able to subtract the 6 ones from the 3-digit number.

Discuss how the exchange method in part a) and the number line in part b) show the same concept in a different way.

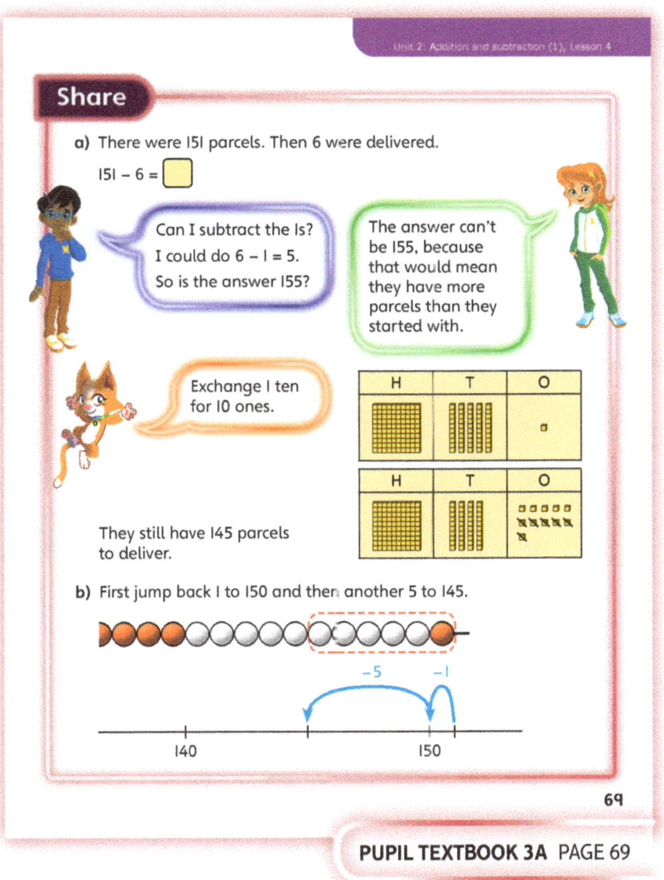

PUPIL TEXTBOOK 3A PAGE 69

Think together

WAYS OF WORKING Whole class teacher led (I do, We do, You do)

ASK

- Questions ❶ and ❷ : *How many 1s do we need to subtract?*
- Questions ❶ and ❷ : *How can you tell if the subtraction will need an exchange?*

IN FOCUS In questions ❶ and ❷ the important thing is for children to recognise how many 1s they need to subtract, and whether or not it requires an exchange. It may be a very useful strategy to model this on a number line, alongside the number in base 10 in a place value grid, so children understand how they can use their knowledge of subtraction within 20 to find the answers efficiently.

STRENGTHEN Children sometimes try to set out a subtraction using two sets of equipment. For example, when trying to represent 256 subtract 7, they may make the number 256 and the number 7, and then become confused about where to take away the 7 from. Ask: *What is the whole? What part are you subtracting?*

DEEPEN In question ❹ children may struggle to see the exchange necessary, when there is a zero in the 10s column. They may not understand that you need to exchange 1 hundred for 10 tens, then 1 ten for 10 ones. Ask: *What happens when you exchange? How many 10s in 1 hundred?*

ASSESSMENT CHECKPOINT Question ❷ requires children to notice that one of the subtractions requires an exchange and the other does not.

ANSWERS

Question ❶ : 15 ones – 8 ones = 7 ones
145 – 8 = 1 hundred + 3 tens + 7 ones
145 – 8 = 137

Question ❷ : Green van: 244 – 9 = 235
Blue van: 239 – 3 = 236
244 – 9 < 239 – 3

The blue van has more parcels left.

Question ❸ : 205 = 211 – 6

Question ❹ : 250 – 7 = 243
205 – 7 = 198

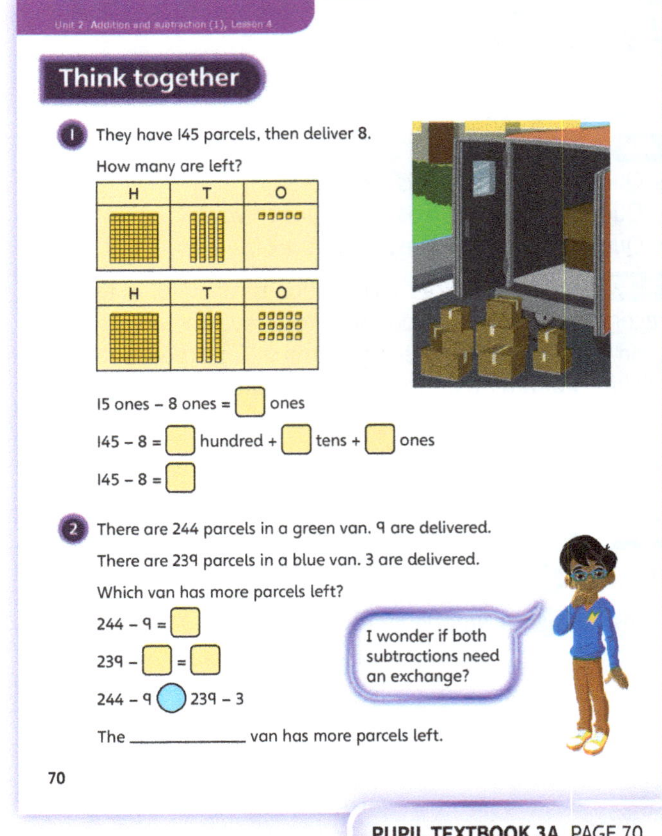

PUPIL TEXTBOOK 3A PAGE 70

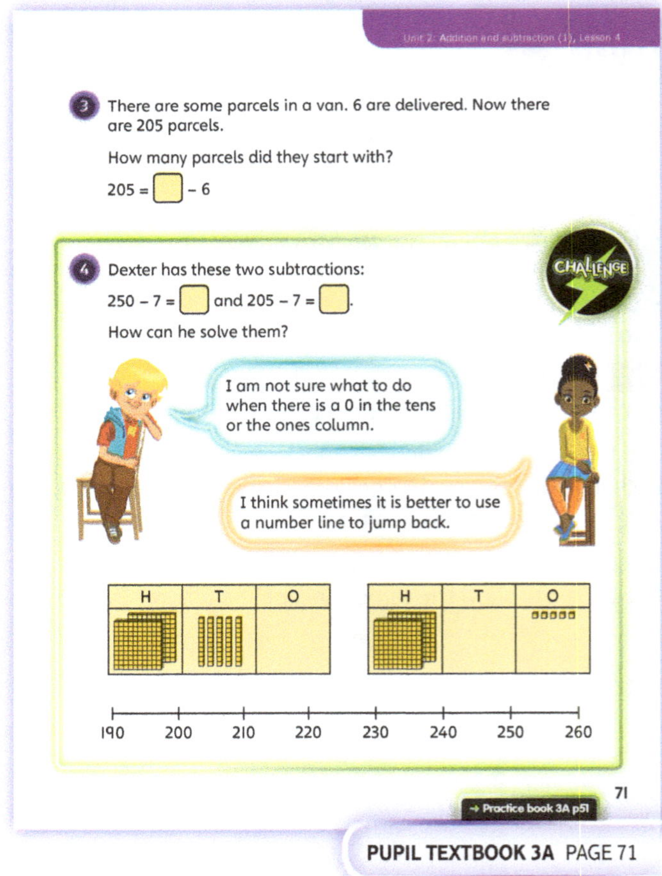

PUPIL TEXTBOOK 3A PAGE 71

Practice

WAYS OF WORKING Independent thinking

IN FOCUS Question **3** requires children first to decide which calculations require an exchange and then to solve them. This is the key – children should be making decisions all the way through about which calculations need an exchange and how best to solve the subtractions efficiently and accurately by choosing the most appropriate method.

STRENGTHEN Children may find it easier to represent their thinking as jumps on a number line, but they should be encouraged to use knowledge of number bonds within 20 to find the answer.

DEEPEN Children should be able to predict when the 10s digit will decrease by 1. Look at question **6**. Challenge children to predict how many subtractions they will perform before they do not need to exchange. They could investigate repeated subtractions of 9 from other starting numbers and explain what they notice.

THINK DIFFERENTLY Question **5** illustrates the very common misconception about reversing the digits to be subtracted. It is vital that children recognise this as a mistake and know to look out for it in their own thinking.

ASSESSMENT CHECKPOINT Answers to question **2** will demonstrate if children have linked the place value with the calculation methods they are using.

ANSWERS Answers for the **Practice** part of the lesson appear in the separate **Practice and Reflect answer guide**.

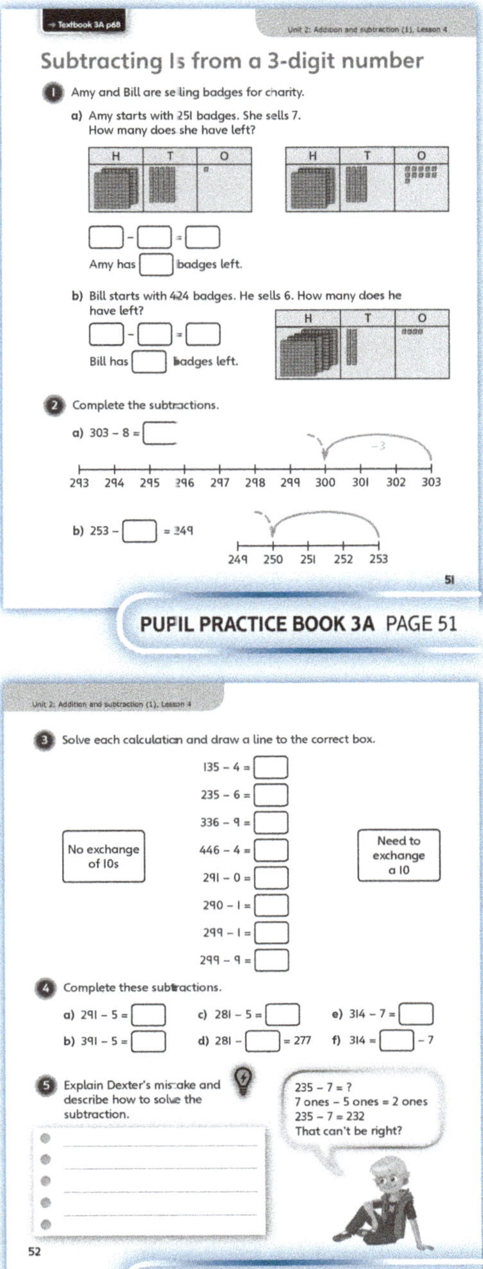

PUPIL PRACTICE BOOK 3A PAGE 51

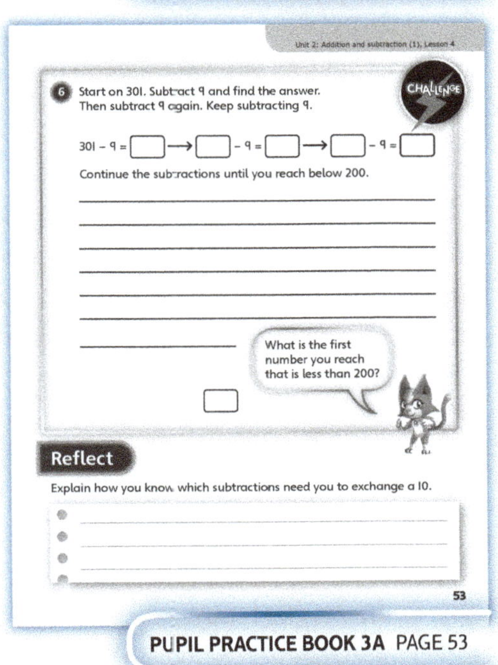

PUPIL PRACTICE BOOK 3A PAGE 52

Reflect

WAYS OF WORKING Group work

IN FOCUS Children should be able to explain how they know which subtractions require them to exchange a 10. Encourage each member of the group to use a different example subtraction to demonstrate their explanation.

ASSESSMENT CHECKPOINT Are children able to create their own examples to demonstrate?

ANSWERS Answers for the **Reflect** part of the lesson appear in the separate **Practice and Reflect answer guide**.

PUPIL PRACTICE BOOK 3A PAGE 53

After the lesson ⏸

- Can children explain when the 10s digit will decrease by 1?
- Can children solve the subtractions by using known number bonds?

Adding and subtracting a 3-digit number and 10s

Learning focus

In this lesson, children will add a multiple of 10 to a 3-digit number by using their knowledge of number bonds to add the 10s digits.

Small steps

→ Previous step: Subtracting 1s from a 3-digit number
→ **This step: Adding and subtracting a 3-digit number and 10s**
→ Next step: Adding a 3-digit number and 10s

NATIONAL CURRICULUM LINKS

Year 3 Number – Addition and Subtraction

- Add and subtract numbers mentally, including: a three-digit number and ones, a three-digit number and tens, a three-digit number and hundreds.
- Solve problems, including missing number problems, using number facts, place value, and more complex addition and subtraction.

ASSESSING MASTERY

Children can explain how to add a multiple of 10 to a 3-digit number in terms of the place value of the digits. Children can explain why, for example, when you add 30 to 654, the 10s digit increases by 3.

COMMON MISCONCEPTIONS

Children may resort to counting in 10s, rather than using knowledge of bonds and place value. Ask:
- *How do you solve 35 + 60 = ___?*

Children may struggle where there is a zero in the 1s or 10s place of the 3-digit number they are adding to. Ask:
- *What is the same and what is different about 302, 312 and 320?*

STRENGTHENING UNDERSTANDING

Encourage children to represent the calculation as jumps on a number line in order to illustrate whether the unknown number is greater or less than the number given. The number line need not be used as a counting strategy, but it can model the structure of the calculations. Display or create part-whole models alongside, to support the use of number bonds.

GOING DEEPER

Challenge children to solve missing number problems, especially when worded as: *20 more than my number is 245. What is my number? or 345 is 30 less than what number?* Encourage children to create and represent these missing number problems on a number line.

KEY LANGUAGE

In lesson: tens (10s), addition, subtraction, part-whole model, zero (0), column, fewer

Other language to be used by the teacher: place value, multiple of 10, ones (1s), hundreds (100s), more, place value grids

STRUCTURES AND REPRESENTATIONS

Number line, place value equipment

RESOURCES

Mandatory: base 10 equipment, place value grids, place value cards

Optional: place value abacus, 100 square

 In the eTextbook of this lesson, you will find interactive links to a selection of teaching tools.

Before you teach

- Can children count in 10s from 0 to 100 and back again?
- Can children count on in 10s from 34? Do they know what stays the same and what changes?

Discover

 Pair work

ASK

- Question **1** a): *What number is represented on the abacus?*
- Question **1** a): *How many beads does Aki add? What number has he added?*
- Question **1** a): *How could he write the calculation?*

IN FOCUS In question **1** a) the key point is that children connect the idea of adding or subtracting 30 with adding or subtracting 3 tens. This will lead to a development in their understanding of more formal column methods over the next series of lessons. Children should experience adding or taking away the 10s by concretely modelling the abacus using base 10 equipment or place value counters.

PRACTICAL TIPS Children could represent the place value abacus context by using base 10 equipment or place value counters placed on three lines, each line labelled 'H', 'T' and 'O' respectively.

Also, children can represent the place value by using a place value grid and representing the numbers with base 10 equipment or place value counters.

ANSWERS

Question **1** a): 351 + 30 = 381

Question **1** b): 381 − 10 = 371

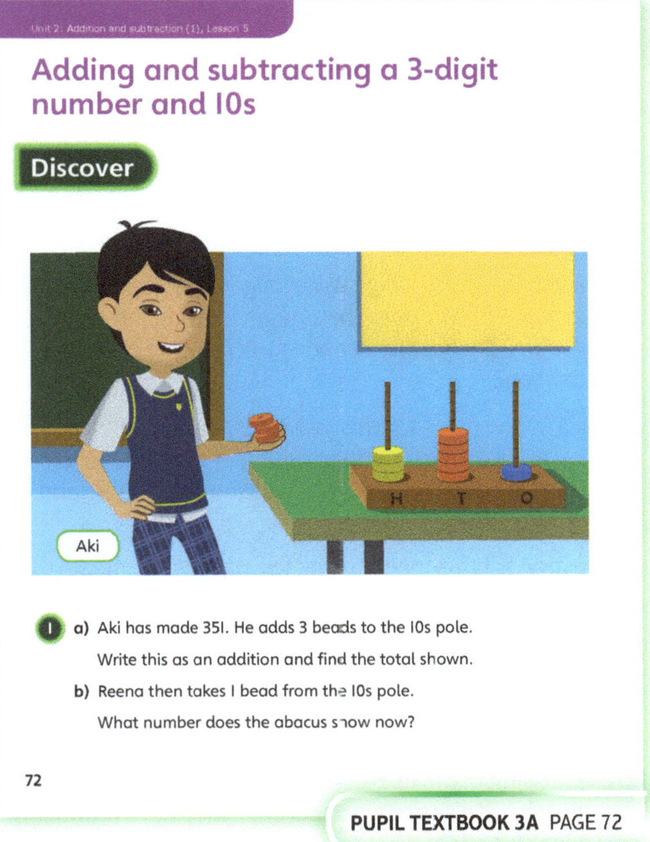

PUPIL TEXTBOOK 3A PAGE 72

Share

 Whole class teacher led

ASK

- Question **1** a): *How many 100s, 10s and 1s are there in Aki's number to start with?*
- Question **1** a): *How does the place value grid represent the addition?*
- Question **1** : *What number bond will calculate how many 10s there are now?*
- Question **1** : *Why does only the 10s digit change?*
- Question **1** b): *Why does the 10s digit decrease by 1 when Reena changes the abacus?*

IN FOCUS The important point is for children to connect the addition and subtraction of 10s with the place value of the numbers being added. Children should discuss why only the 10s digit changes for these calculations, and how they can use their knowledge of bonds within 10 to solve the calculations accurately.

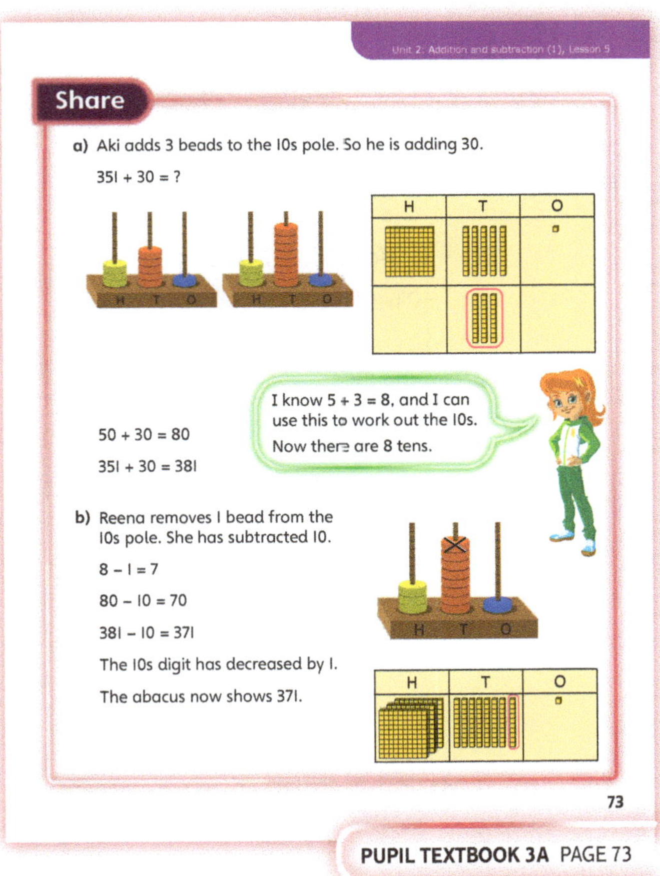

PUPIL TEXTBOOK 3A PAGE 73

Think together

WAYS OF WORKING Whole class teacher led (I do, We do, You do)

ASK

- Questions ❶ and ❷: *Can you explain when the 10s digit increases or decreases?*
- Question ❶: *Which digit will change? How do you know?*
- Question ❷: *Why does the 1s digit not change?*
- Question ❸ *What number bond will help you solve this calculation?*

IN FOCUS The focus for this section is for children to develop a fluency and understanding of why and how the 10s digit changes, by thinking about the place value and the number bonds they can use.

In question ❸ they need to link the part-whole model to the calculation, which requires them to think deeply about the 10s being added or subtracted.

STRENGTHEN It may help some children to use a 100 square, to be able to visualise the effect of adding or subtracting 10s.

DEEPEN Ask children to think of several different solutions to the missing digits in 2___4 − ___0 = 204. Can they convince themselves or a partner that they have found all possible solutions?

ASSESSMENT CHECKPOINT Question ❸ requires children to explain the calculations in terms of addition and subtraction of 10s.

ANSWERS

Question ❶: 8 tens - 5 tens = 3 tens
582 - 50 = 532

Question ❷: A: 451 − 40 = 411 Abacus A shows 411.
B: 451 + 40 = 491 Abacus B shows 491.

Question ❸: 414 + 70 and 280 − 10 both use 1 + 7 = 8 part-whole model

575 − 60 uses 1 + 6 = 7 part-whole model

124 + 60 uses 2 + 6 = 8 part-whole model

382 + 10 and 990 − 80 use 1 + 8 = 9 part-whole model

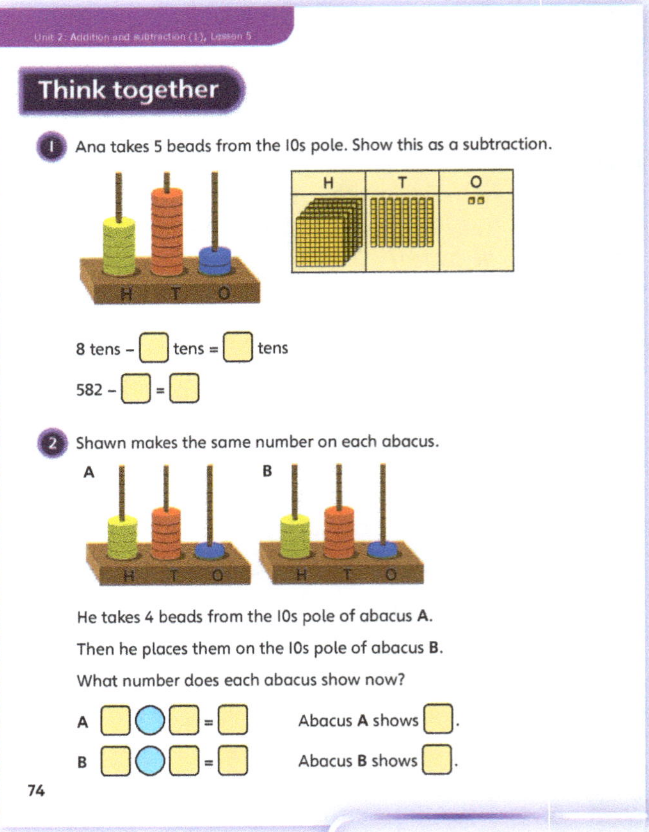

PUPIL TEXTBOOK 3A PAGE 74

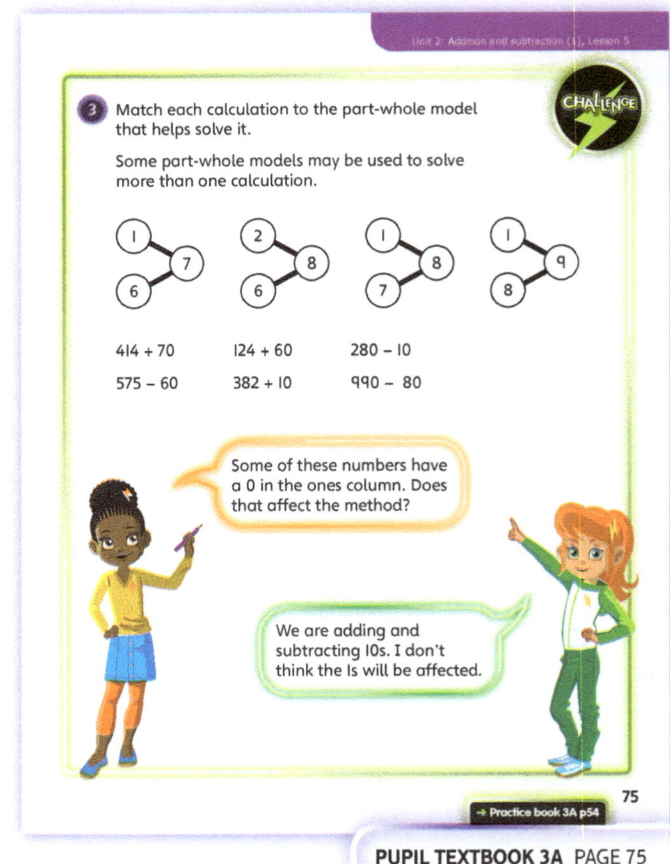

PUPIL TEXTBOOK 3A PAGE 75

Practice

WAYS OF WORKING Independent thinking

IN FOCUS The focus in this section is to develop fluency with mental methods for adding and subtracting 10s. In questions ❶ and ❷ the key points are that children recognise how to use their number bonds and various representations to explore the concept.

STRENGTHEN It may help children to represent the additions and subtractions on a number line in order to recognise what they have to do to find the missing information.

DEEPEN Questions ❸ and ❺ both require children to find missing numbers. Children may struggle to see whether they need to add or subtract to find the missing number. Encourage children to represent the calculation as jumps on a number line, and to reread the calculation to check their final answer works.

THINK DIFFERENTLY Question ❺ contains equations with calculations on either side of the equals sign. Give children a calculation with two missing numbers, such as 225 + _0 = 295 − _0, and ask if they can find all the possible solutions.

ASSESSMENT CHECKPOINT Question ❸ should demonstrate if children have understood the idea of adding or subtracting a multiple of 10 to or from a 3-digit number accurately.

ANSWERS Answers for the **Practice** part of the lesson appear in the separate **Practice and Reflect answer guide**.

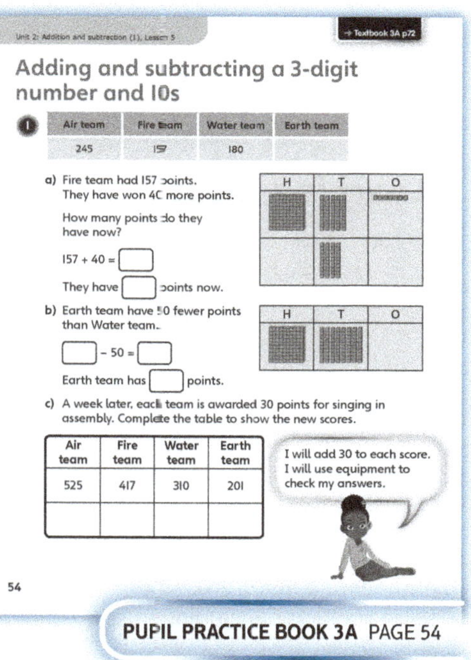

PUPIL PRACTICE BOOK 3A PAGE 54

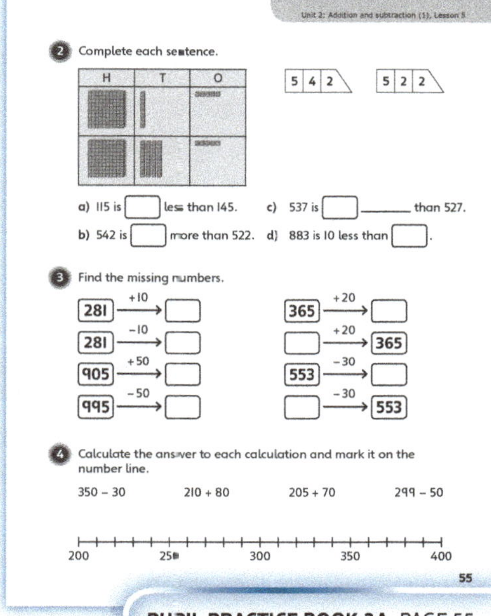

PUPIL PRACTICE BOOK 3A PAGE 55

Reflect

WAYS OF WORKING Pair work

IN FOCUS Children should focus on convincing their partner what the 10s digits will be, without having to work out the full answer as proof. This way, they will have to give reasons about the place value in the calculation, rather than just the process.

ASSESSMENT CHECKPOINT Can children justify their decision using the relevant number bonds?

ANSWERS Answers for the **Reflect** part of the lesson appear in the separate **Practice and Reflect answer guide**.

After the lesson 

- Do children understand that the 10s digit increases or decreases depending on the 10s being added or subtracted?
- Can children use number bonds to calculate accurately?
- Can children also solve missing number problems?

PUPIL PRACTICE BOOK 3A PAGE 56

Adding a 3-digit number and 10s

Learning focus

In this lesson, children will develop their understanding of adding 10s to a 3-digit number, including examples which require exchange of 10 tens for 1 hundred.

Small steps

→ Previous step: Adding and subtracting a 3-digit number and 10s
→ **This step: Adding a 3-digit number and 10s**
→ Next step: Subtracting 10s from a 3-digit number

NATIONAL CURRICULUM LINKS

Year 3 Number – Addition and Subtraction

- Add and subtract numbers mentally, including: a three-digit number and ones, a three-digit number and tens, a three-digit number and hundreds.
- Solve problems, including missing number problems, using number facts, place value, and more complex addition and subtraction.

ASSESSING MASTERY

Children can add multiples of 10 and recognise when they need to exchange 10 tens for 1 hundred.

COMMON MISCONCEPTIONS

Children may struggle with the flexible partitioning of the 10s required to bridge the hundred. Ask:
- *How many different ways can you partition 80 using 10s?*

Bridging the hundreds can cause children to struggle with exchange. They may often forget to include the exchanged digit or, for example, misconstrue 13 tens as 103. Ask:
- *What number is 13 tens?*

STRENGTHENING UNDERSTANDING

Children should explore the concept of exchange using place value equipment. It may also help to represent the additions on a number line, and justify why the addition bridges into the next hundred.

GOING DEEPER

Now that children will be able to add a multiple of 10 to any 3-digit number, challenge children to explore such conjectures as: *When you add 90 to a number, you always need to exchange 10 tens; When you add 50, you sometimes do not need to exchange.*

KEY LANGUAGE

In lesson: addition, tens (10s), number line, exchange, calculations, 3-digit number

Other language to be used by the teacher: hundreds (100s), ones (1s)

STRUCTURES AND REPRESENTATIONS

Number lines, place value equipment

RESOURCES

Mandatory: base 10 equipment, place value grids, place value cards, place value counters

Optional: 0–9 digit cards

 In the eTextbook of this lesson, you will find interactive links to a selection of teaching tools.

Before you teach 

- Can children answer questions such as: *What number is 17 tens?*
- Can children count in 10s, for example, from 65 up to 165?

Discover

WAYS OF WORKING Pair work

ASK

- Question **1** a): *What is the same and what is different about the calculations?*
- Question **1** a): *Which calculation requires more steps of thinking?*
- Question **1** a): *How could you represent the additions?*

IN FOCUS In question **1** a) the important point is for children to notice that they can employ the method of adding 10s for one of the calculations, but also recognise that they will need to adapt their thinking for the other calculation.

PRACTICAL TIPS Represent the age of the beech tree using place value equipment. Also locate its age on a number line. Then represent the numbers to be added using equipment and jumps on a number line.

ANSWERS

Question **1** a): 184 + 10 = 194.
The birch tree is 194 years old.
184 + 20 = 204.
The horse chestnut tree is 204 years old.

Question **1** b): Number line from 184 with a jump of 20 to 204. This jump may be broken up, for example, into two 10s, or a jump of 6, then 10 and finally 4.

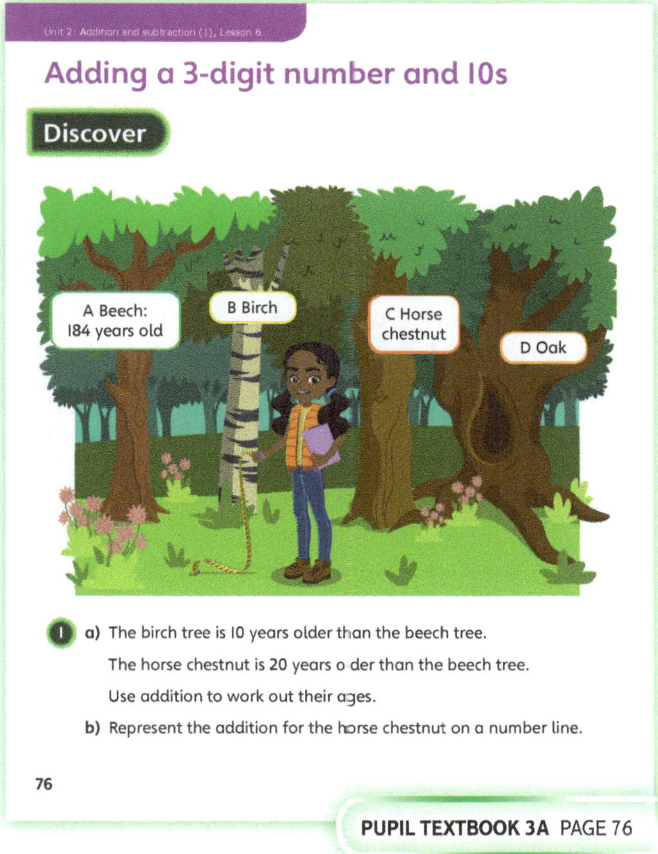

PUPIL TEXTBOOK 3A PAGE 76

Share

WAYS OF WORKING Whole class teacher led

ASK

- Question **1** a): *In the horse chestnut part of question **1** a) how do you know that the age will be greater than 200? Are you exchanging 10 ones or 10 tens?*
- Question **1** a): *Could you represent the second calculation as 184 + 10 + 10? What would this look like?*

IN FOCUS The idea of exchange is key to this lesson. Here we see that one of the calculations bridges into the next hundred, and children should discuss and explore how this is represented in different ways by the place value grids and by the number line. Although children should gain developmental fluency, the concept of exchange will support later development of more formal written methods.

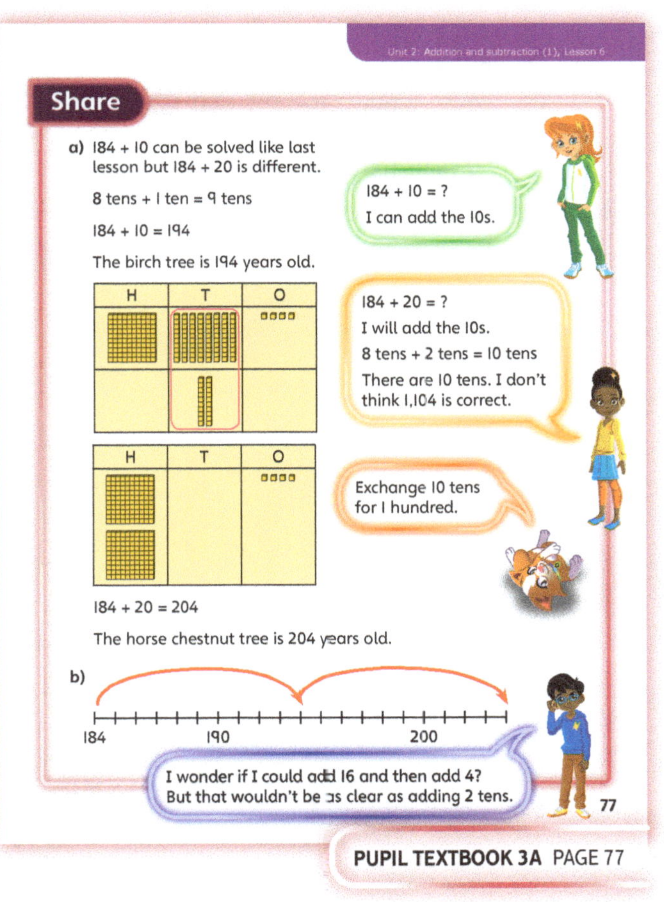

PUPIL TEXTBOOK 3A PAGE 77

Think together

WAYS OF WORKING Whole class teacher led (I do, We do, You do)

ASK

- Questions **1** and **2** : *How many 10s are there altogether?*
- Questions **1** and **2** : *What would this calculation look like on a number line?*
- Question **4** : *How can you predict whether the addition will cross into the next hundred, even before calculating the full answer?*

IN FOCUS The important aspect of this part of the lesson is that children develop mental fluency alongside an understanding of the concept of exchange.

STRENGTHEN Represent the numbers using place value equipment and show the structure of the addition on a number line. Discuss how the crossing into the next hundred is related to the exchange.

DEEPEN Question **3** contains a missing number problem. Ask children to consider different ways to solve this and to justify their answer.

ASSESSMENT CHECKPOINT Can children answer and justify their reasoning for question **3**? What methods do they choose to use to represent their thinking? If children are able to justify in terms of exchange, then they have understood the concept deeply.

ANSWERS

Question **1** : 184 + 50 = 234 The oak tree is 234 years old.

Question **2** : 260 + 90 = 350 The giant redwood will be 350 years old.

Question **3** : Present day: 385 + 0, 385 years;
30 years from now: 385 + 30, 415 years;
60 years from now: 385 + 60, 445 years;
90 years from now: 385 + 90, 475 years.

Question **4** : You need to exchange 10 tens for 1 hundred to solve: 60 + 365, 130 + 70 and 181 + 30 but not for 40 + 401.

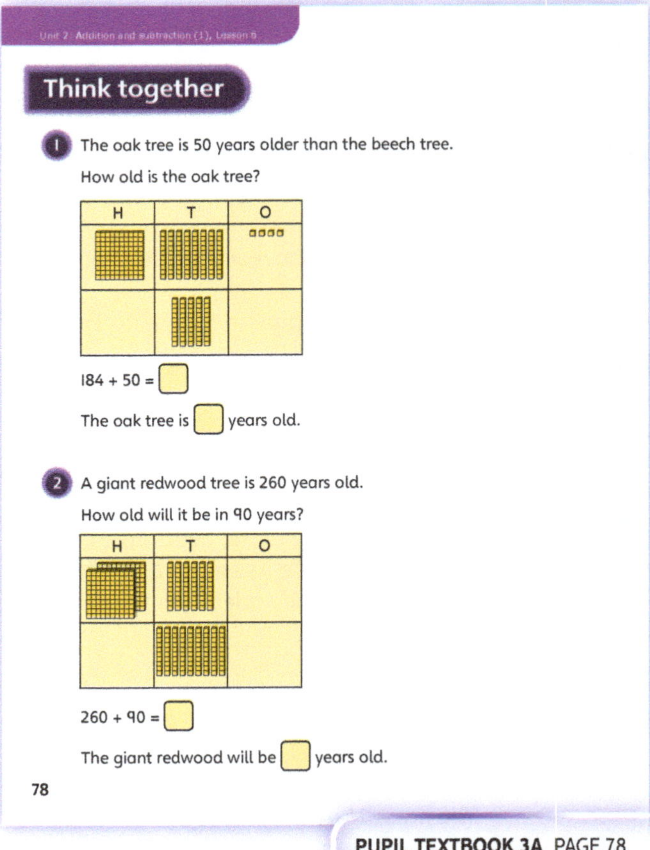

PUPIL TEXTBOOK 3A PAGE 78

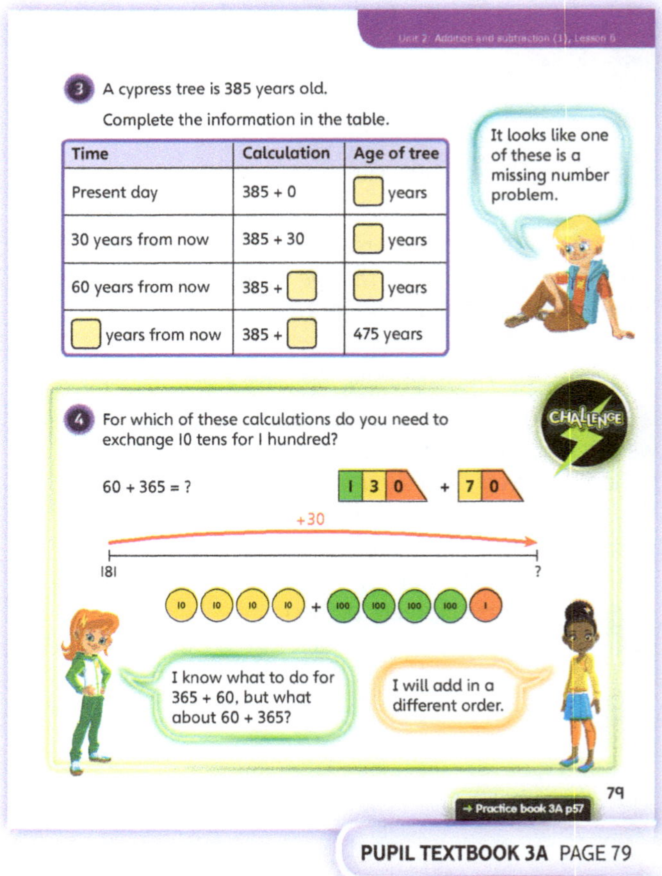

PUPIL TEXTBOOK 3A PAGE 79

Practice

WAYS OF WORKING Independent thinking

IN FOCUS Children should develop their fluency and mental addition through this practice. In questions **1**, **2**, **3**, and **5** they may select equipment to represent their thinking, but should also be able to justify their reasoning based on number bonds within 20, and the concept of exchange.

STRENGTHEN Some children may find it more natural to represent the additions on a number line, and it may also help to display relevant bonds within 20 on part-whole models.

DEEPEN In question **6**, there are multiple different solutions for the missing digits. Challenge children to find all the possible solutions and to convince their partner that they have found them all.

THINK DIFFERENTLY Question **4** contains a mistake where Isla has failed to exchange the 10s for another hundred. Encourage children to check their own work to see if they have made this mistake anywhere.

ASSESSMENT CHECKPOINT Finding the missing numbers in questions **3** and **5** requires a deep understanding of the concept of exchange.

ANSWERS Answers for the **Practice** part of the lesson appear in the separate **Practice and Reflect answer guide**.

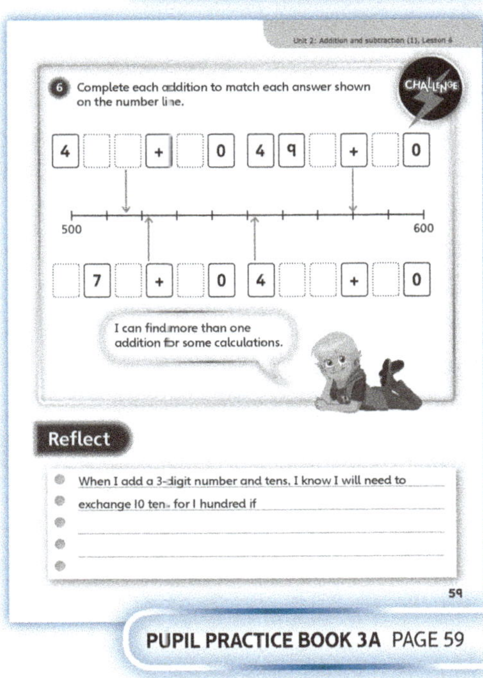

PUPIL PRACTICE BOOK 3A PAGE 57

PUPIL PRACTICE BOOK 3A PAGE 58

Reflect

WAYS OF WORKING Independent thinking

IN FOCUS The sentence stem prompts children to really think about the relationship between the additions and the concept of place value. Encourage children to show examples of when exchange is necessary and when it is not necessary.

ASSESSMENT CHECKPOINT This section will show if children engage decision making processes, rather than simply performing rote calculations.

ANSWERS Answers for the **Reflect** part of the lesson appear in the separate **Practice and Reflect answer guide**.

After the lesson 

- Can children decide when exchange is necessary?
- Can children justify their answers to themselves and to others by invoking number bonds?
- Can children choose a representation to illustrate their mental methods?

PUPIL PRACTICE BOOK 3A PAGE 59

Subtracting 10s from a 3-digit number

Learning focus

In this lesson, children will subtract a multiple of 10 from a 3-digit number, including where they have to exchange 1 hundred for 10 tens.

Small steps

→ Previous step: Adding a 3-digit number and 10s
→ **This step: Subtracting 10s from a 3-digit number**
→ Next step: Adding and subtracting a 3-digit and 2-digit number

NATIONAL CURRICULUM LINKS

Year 3 Number – Addition and Subtraction

- Add and subtract numbers mentally, including: a three-digit number and ones, a three-digit number and tens, a three-digit number and hundreds.
- Solve problems, including missing number problems, using number facts, place value, and more complex addition and subtraction.

ASSESSING MASTERY

Children can recognise when they need to exchange 10s when subtracting a multiple of 10 from a 3-digit number. Children can perform the calculations accurately and fluently with mental methods, often supported by jottings.

COMMON MISCONCEPTIONS

A common misconception for children to have is thinking they can transpose the 10s digits. For example, in 230 – 70 they may perform 70 – 30 = 40 to give a 10s digit of 4. Ask:
- *How many 10s are being subtracted?*

STRENGTHENING UNDERSTANDING

Children need to see the number line and the exchange side by side, to recognise why the subtraction requires a part taken from a whole.

GOING DEEPER

This lesson models a number of different ways of representing the subtractions. Challenge children to explore and adapt the alternative representations. Can they think of a number of different ways of presenting the concept, and can they decide which shows the method required most effectively?

KEY LANGUAGE

In lesson: subtraction, exchange, difference, calculation, number line, part-whole, method, more, metres (m)
Other language to be used by the teacher: 3-digit number

STRUCTURES AND REPRESENTATIONS

Number lines, part-whole model, place value equipment, bar model

RESOURCES

Mandatory: base 10 equipment, place value grids, place value cards
Optional: 0–9 digit cards

 In the eTextbook of this lesson, you will find interactive links to a selection of teaching tools.

Before you teach 

- Can children identify what is the same and what is different about 5 – 4, 15 – 4, and 15 – 8?
- Can children partition 43 in 3 different ways? What about 143?

Discover

 Pair work

- Question ❶ : *What kind of problem is this? What calculation is required?*
- Question ❶ a): *What is the whole? What is the part to take away?*
- Question ❶ a): *How could you represent 210 m of fabric?*

 The focus of this lesson is for children to understand that in question ❶ a) this subtraction crosses a hundred boundary and so requires an exchange, in a similar way to the previous lesson on addition.

 This could be modelled with a ball of string or wool to represent the fabric. Children can represent the length using a number line, or using a bar model.

Question ❶ a): 210 m – 20 m = 190 m

190 m are left.

Question ❶ b): 190 m – 140 m = 50 m

Jen sold 50 m.

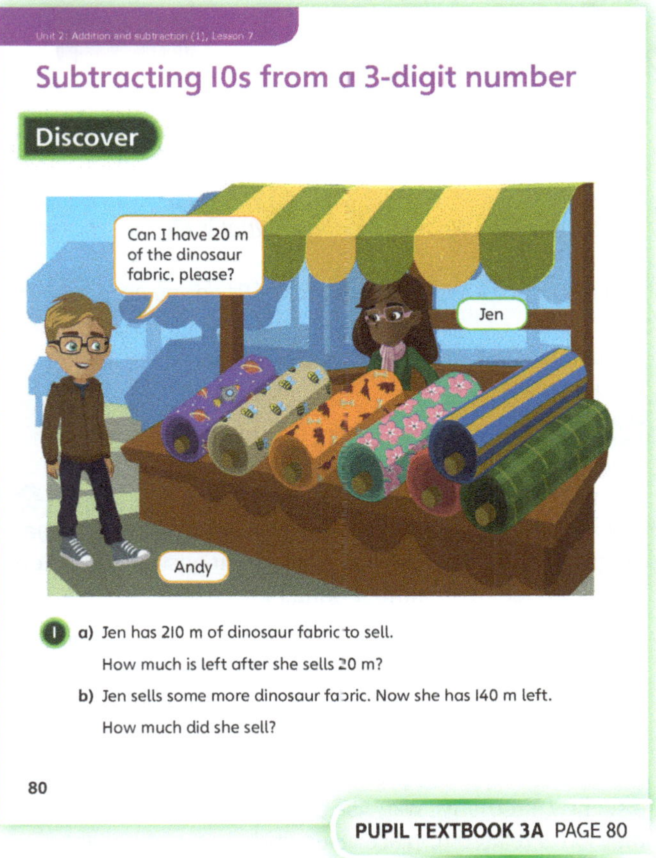

PUPIL TEXTBOOK 3A PAGE 80

Share

 Whole class teacher led

- Question ❶ a): *How does Astrid know that her first answer is incorrect?*
- Question ❶ a): *What mistake has Astrid made?*
- Question ❶ a): *What exchange has been made?*
- Question ❶ a): *How does this link to the previous lesson about adding 10s? What is the same? What is different?*
- Question ❶ a): *Could you represent this question using a number line?*
- Question ❶ b): *How does the number line help solve this problem?*

 In question ❶ a): the important point for children is to recognise the need to exchange 1 hundred for 10 tens, so that the subtraction can be made. The calculations bridge into the previous hundred, and children should discuss and explore how this can be represented using part-whole models and number lines.

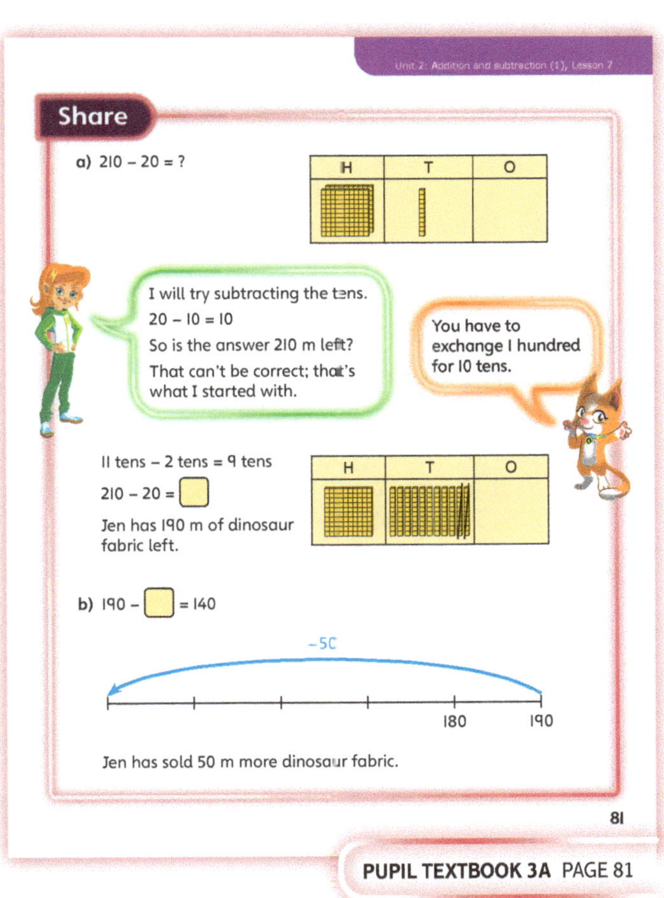

PUPIL TEXTBOOK 3A PAGE 81

Think together

WAYS OF WORKING Whole class teacher led (I do, We do, You do)

ASK

- Question **1** : *What is the whole? What is the part that you know?*
- Questions **1** and **2** : *How many 10s do you need to subtract?*
- Questions **1** and **2** : *What would this look like on a number line?*

IN FOCUS The important thing for children here is to develop fluency with the mental subtractions, alongside an understanding of the exchange required. In question **4** children should be able to explain their thinking using the language of place value, and justify their answers based on their knowledge of number bonds within 20.

STRENGTHEN Ask children to represent the whole as place value equipment, and then enact the exchange using the equipment. They always need to remember to regroup and rename 1 hundred for 10 tens.

DEEPEN Question **4** presents the concept of exchange in a slightly alternative way. Ask children to discuss how this method is related to place value grids and number lines. Children should discuss the different methods and make judgements about which shows the concept most clearly, in their opinion, and ask them to justify with reasons why we use place value and how it helps develop mental methods of subtraction.

ASSESSMENT CHECKPOINT Question **3** requires children to reason in a way that requires them to solve a missing number subtraction, which should ensure deep understanding.

ANSWERS

Question **1** : 13 tens – 5 tens = 8 tens
335 – 50 = 285
There is 285 m of space fabric left.

Question **2** : 213 – 80 = 133 m
Jen has 133 m of bee fabric more than Toshi.

Question **3** : 425 – 50 = 375

Question **4** : Flo is solving 235 – 60 by exchanging 1 hundred for 10 tens, which gives her 13 tens. 13 tens – 6 tens = 7 tens. 235 – 60 = 175.

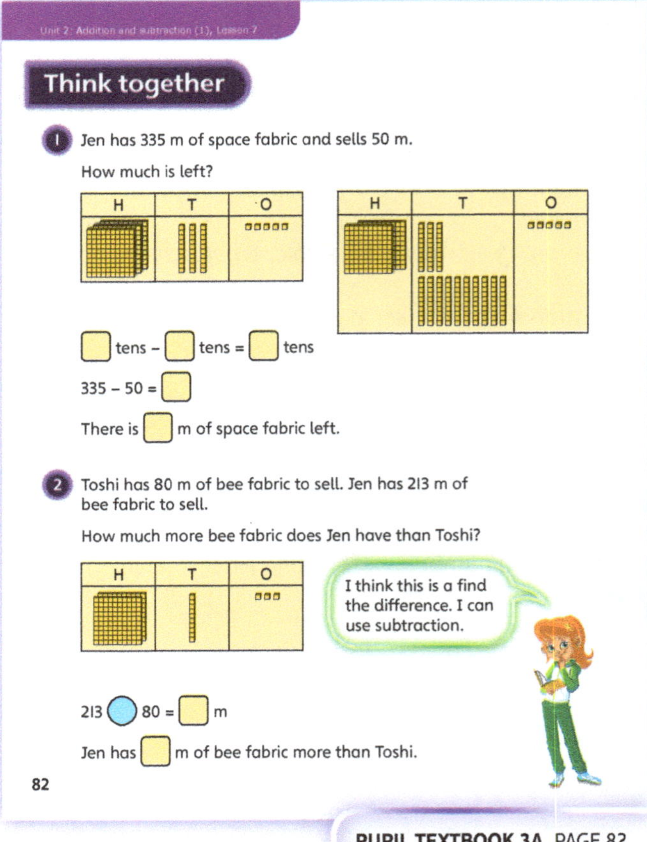

PUPIL TEXTBOOK 3A PAGE 82

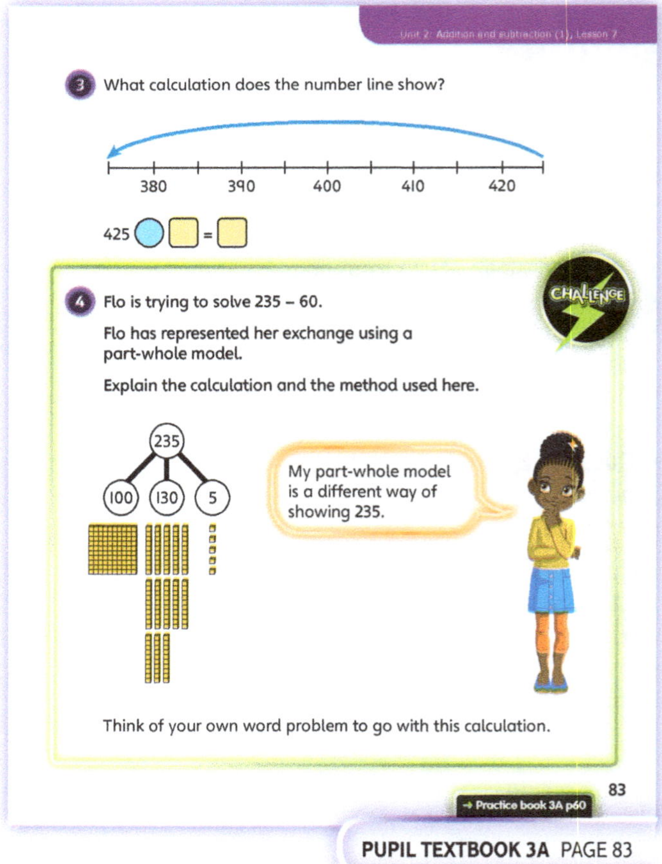

PUPIL TEXTBOOK 3A PAGE 83

Practice

WAYS OF WORKING Independent thinking

IN FOCUS Children develop their fluency through questions which gradually become more abstract. Question ❶ prompts the use of base 10 equipment to support the idea of exchange.

STRENGTHEN The use of base 10 equipment should prompt children to use exchange, but some children may find it helps to also represent the subtractions as jumps on a number line. Encourage children to use place value equipment and drawings to structure their thinking, and then use their knowledge of number bonds to find the answers.

DEEPEN There are multiple solutions for the missing digits in question ❺. Although question ❻ appears to have multiple solutions, in fact, there are only two solutions. Challenge children to explain this and think about inverse operations.

THINK DIFFERENTLY Question ❷ represents the exchange in a different way. This representation requires children to think flexibly and be able to focus on the effect of place value. Encourage children to explore alternative representations.

ASSESSMENT CHECKPOINT Questions ❸ and ❹ will demonstrate whether children can complete the abstract calculations accurately and with deep understanding.

ANSWERS Answers for the **Practice** part of the lesson appear in the separate **Practice and Reflect answer guide**.

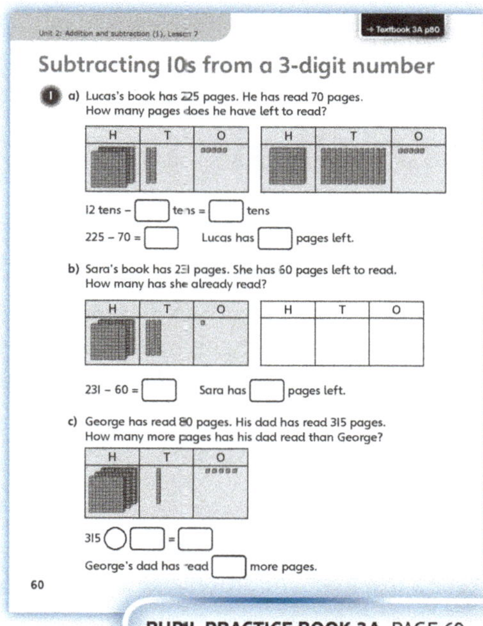

PUPIL PRACTICE BOOK 3A PAGE 60

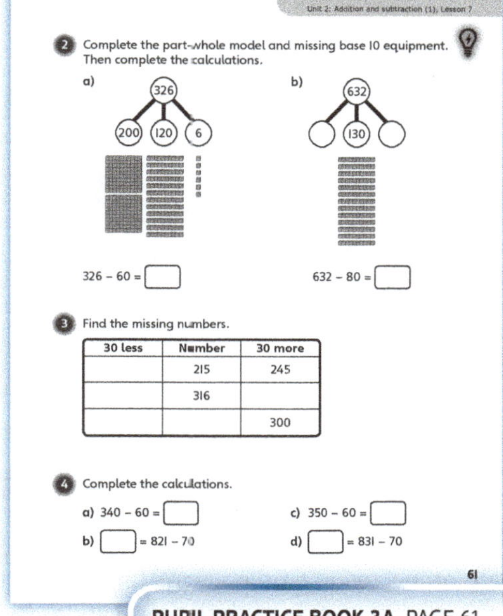

PUPIL PRACTICE BOOK 3A PAGE 61

Reflect

WAYS OF WORKING Pair work

IN FOCUS Children should rehearse their answers and see if they have completed the sentence in terms of the reasoning required. Some children may just work out the answer, but the focus is on explaining the link with place value and exchange.

ASSESSMENT CHECKPOINT If children can explain the reasoning, then they will be able to understand if their own subtractions are accurate.

ANSWERS Answers for the **Reflect** part of the lesson appear in the separate **Practice and Reflect answer guide**.

After the lesson 

- Can children represent the exchange using place value equipment?
- Do children understand how to use number bonds to complete the subtractions?
- What different representations can children interpret?

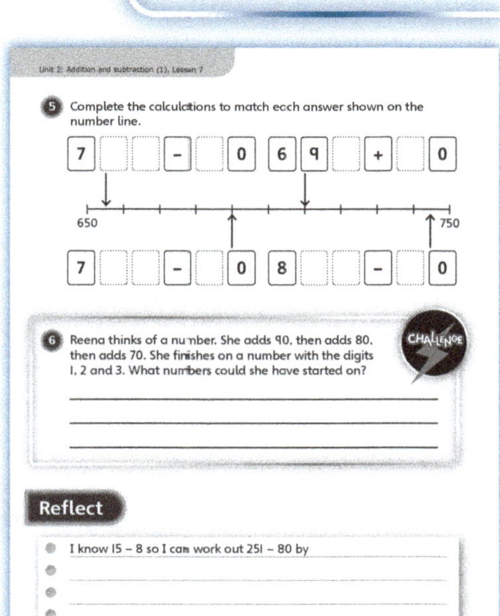

PUPIL PRACTICE BOOK 3A PAGE 62

Adding and subtracting a 3-digit and a 2-digit number

Learning focus

In this lesson, children will develop some more formal written methods, by considering addition and subtraction of a 3-digit and a 2-digit number where no exchange is required.

Small steps

→ Previous step: Subtracting 10s from a 3-digit number
→ **This step: Adding and subtracting a 3-digit and a 2-digit number**
→ Next step: Adding a 3-digit and a 2-digit number

NATIONAL CURRICULUM LINKS

Year 3 Number – Addition and Subtraction

- Add and subtract numbers with up to three digits, using formal written methods of columnar addition and subtraction.
- Add and subtract numbers mentally, including: a three-digit number and ones, a three-digit number and tens, a three-digit number and hundreds.

ASSESSING MASTERY

Children can add and subtract a 3-digit number and a 2-digit number, and represent the addition using a column method, without the need for exchange.

COMMON MISCONCEPTIONS

Children are beginning to represent calculations more formally. This abstract jump can cause children some confusion. Ask:
- *How does this way of writing the calculation match the equipment?*

When representing an addition using place value equipment, both numbers are represented. However, when subtracting, only the whole is represented. This can be confusing for children, as the column subtraction shows both the whole and the part as separate numbers. Ask:
- *What is the whole and what are the parts in this calculation?*

STRENGTHENING UNDERSTANDING

Using a 'before, then, now' structure can help children visualise the calculation and understand how the parts and the whole are connected in different ways for a subtraction and an addition. Using base 10 equipment and number lines may help children get to grips with the calculations they are being asked to make.

GOING DEEPER

Challenge children to make decisions about different calculations. For example, solving 205 + 50 may be easier as a mental calculation, whereas others may be better solved using jottings or written methods.

KEY LANGUAGE

In lesson: addition, subtraction, tens (10s), ones (1s), hundreds (100s), total, method, altogether, calculation, digits, miles

Other language to be used by the teacher: columns

STRUCTURES AND REPRESENTATIONS

Column method, place value grids, number lines

RESOURCES

Mandatory: base 10 equipment, place value grids, place value cards, place value counters

Optional: 0–9 digit cards

 In the eTextbook of this lesson, you will find interactive links to a selection of teaching tools.

Before you teach 

- Do children know what a 3-digit number means? What about a 2-digit number?
- What different ways have children written additions and subtractions?

Discover

ASK

- Question **1** a): *What are the parts of the journey?*
- Question **1** a): *What calculation would find the whole?*
- Question **1** b): *What makes this calculation different from the ones you have solved so far in this unit?*

IN FOCUS The most important aspect is that children recognise how they will require more steps for the calculations, as they are adding or subtracting a 2-digit number. Children should recognise the importance of all the work they have previously done to embed a strong understanding of place value.

PRACTICAL TIPS Children could represent the 'maps' as journeys by tracing their finger along the paths shown on the maps. They could use a counter or a cube to represent the vehicle on its journey.

ANSWERS

Question **1** a): 141 + 47 = 188 miles

Question **1** b): 141 − 11 = 130 miles

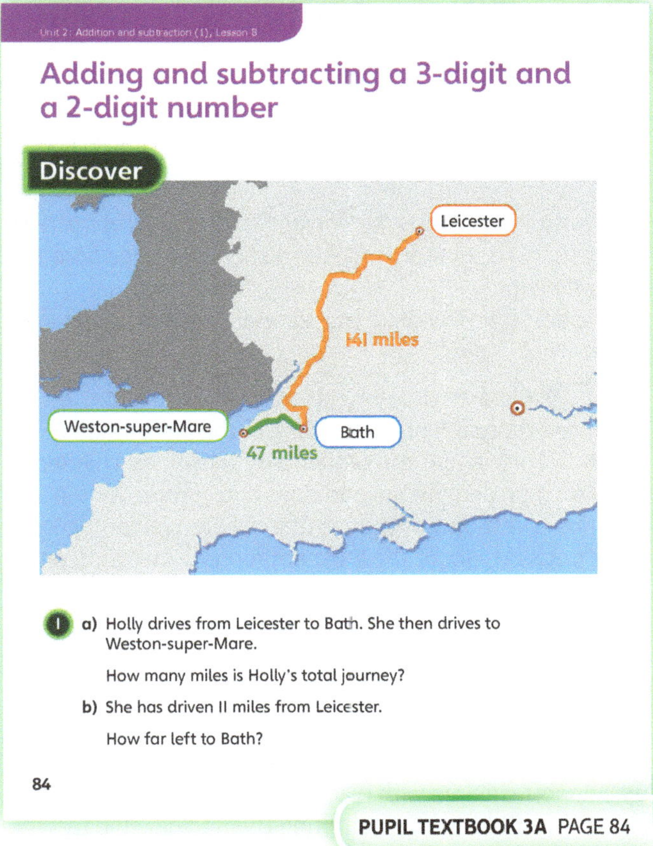

PUPIL TEXTBOOK 3A PAGE 84

Share

ASK

- Question **1** a): *How has the equipment been arranged to make the calculation clearer?*
- Question **1** a): *Can you see how the column addition matches the place value method?*
- Question **1** a): *What number bonds help add the 1s and the 10s?*
- Question **1** a): *What do you notice about adding the 100s?*

IN FOCUS It is important that children recognise how the arrangement of the place value equipment is mirrored in the written column method in question **1** a). Children are now starting to develop more formal written methods, and the work has built on a thorough understanding of place value which has developed in all of the lessons so far in Year 3.

Children should use known number bonds rather than counting the equipment.

In question **1** b) children have to set out the column method themselves.

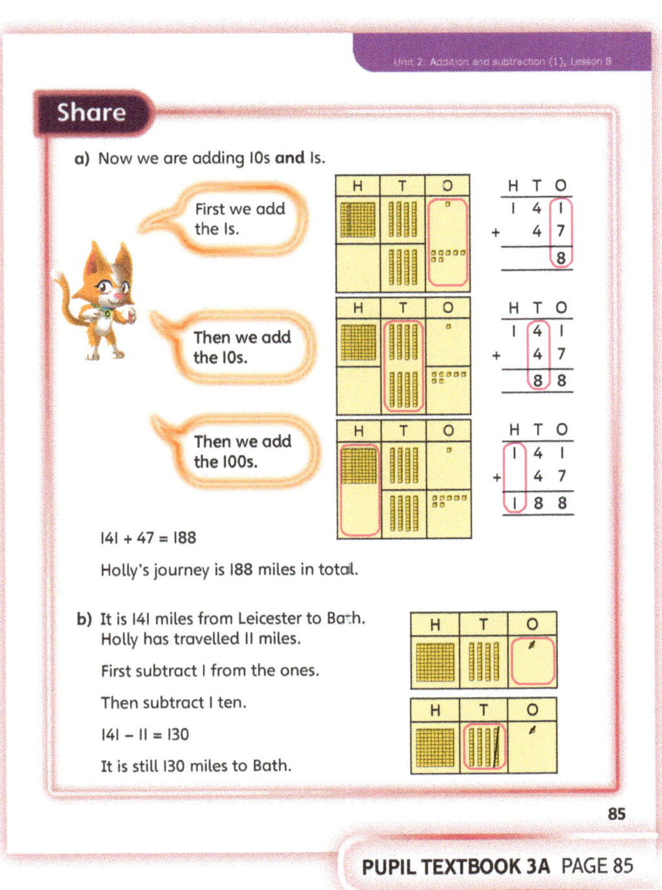

PUPIL TEXTBOOK 3A PAGE 85

Think together

WAYS OF WORKING Whole class teacher led (I do, We do, You do)

ASK

- Question **1** : *What is the whole? What is being subtracted?*
- Question **2** : *How does the calculation match the base 10 equipment?*
- Question **3** : *How could you represent this calculation in a part-whole model?*

IN FOCUS Children develop their written calculations by building a thorough understanding of the place value columns. Throughout the sequence of questions, children focus on representing the additions and subtractions in columns. Question **4** requires that children remember to make decisions about the most appropriate methods, rather than always reverting to one method. It is important that children remember to make decisions such as these before any calculation.

STRENGTHEN Some children may find that writing their own column methods without a scaffold can be difficult, and they may struggle to ensure columns line up. Support children with place value grids, and column method scaffolds while prompting them to develop the transcription skills needed.

DEEPEN Question **1** has an answer that is a multiple of 100. Ask children to explain how they could generate any number of subtractions where the answer is a multiple of 100. Can they adapt their thinking to produce subtractions with a given answer, such as 201 or 410?

ASSESSMENT CHECKPOINT Question **4** prompts children to analyse the number of steps needed to complete an addition or a subtraction. Children who understand the differences in complexity for different calculations will demonstrate a deep understanding.

ANSWERS

Question **1** : 274 – 74 = 200
There are 200 miles left of the journey.

Question **2** : 133 – 21 = 112
112 miles are left.

Question **3** : 209 + 52 = 261
The total journey is 261 miles.

Question **4** : Children should recognise that some of the calculations, such as 609 – 7 suit a mental method, but some, such as 234 + 52 are better suited to written methods.

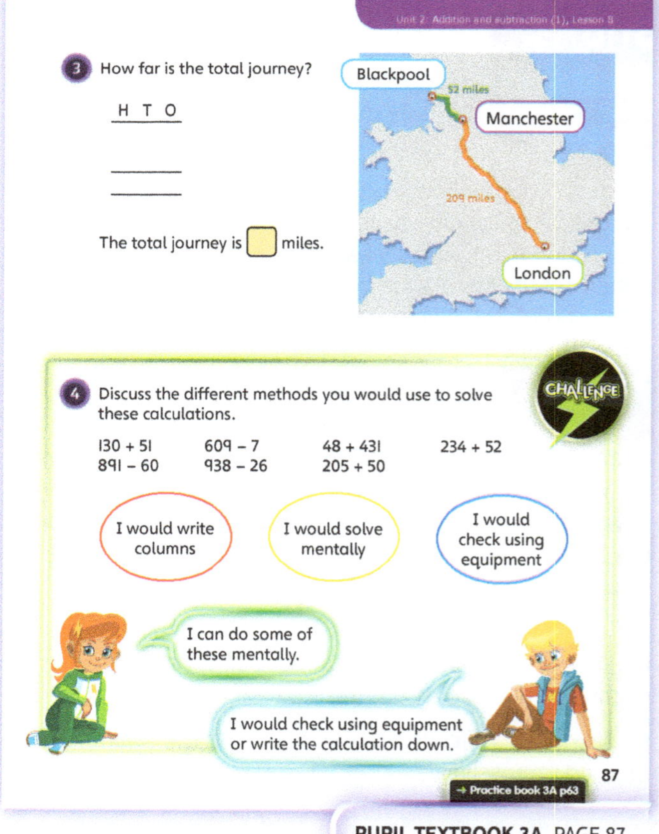

PUPIL TEXTBOOK 3A PAGE 86

PUPIL TEXTBOOK 3A PAGE 87

Practice

WAYS OF WORKING   Independent thinking

IN FOCUS   In the course of this lesson, children focus on developing fluency in the column methods. As they develop confidence, they should be able to explain in questions ① a) and b) how the number bonds they use to find the 10s digits and the 1s digits are related to the place value.

STRENGTHEN   Question ③ uses variation to support the understanding of place value, and the number bonds required are straightforward. Use this question to build confidence and build on the patterns uncovered to show the relationships in place value in column methods.

DEEPEN   Question ⑤ requires an understanding of inverse operations. It requires children to fully understand the inverse relationships of parts and wholes in additions as compared with subtractions. Challenge children to find multiple solutions to _ _ _ + _ _ = 980 using the column method.

ASSESSMENT CHECKPOINT   Question ④ requires children to complete additions and subtractions accurately by asking them to apply the column method for addition and subtraction. Children who can solve these calculations are showing that they are able to use written methods successfully.

ANSWERS   Answers for the **Practice** part of the lesson appear in the separate **Practice and Reflect answer guide**.

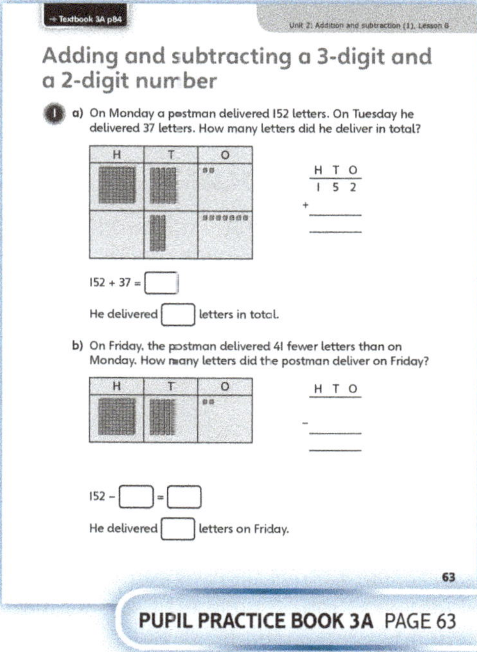

PUPIL PRACTICE BOOK 3A PAGE 63

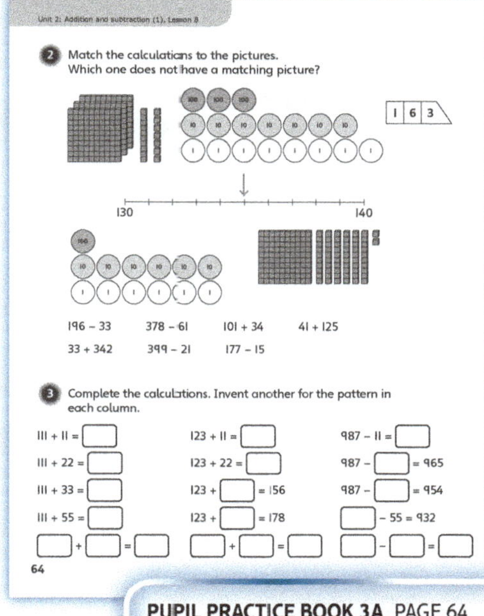

PUPIL PRACTICE BOOK 3A PAGE 64

Reflect

WAYS OF WORKING   Pair work

IN FOCUS   For this, children are asked to justify their calculations. They may choose to represent the calculations using diagrams or place value equipment. It may be that they justify their calculations by using column methods.

ASSESSMENT CHECKPOINT   Can children accurately assess the accuracy of their calculations?

ANSWERS   Answers for the **Reflect** part of the lesson appear in the separate **Practice and Reflect answer guide**.

After the lesson 

- Can children solve additions and subtractions between a 3-digit number and a 2-digit number involving exchange?
- Can children use written methods to complete calculations?

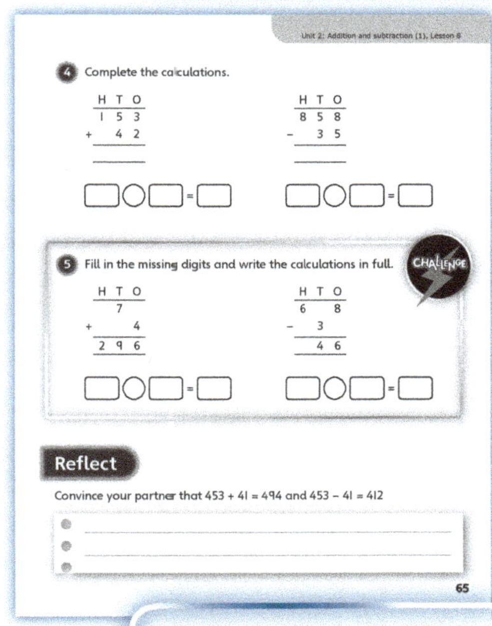

PUPIL PRACTICE BOOK 3A PAGE 65

Adding a 3-digit and a 2-digit number

Learning focus

In this lesson, children will develop written methods for addition, including exchange of 10s and 1s.

Small steps

→ Previous step: Adding and subtracting a 3-digit and a 2-digit number
→ **This step: Adding a 3-digit and a 2-digit number**
→ Next step: Subtracting a 2-digit number from a 3-digit number

NATIONAL CURRICULUM LINKS

Year 3 Number – Addition and Subtraction

- Add and subtract numbers with up to three digits, using formal written methods of columnar addition and subtraction.
- Add and subtract numbers mentally, including: a three-digit number and ones, a three-digit number and tens, a three-digit number and hundreds.
- Solve problems, including missing number problems, using number facts, place value, and more complex addition and subtraction.

ASSESSING MASTERY

Children can add a 3-digit and a 2-digit number accurately using a written column method.

COMMON MISCONCEPTIONS

Children may find it confusing where an exchange has a 'knock-on' effect, such as in 128 + 73, where the exchange of ones also causes an exchange of tens. Ask:

- *Will this need an exchange of 1s, 10s, or both?*

Children may become confused where they have to exchange both tens and ones in the same calculation. Ask:
- *What will you need to exchange in this calculation: 123 + 98?*

STRENGTHENING UNDERSTANDING

Encourage children to use a range of place value equipment to explore the idea of exchange of 10 ones for 1 ten, and 10 tens for 1 hundred. Allow children to experience the exchange by physically combining and regrouping the equipment.

GOING DEEPER

Challenge children to create different additions with the same total. Ask them to explore the relationship between calculations that create the same totals.

KEY LANGUAGE

In lesson: addition, 3-digit number, 2-digit number, altogether, total, exchange, ones (1s), tens (10s), hundreds (100s), digit, columns, calculation, centimetre (cm), 10 ones, 10 tens

Other language to be used by the teacher: column method, column addition

STRUCTURES AND REPRESENTATIONS

Place value equipment, column method

RESOURCES

Mandatory: place value grids, base 10 equipment, place value cards, place value counters, 0–9 digit cards

 In the eTextbook of this lesson, you will find interactive links to a selection of teaching tools.

Before you teach 

- Can children create different additions with a total of, for example, 21 or 101?
- Can children recognise whether an addition of single digits is greater than 10?

Discover

 Pair work

ASK

- Question **1** : *Can you represent the prices using equipment?*
- Question **1** : *How many 1s in the parts? How many 10s?*
- Question **1** : *What is the best way to represent the additions?*
- Question **1** : *Which order should you add the parts?*

IN FOCUS The important aspect for children to recognise in question **1** : is that the calculations require exchange, and so there will be more steps than for the method in the previous lesson.

PRACTICAL TIPS Although the context uses money, it is best not to represent the values using coins or notes. Instead, use place value equipment to represent the different amounts shown in the shop.

ANSWERS

Question **1** a): 275 + 16 = £291

Question **1** b): 45 + 61 = £106

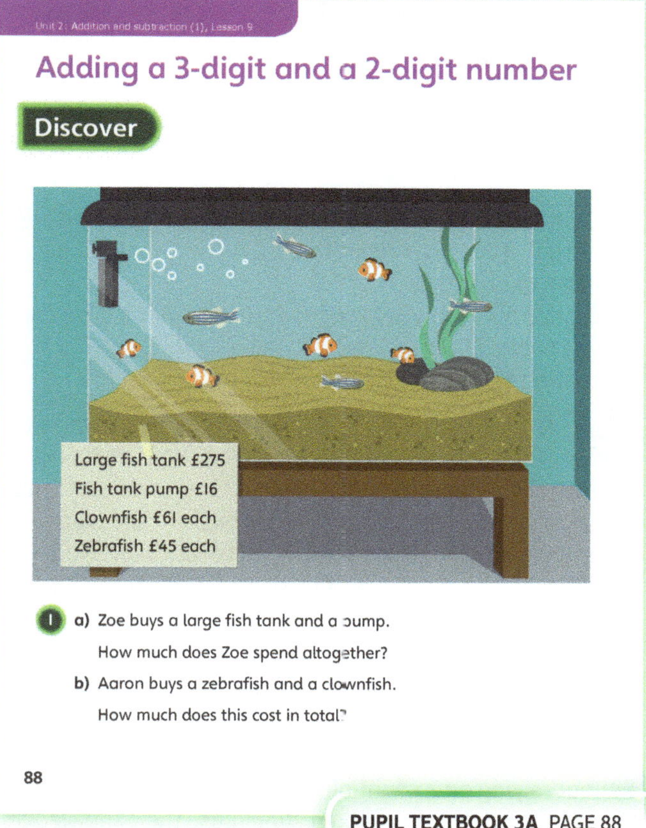

PUPIL TEXTBOOK 3A PAGE 88

Share

 Whole class teacher led

ASK

- Question **1** a): *Can you see the exchange of 10 ones?*
- Question **1** a): *How is this represented in the column method?*
- Question **1** a): *Where is the exchanged ten shown?*
- Question **1** a): *Why do we add the 10s first and not the 100s?*

IN FOCUS There are three important aspects to question **1** a):

- How to represent the exchange in written column methods.
- Why we add the 10s first, in order not to miss the exchanged 10.
- How to make sure children remember to add an exchanged digit.

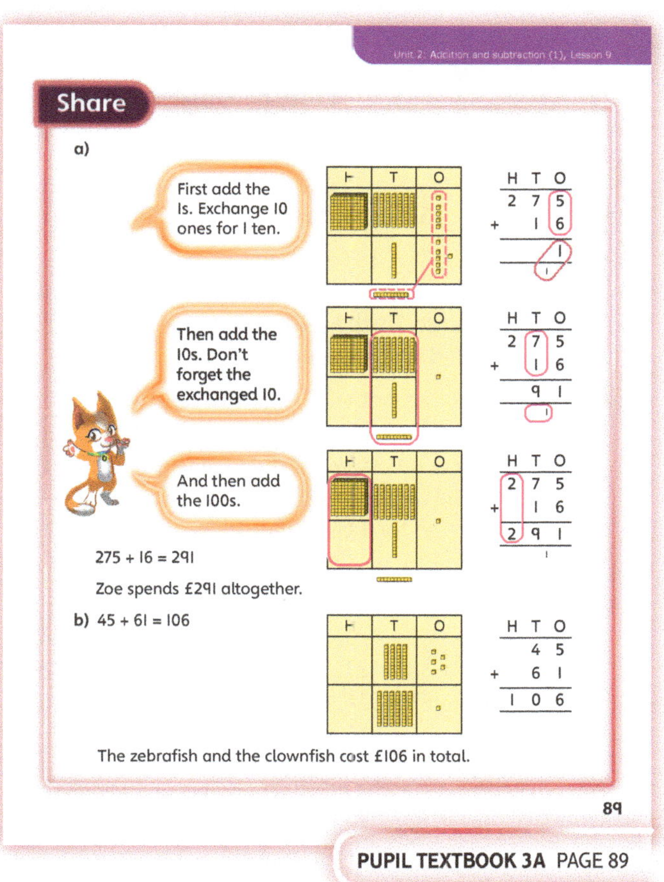

PUPIL TEXTBOOK 3A PAGE 89

Think together

WAYS OF WORKING Whole class teacher led (I do, We do, You do)

ASK

- Question ❶ : *Can you spot if an exchange will be needed?*
- Question ❷ : *How do you represent the exchange in the column method?*
- Question ❷ : *Can you explain why the additions have the same answer?*
- Question ❸ : *What mistakes should we look out for?*

IN FOCUS The focus of this lesson is for children to develop confidence and fluency in the column methods involving exchange, while continuing to deepen their understanding of exchange by using equipment to represent the place value. In questions ❶ and ❷ it is vital that children are able to explain the process of the calculation method in terms of the exchanges needed and known number bonds.

STRENGTHEN Children may find it very useful to use equipment to model each calculation alongside writing the column method. They should complete the calculation in stages, and with the equipment to model the process, without relying on a counting strategy. For example, if there are 3 ones and 9 ones, children should use their knowledge of number bonds to work out 3 + 9 = 12, rather than counting the equipment.

DEEPEN Question ❷ and question ❹ contain prompts to consider different additions which produce the same answers. Challenge children to explore the effect of transposing digits.

ASSESSMENT CHECKPOINT If children can find and explain the mistakes in question ❸, then they have developed a good understanding of the column method.

ANSWERS

Question ❶ : 275 + 61 = 336 Tia spends £336 in total.

Question ❷ : 126 + 57 = 183 and 156 + 27 = 183. They have the same answer. The number of 1s added is the same and the number of 10s added is the same.

Question ❸ : Mark has not exchanged 10 tens for 1 hundred: 154 + 72 is not 126 but 226. Poppy has shown 2 digits in the 1s column of the answer, rather than exchanged the 10 ones for 1 ten. 164 + 27 is not 1,811 but instead 191.

Question ❹ : Nine additions as shown below:
333 + 88 = 421, 338 + 38 = 376, 338 + 83 = 421, 383 + 38 = 421, 383 + 83 = 466, 833 + 38 = 871, 833 + 83 = 916, 838 + 33 = 871, 883 + 33 = 916

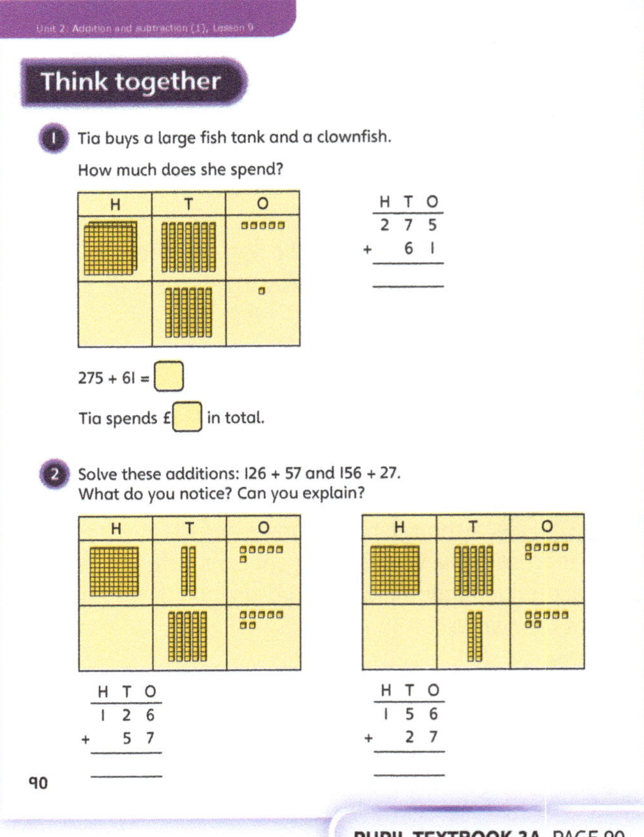

PUPIL TEXTBOOK 3A PAGE 90

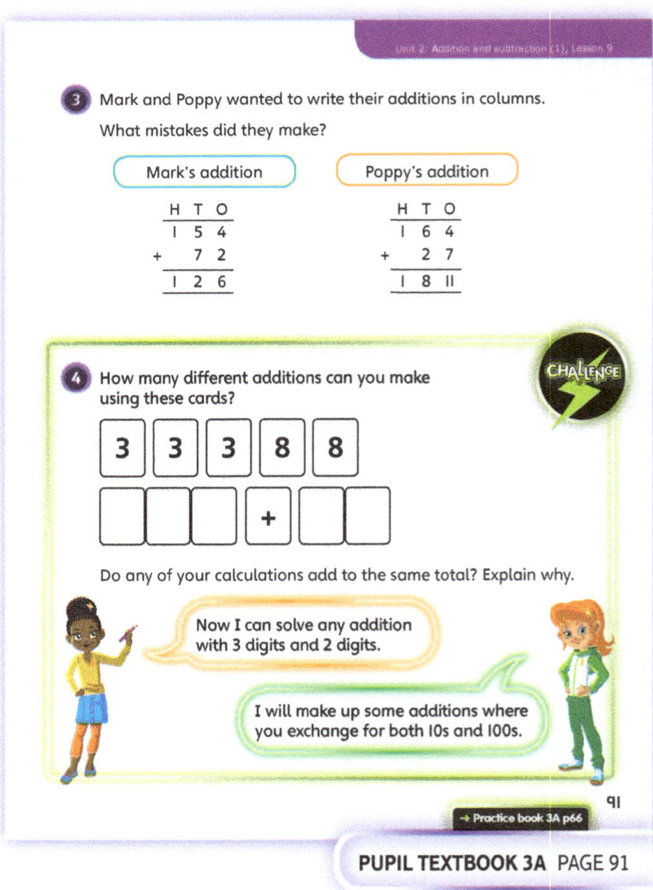

PUPIL TEXTBOOK 3A PAGE 91

Practice

WAYS OF WORKING Independent thinking

IN FOCUS Children develop their fluency and also make decisions about which parts need exchange. They will need to understand how to use the column methods accurately in order to solve calculations, but in questions **5** and **6** they will also need to reason some answers based on their understanding of inverse operations for missing number and missing digit problems.

STRENGTHEN Support children to write their own column methods by using place value grids or column scaffolds.

DEEPEN Question **6** requires children to reason about additions which have to exchange both 10s and 1s. They will have to work logically, by considering additions where the exchange has a knock-on effect, such as with 189 + 12, which appears to only require exchange of 1s, but that exchange causes an exchange of 10s as well.

THINK DIFFERENTLY Question **5** prompts children to reason the missing digits, rather than simply to rely on following the calculation method by rote.

ASSESSMENT CHECKPOINT Accurate completion of question **4** will demonstrate fluency in children's understanding of how to add a 3-digit number and a 2-digit number.

ANSWERS Answers for the **Practice** part of the lesson appear in the separate **Practice and Reflect answer guide**.

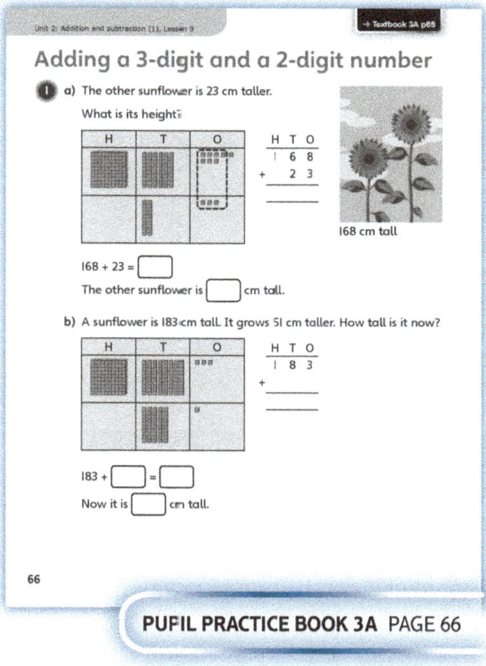

PUPIL PRACTICE BOOK 3A PAGE 66

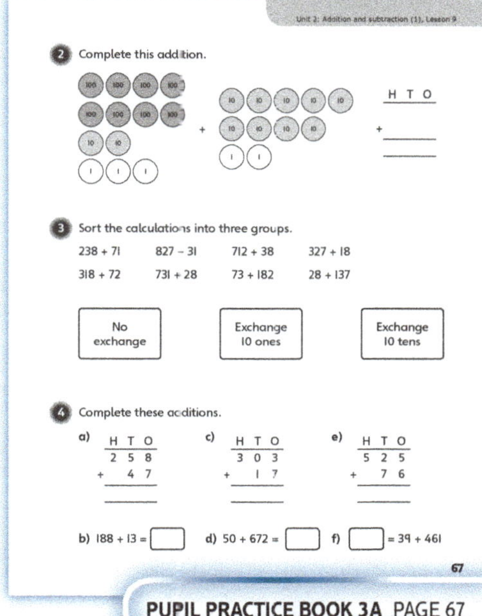

PUPIL PRACTICE BOOK 3A PAGE 67

Reflect

WAYS OF WORKING Pair work

IN FOCUS Children should discuss the different steps they go through and discuss what the most important decisions to make are as they complete an addition.

ASSESSMENT CHECKPOINT Can children explain how to work logically, and to check for exchange at each stage?

ANSWERS Answers for the **Reflect** part of the lesson appear in the separate **Practice and Reflect answer guide**.

After the lesson

- Are children able to represent the additions using place value equipment?
- Can children explain how to represent an exchanged 10 or 100 in a written column method?
- Do children use their knowledge of bonds to complete the addition of digits?

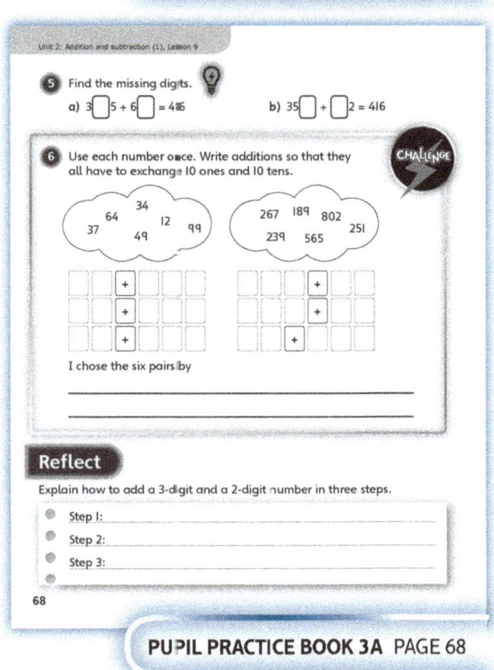

PUPIL PRACTICE BOOK 3A PAGE 68

Subtracting a 2-digit number from a 3-digit number

Learning focus

In this lesson, children will subtract using column methods with exchange where necessary.

Small steps

→ Previous step: Adding a 3-digit and a 2-digit number
→ **This step: Subtracting a 2-digit number from a 3-digit number**
→ Next step: Addition and subtraction patterns

NATIONAL CURRICULUM LINKS

Year 3 Number – Addition and Subtraction

- Add and subtract numbers with up to three digits, using formal written methods of columnar addition and subtraction.
- Add and subtract numbers mentally, including: a three-digit number and ones, a three-digit number and tens, a three-digit number and hundreds.
- Solve problems, including missing number problems, using number facts, place value, and more complex addition and subtraction.

ASSESSING MASTERY

Children can use column subtraction written methods to complete subtractions of a 2-digit number from a 3-digit number where exchange is needed.

COMMON MISCONCEPTIONS

Representing the exchange in written column methods can be challenging for children, as they may struggle to understand why certain digits are crossed out and small digits are written next to other digits. Ask:
- *Why is this number crossed out? What does this '1' mean that has been added next to the 5 ones?*

Children may struggle where there is an exchange required but a zero in the 10s column. Ask:
- *Can you exchange tens in this calculation: 305 – 58?*

STRENGTHENING UNDERSTANDING

Children will need plenty of experience of regrouping the place value equipment and discussion of how this exchange and regrouping is connected to the way column subtractions are written. Make sure there are plenty of opportunities for children to explore this method alongside the equipment.

GOING DEEPER

Encourage children to think about whether a written column method is the most efficient way of solving each subtraction. It may be that certain subtractions are better suited to a mental method, or one represented on a number line. Challenge children to scan the questions for where alternative methods may be more appropriate.

KEY LANGUAGE

In lesson: subtraction, 2-digit number, 3-digit number, ones (1s), tens (10s), hundreds (100s), order, exchange, columns, zero (0), addition, symbol

Other language to be used by the teacher: regroup, column method, column subtraction, partition

STRUCTURES AND REPRESENTATIONS

Number line, column method, place value equipment

RESOURCES

Mandatory: base 10 equipment, place value grids, place value cards, 0–9 digit cards

Optional: place value counters

 In the eTextbook of this lesson, you will find interactive links to a selection of teaching tools.

Before you teach ⏸

- What is the same and what is different about the methods used for addition and subtraction?

Discover

WAYS OF WORKING Pair work

ASK

- Question ① a): *How do you know what operation is required? What is it about the problem that suggests subtraction?*
- Question ① a): *What do you notice about the digits of the two numbers?*
- Question ① a): *Would you need to represent the 38 as well as the 175?*

IN FOCUS The important aspect of question ① a): is that children recognise that an exchange is required, as the 8 ones needs to be subtracted, but 175 has 5 ones as it is normally partitioned.

PRACTICAL TIPS Represent the number of trees using place value equipment.

ANSWERS

Question ① a): Luis subtracted the ones in the wrong order.

Question ① b): 175 − 38 = 137

PUPIL TEXTBOOK 3A PAGE 92

Share

WAYS OF WORKING Whole class teacher led

ASK

- Question ① a): *Can you explain why what Luis did is a mistake?*
- Question ① a): *Can you see how the exchange is represented by the equipment?*
- Question ① b): *What does this exchange look like in the written column method?*

IN FOCUS The most important aspect of teaching this question is to help children to understand how to represent the exchange in written columns. In question ① b) they will need to understand why the digit 7 is crossed out and replaced by a 6, and what the 1 written next to the 5 represents. The danger is that children simply learn this as a process, rather than understanding why these changes are made.

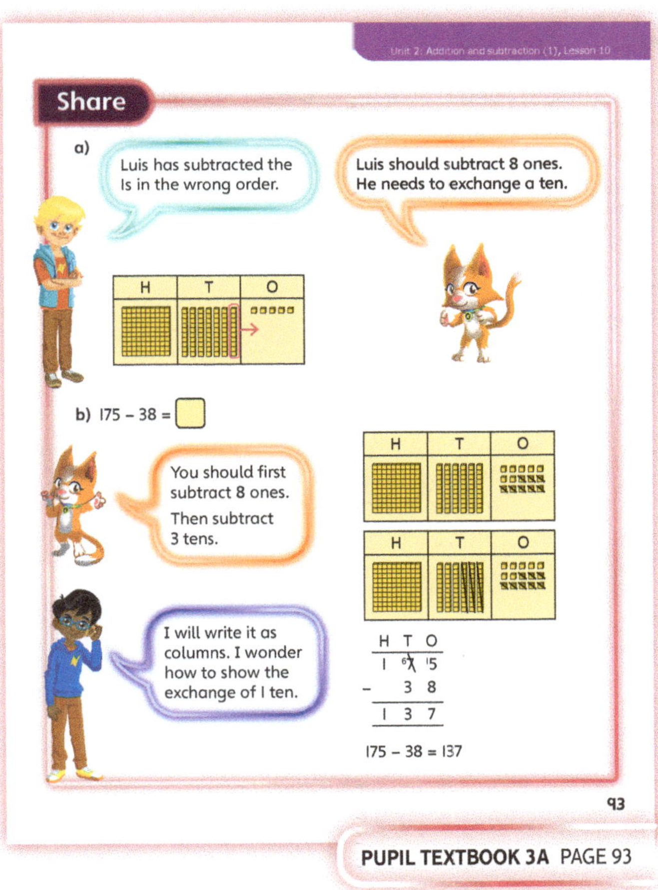

PUPIL TEXTBOOK 3A PAGE 93

125

Think together

WAYS OF WORKING Whole class teacher led (I do, We do, You do)

ASK

- Question **1** : *Do you need to exchange a ten? Or a hundred?*
- Question **1** : *How do you spot if you need to exchange?*
- Question **1** : *Why do you start by subtracting the ones?*
- Question **2** : *How can you represent the exchange using equipment?*
- Question **2** : *How can you represent the exchange in the written column method?*

IN FOCUS Children will be developing confidence and fluency in how to represent the exchange required. In question **3** they should represent the calculation with equipment alongside the written method, and discuss how the two ways of showing the exchange are related.

STRENGTHEN Again, in this lesson a common misconception is to transpose digits, and simply subtract the smaller digit from the larger, without considering which is the whole and which the part. Use a number line and equipment to demonstrate clearly the parts that are being subtracted, in comparison with the whole.

DEEPEN Question **4** challenges children to understand how to exchange where there are zero tens. Children should discuss how to solve with exchange across two columns, but may also recognise that these questions may be better represented with a number line or some other method. Challenge children to compare and contrast different methods.

ASSESSMENT CHECKPOINT Question **3** requires exchange of both first ten and then a hundred. Successfully completing this and explaining the exchange will indicate a good understanding.

ANSWERS

Question **1** : 246 – 63 = 183
183 trees are left.

Question **2** : 191 – 55 = 136
They planted 136 more birch trees.

Question **3** : 221 – 43 = 178

Question **4** : 302 – 65 = 237
To solve the subtraction Flo needs to first exchange from the hundreds to the tens, and then exchange from the tens to the ones.

PUPIL TEXTBOOK 3A PAGE 94

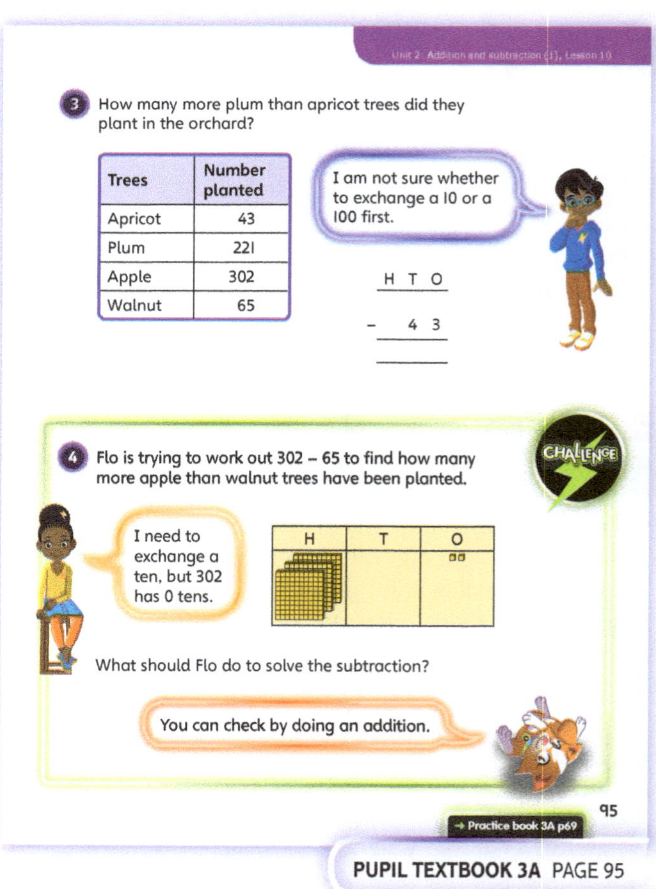

PUPIL TEXTBOOK 3A PAGE 95

Practice

WAYS OF WORKING Independent thinking

IN FOCUS The key point of this practice in questions ❶ and ❷ is to allow children the chance to build confidence and fluency alongside an understanding of the place value.

STRENGTHEN Build confidence by working on subtractions which only require exchange of one column at a time, until children are confident with this aspect. Begin with examples where the ones require an exchange of 1 ten, and then move to where the tens require an exchange of 1 hundred.

DEEPEN The challenge in question ❻ requires significant logical deduction. Children will need to reason from the fact that the 100s digit changes, and so the subtraction of the 10s digit does require an exchange. This is a very logical puzzle, and children may have to explore it using some trial and error approaches first.

THINK DIFFERENTLY The mistake in question ❺ is to tackle the misconception where the digits from the whole and the part have been transposed.

ASSESSMENT CHECKPOINT Question ❸ requires a good level of understanding of the method, and of how to exchange in one or two columns.

ANSWERS Answers for the **Practice** part of the lesson appear in the separate **Practice and Reflect answer guide**.

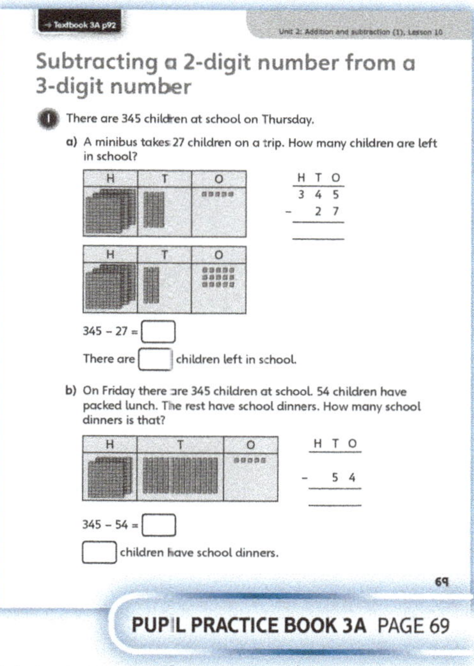

PUPIL PRACTICE BOOK 3A PAGE 69

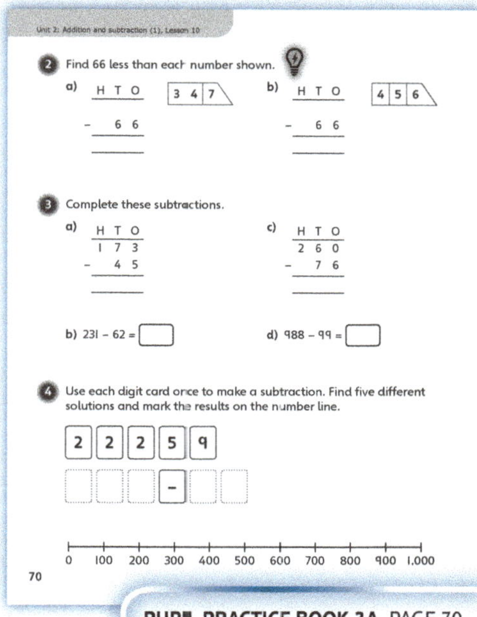

PUPIL PRACTICE BOOK 3A PAGE 70

Reflect

WAYS OF WORKING Group work

IN FOCUS Children should look back over all 10 lessons in this unit as a group, and discuss the most important learning points. They may not all agree, but they should listen to one another and decide afterwards what are the three main points for themselves.

ASSESSMENT CHECKPOINT Have children retained the learning from across the whole unit of ten lessons?

ANSWERS Answers for the **Reflect** part of the lesson appear in the separate **Practice and Reflect answer guide**.

After the lesson 

- Can children represent the exchange in subtractions?
- Can children understand how to write the exchange using written column methods?

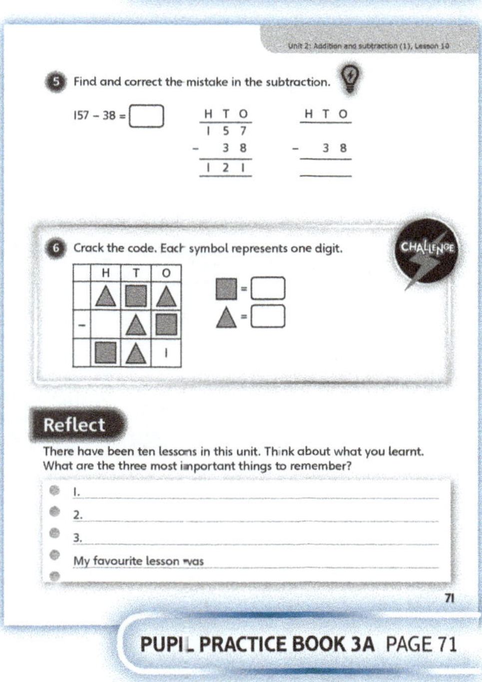

PUPIL PRACTICE BOOK 3A PAGE 71

End of unit check

Don't forget the *Power Maths* unit assessment grid on p26.

WAYS OF WORKING Group work – adult led

IN FOCUS Question **1** focuses on children being able to calculate a subtraction of one 3-digit number from another.

Question **2** focuses on children's understanding of where exchange of 10 ones for 1 ten is necessary.

Questions **3**, **4** and **5** focus on children being able to calculate addition and subtraction involving an exchange accurately.

Question **6** focuses on children's ability to solve missing digit calculations accurately.

ANSWERS AND COMMENTARY Children who have mastered this unit will be able to use mental methods, diagrams and place value grids to add and subtract single-digit and 2-digit numbers to and from 3-digit numbers. Children will be able to explain where an exchange was necessary, and how this relates to bridging a 10 or 100.

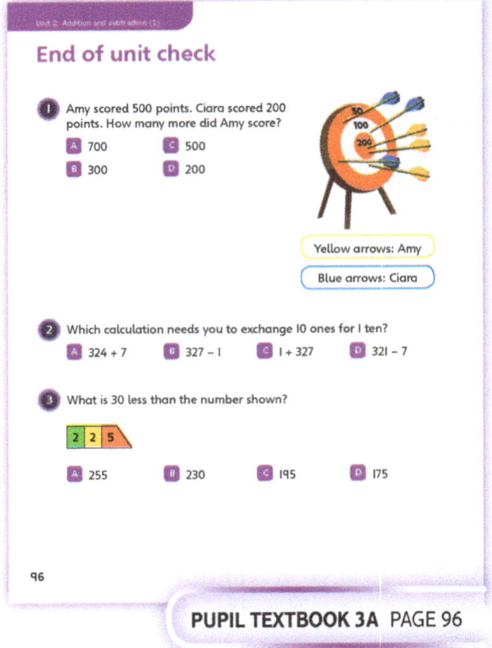

PUPIL TEXTBOOK 3A PAGE 96

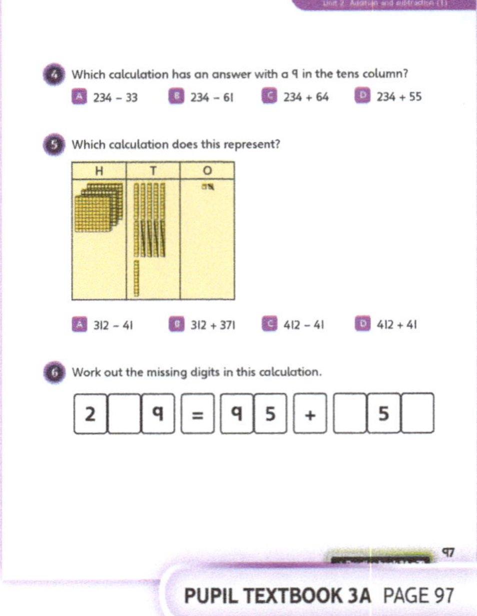

PUPIL TEXTBOOK 3A PAGE 97

Q	A	WRONG ANSWERS AND MISCONCEPTIONS	STRENGTHENING UNDERSTANDING
1	B	A suggests children have not understood parts and wholes.	Encourage children to draw number lines or manipulate place value equipment to justify or explore their own answers further, or to disprove errors.
2	A	D suggests children have not understood the different exchanges required for additions and subtractions.	
3	C	D suggests children have not accurately bridged the 100.	
4	C	D suggests children are not confident using the language of columns in preparation for the next unit.	
5	C	A suggests children have not developed a full understanding of the place value columns.	
6	249 = 95 + 154	Incorrect answers may suggest confusing place value columns.	

My journal

WAYS OF WORKING Independent thinking

ANSWERS AND COMMENTARY

Children should describe and explain the different methods represented, and highlight the similarities and differences. They will have to think very carefully about the action being shown, and the order of the different parts of the subtraction. Children will need to be able to use the language of partitioning and place value to explain what they notice.

If children need support in unpicking the methods, ask:
- *Which method would you use?*
- *What happens first in each method?*

If children are struggling to find similarities or differences, ask:
- *Do all the methods produce the same answer?*
- *How many 10s and 1s are being subtracted?*

Power check

WAYS OF WORKING Independent thinking

ASK

- *What numbers could you add and subtract before this unit?*
- *Do you know how to use equipment or drawings to support or explain your calculations?*
- *Have you built more confidence in any areas?*
- *Do you feel confident knowing when and how to use exchange?*
- *Can you show your answers in different ways?*

Power play

WAYS OF WORKING Pair work

IN FOCUS Use this power play to encourage children to explore properties of calculations, and to make predictions and conjectures that they can then check.

ANSWERS AND COMMENTARY Children will most likely need to begin by using a trial and error, then a trial and improve method, to explore the context. They should start to make conjectures about what digits could be involved, and should then test these and modify their predictions as appropriate.

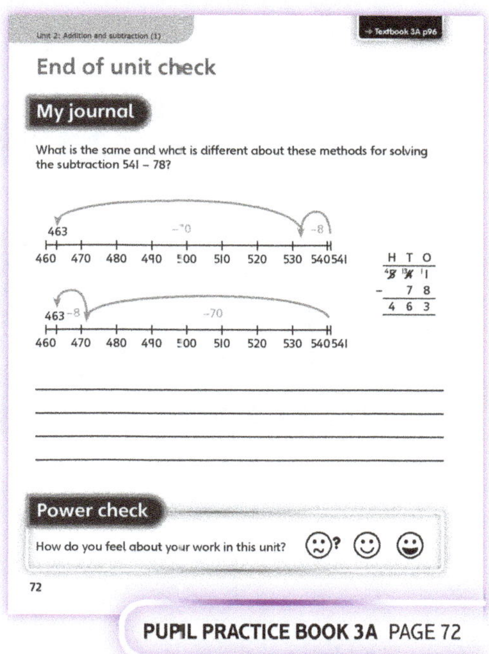

PUPIL PRACTICE BOOK 3A PAGE 72

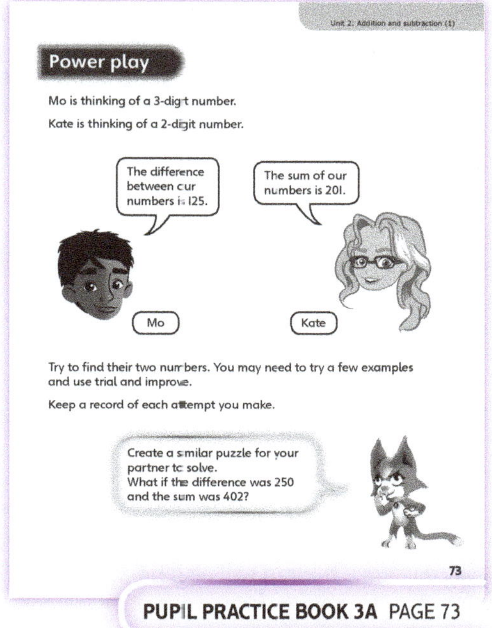

PUPIL PRACTICE BOOK 3A PAGE 73

After the unit ⏸

- Can children decide when to use a mental method, a drawing, equipment or a written method for different calculations?
- Can children use deep understanding to solve missing number and missing digit problems?

Strengthen and **Deepen** activities for this unit can be found in the *Power Maths* online subscription.

Unit 3
Addition and subtraction ②

Mastery Expert tip! "When teaching this unit, I would encourage children to use their new skills to explore properties of numbers, such as odds and evens, in even greater depth, by asking them to find patterns and make conjectures about what happens when you add or subtract odd and even numbers up to 1,000."

Don't forget to watch the Unit 3 video!

WHY THIS UNIT IS IMPORTANT

This unit develops a depth of understanding of the key skills of formal addition and subtraction through place value, checking strategies and mental methods. Children will use their growing understanding to explore calculations which do or do not require exchange, developing fluency, accuracy and confidence in their ability to perform these calculations. They will be able to apply checking strategies to decide for themselves whether their answer is reasonable or likely to be an error.

WHERE THIS UNIT FITS

→ Unit 2: Addition and subtraction (1)

→ **Unit 3: Addition and subtraction (2)**

→ Unit 4: Multiplication and division (1)

In this unit, children build on previous learning to add and subtract numbers with up to three digits. They begin by using place value equipment and formal written methods of columnar addition and subtraction.

Before they start this unit, it is expected that children:

• know how to partition numbers to 1,000 flexibly
• understand the concept of exchange in addition and subtraction
• know how to represent additions and subtractions using place value equipment and a place value grid.

ASSESSING MASTERY

Children who have mastered this unit will be able to add or subtract numbers with up to 3 digits within 1,000. They will be able to justify whether or not an exchange was necessary and be able to explain the effect of doing an exchange in terms of place value. Children will also be able to justify an answer through checking strategies of approximation, estimation and the use of inverse operations.

COMMON MISCONCEPTIONS	STRENGTHENING UNDERSTANDING	GOING DEEPER
Children may resort to counting strategies rather than using known number bonds to add digits within an addition.	Use base 10 equipment to model an addition or subtraction, but encourage children to visualise the known number bond that finds the total or the part required to complete the parts of the calculation.	Explore the properties of numbers which result from adding odds and evens, or numbers that end in a 5. Discuss problems with multiple solutions (for example, how many subtractions can you find that end in 1?).
Children may learn the process of column methods without fully understanding how it relates to place value and flexible partitioning.	Encourage children to use the language of 1s, 10s and 100s flexibly. Use a part-whole model alongside a column method to represent how the number has been partitioned.	There are many opportunities for all children to engage in missing number or missing digit problems, which dig deep into the mechanics of the methods. This requires children to explore how and why the process works.

Unit 3: Addition and subtraction ❷

Use these pages to introduce the focus to children. You can use the characters to explore different ways of working too!

STRUCTURES AND REPRESENTATIONS

Place value grid: Use this to help children to represent the partitions and exchanges required when adding and subtracting. It should support the understanding of column methods.

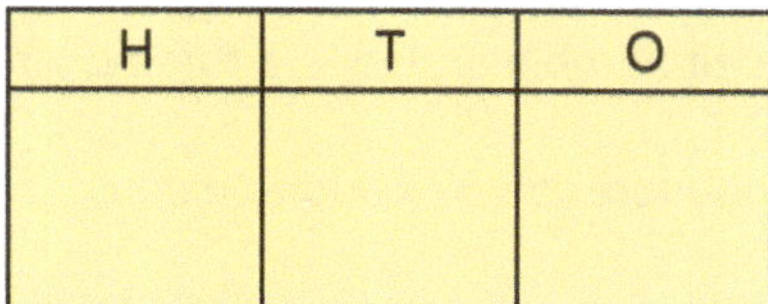

H	T	O

Base 10 equipment: These can be manipulated by children to model the differences between addition and subtraction (by taking away or by comparing). Base 10 equipment can be used in conjunction with place value grids to give structure.

H	T	O

Column methods: Column methods will be used to present efficient and accurate addition and subtraction. Children should practise by using the scaffolded examples provided in the book, but should also experience writing their own calculations in columns.

```
  H  T  O

+ _______

  _______
```

KEY LANGUAGE

There is some key language that children will need to know as part of the learning in this unit.

- ➜ add, addition
- ➜ subtract, subtraction
- ➜ total, altogether
- ➜ exchange
- ➜ part-whole, whole, part
- ➜ place value
- ➜ hundreds (100s), tens (10s), ones (1s)
- ➜ column method
- ➜ mental method, mentally
- ➜ estimate, estimation
- ➜ approximate, approx., approximation, approximately, about
- ➜ fact family
- ➜ bar model
- ➜ digits
- ➜ multiple
- ➜ logically
- ➜ function machine

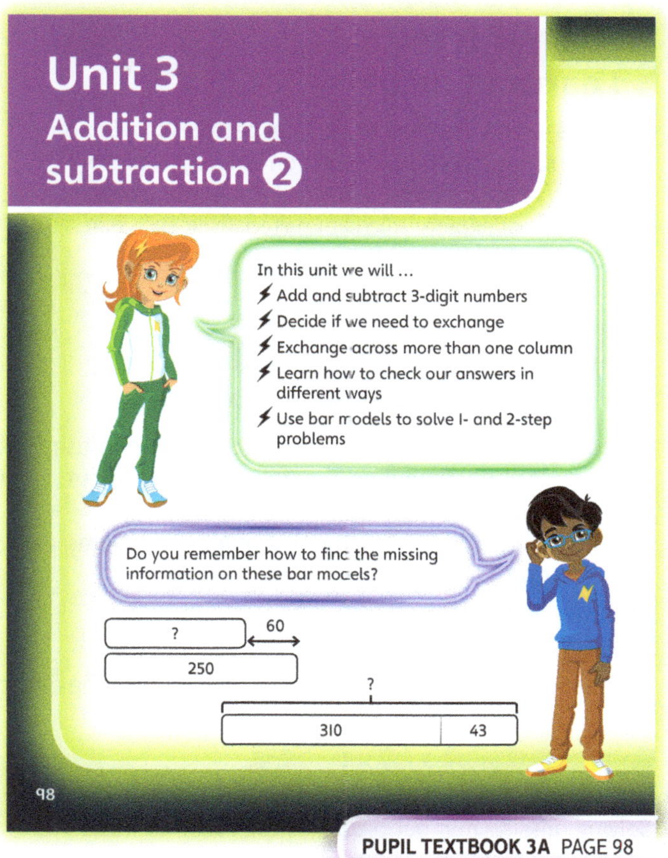

PUPIL TEXTBOOK 3A PAGE 98

PUPIL TEXTBOOK 3A PAGE 99

Addition and subtraction patterns

Learning focus

In this lesson, children will explore patterns in addition and subtraction and the effect on different digits of adding or subtracting 1s, 10s or 100s. At this stage, they do not explore calculations requiring exchange.

Small steps

→ Previous step: Subtracting a 2-digit number from a 3-digit number
→ **This step: Addition and subtraction patterns**
→ Next step: Adding two 3-digit numbers (1)

NATIONAL CURRICULUM LINKS

Year 3 Number – Addition and Subtraction

- Add and subtract numbers with up to three digits, using formal written methods of columnar addition and subtraction.
- Add and subtract numbers mentally, including: a three-digit number and ones, a three-digit number and tens, a three-digit number and hundreds.
- Solve problems, including missing number problems, using number facts, place value, and more complex addition and subtraction.

ASSESSING MASTERY

Children can identify similarities and differences in the effect of adding or subtracting 1s, 10s or hundreds, for example, 2, 20 or 200.

COMMON MISCONCEPTIONS

Solving missing number problems can cause confusion as to which operation to use. Discuss whether children actually need to add or subtract in order to identify the missing number. What strategies do they use? Ask:

- *How could you solve __ + 20 = 220?*
- *What if the missing number is in a different position, for example 200 + __ = 220?*

STRENGTHENING UNDERSTANDING

Using a part-whole model to represent the missing number problems can help support children as they develop their understanding of the calculations required.

GOING DEEPER

The function machines can be used for exploring missing number problems and for exploring hidden operations, too. Being able to investigate missing operations requires two steps of thinking.

KEY LANGUAGE

In lesson: subtraction, function machine, ones (1s), tens (10s), hundreds (100s), part-whole models, **digit**

Other language to be used by the teacher: operation

STRUCTURES AND REPRESENTATIONS

Function machines, part-whole model, place value grid

RESOURCES

Mandatory: base 10 equipment or place value equipment

Optional: part-whole models

 In the eTextbook of this lesson, you will find interactive links to a selection of teaching tools.

Before you teach 

- Can children identify which digits change when adding 5?
- Can they explain what is the same and what is different about these calculations: 11 + 2, 11 + 20, 20 + 11, 2 + 11?

Discover

 Pair work

ASK

- Question ① a): *What is the same and what is different about the three machines?*
- Question ① a): *What do you notice about the numbers being inputted into each machine?*
- Question ① a): *Can you predict anything about how each number will change?*
- Question ① b): *What is different about how you work out the answer to question ① b) compared to ① a)?*

IN FOCUS   The important aspect is to notice the similarities and differences in the functions in question ① a). The aim is for children to notice that they are linked, but that the place value of the digits may have a noticeable and predictable effect on how the input number is changed.

PRACTICAL TIPS   The function machines could be role-played by the teacher or children. The children 'input' a number by handing the person a number written on a whiteboard, or by telling them the number. The function machine would then output a number in a similar way. (After having made some machine-like noises, say: 'The output is 156.')

ANSWERS

Question ① a): 154 + 2 = 156, 154 + 20 = 174,
154 + 200 = 354

Question ① b): 797 − 200 = 597,
597 + 200 = 797; Jamie's number was 597.

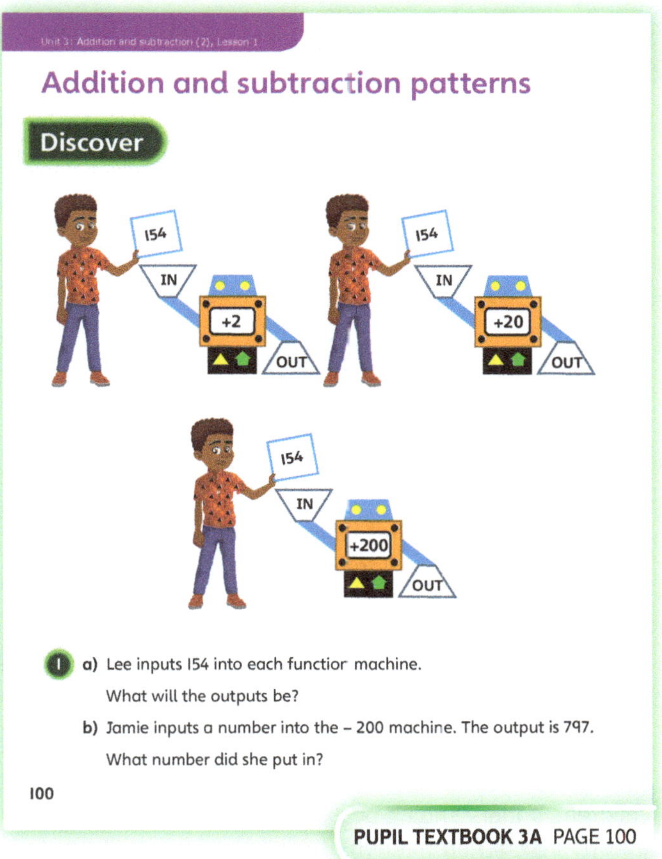

PUPIL TEXTBOOK 3A PAGE 100

Share

 Whole class teacher led

ASK

- Question ① a): *Which digit changes when you add 2, 20 or 200?*
- Question ① a): *Why do the other digits not change?*
- Question ① a): *Which number bonds help you solve each addition?*
- Question ① b): *Why is this a missing number problem?*
- Question ① b): *How does the part-whole model show how to find the missing number?*

IN FOCUS   For question ① a), children should discuss and develop an awareness that the effect of adding 100s, 10s or 1s in the given examples is that only one digit of the input number changes. They need to understand that this is due to the place value of the numbers being added.

In question ① b), children should recognise that this is like a missing number problem. They need to be able to explain how the part-whole model represents this and support the reasons for choosing the operation required.

PUPIL TEXTBOOK 3A PAGE 101

Think together

WAYS OF WORKING Whole class teacher led (I do, We do, You do)

ASK

- Questions **1**, **2**, **3** a) and **3** b): *What operation will you need to perform?*
- Questions **1** and **2**: *Can you predict how the digits will change?*
- Questions **1**, **2** and **3** b): *What do you notice about the effect on the digits?*

IN FOCUS In questions **1** and **2**, children develop their expertise so that they understand whether they need to add or subtract. They then extend this in question **3** a) to reasoning about the effect on the different digits. By the end of the sequence, children should be able to explain how to reason based on the effect on the 1s, 10s or 100s digits.

STRENGTHEN Ensure that children represent their thinking using place value equipment and on a place value grid. Discuss how to use number bonds to work out the answers.

DEEPEN At this stage, we are not considering calculations which require exchange or that bridge 10s or 100s. Deepen children's understanding by exploring the patterns in question **3** b). Can children develop their own patterns to create a family of puzzles like these? Ask children to explain how their thinking has to change in order to invent the puzzles, as opposed to simply solving the question.

ASSESSMENT CHECKPOINT Can children explain the missing parts of the calculations in question **3** b) by looking at which of the digits change?

ANSWERS

Question **1** : 321 + 5 = 326, 321 + 50 = 371, 321 + 500 = 821

Question **2** : 546 − 3 = 543, 546 − 30 = 516, 546 − 300 = 246

Question **3** a): 259 − 253 = 6, 253 + 6 = 259, so + 6
953 − 253 = 700, 253 + 700 = 953, so + 700
253 − 203 = 50, 253 − 50 = 203, so − 50

Question **3** b): 113 = 111 + 2   555 = 755 − 200
131 = 111 + 20   555 = 557 − 2
311 = 111 + 200   555 = 575 − 20

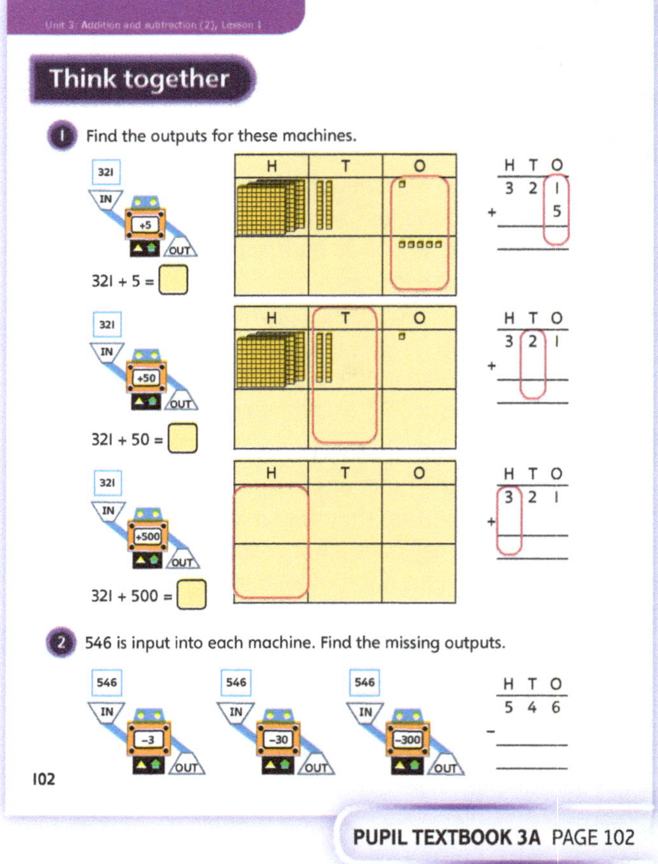

PUPIL TEXTBOOK 3A PAGE 102

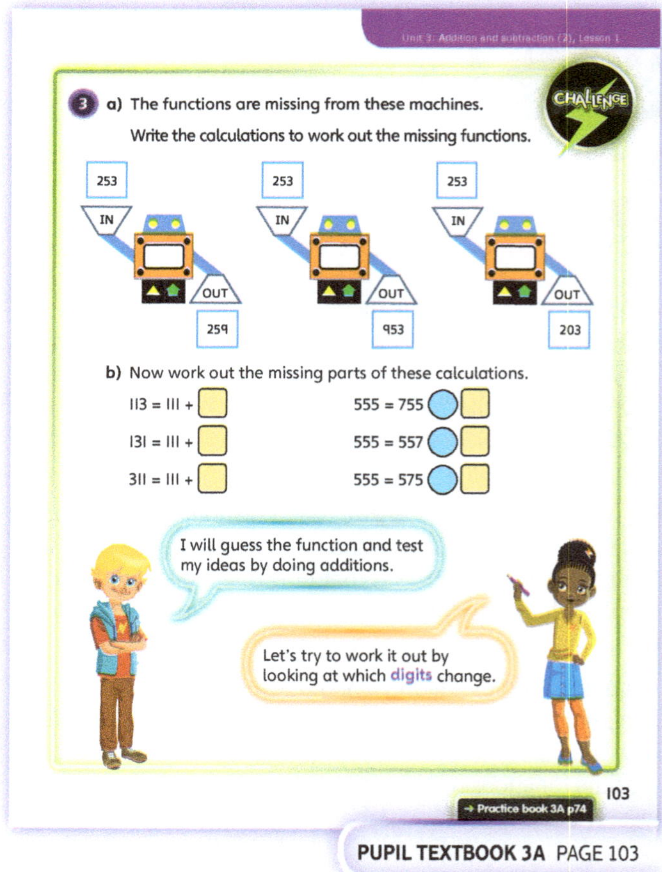

PUPIL TEXTBOOK 3A PAGE 103

Practice

WAYS OF WORKING Independent thinking

IN FOCUS The focus of question ❶ is on developing fluency in adding 1s, 10s or 100s. From question ❷ onward, children develop their reasoning around the problems. They apply these fluency and reasoning skills to word problems and to a variety of missing number problems presented in slightly different contexts.

STRENGTHEN Base 10 equipment is used to support thinking in the earlier questions, but children should feel they can continue to use the equipment, even when not directly prompted. It may also help children to represent the calculations on part-whole models, in order to understand if they need to add to find the whole, or subtract to find a part.

DEEPEN The patterns of variation in question ❹ are designed to push children to a deeper understanding. They will require a strong understanding of the structure of the patterns, in order to spot the different reasoning required. Question ❺ prompts children to notice that the technique they have been using does not apply to all situations. Children will start to consider which of these calculations require exchange.

ASSESSMENT CHECKPOINT Question ❸ requires a good understanding of how the calculations work, in order to identify which output matches which function.

ANSWERS Answers for the **Practice** part of the lesson appear in the separate **Practice and Reflect answer guide**.

Reflect

WAYS OF WORKING Independent thinking

IN FOCUS It is important that children show how they work through the calculations, not simply give an answer. They may draw base 10 equipment or number lines or show the calculation as a part-whole diagram. Ask children to compare and discuss their representations with other children.

ASSESSMENT CHECKPOINT Can children explain their representations in terms of place value?

ANSWERS Answers for the **Reflect** part of the lesson appear in the separate **Practice and Reflect answer guide**.

After the lesson 

- Do children know how to work out the calculations by considering which digits change?
- Can children solve missing number problems?
- Can children represent their thinking with equipment or models to justify their answers?

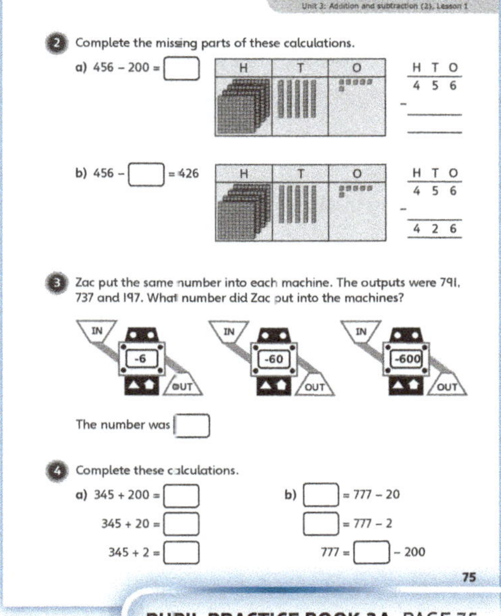

PUPIL PRACTICE BOOK 3A PAGE 74

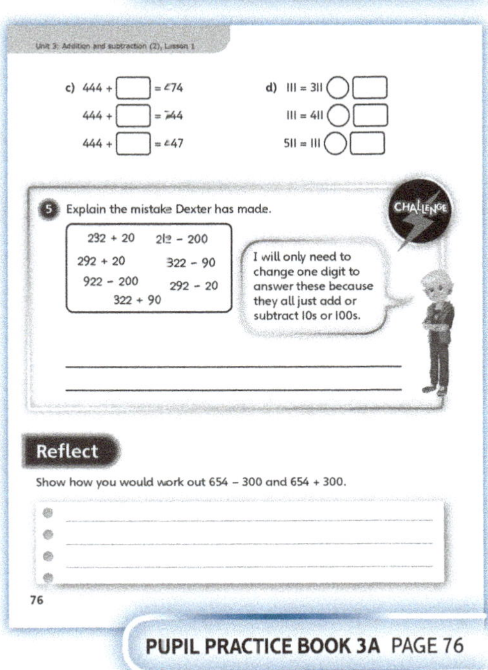

PUPIL PRACTICE BOOK 3A PAGE 75

PUPIL PRACTICE BOOK 3A PAGE 76

Adding two 3-digit numbers 

Learning focus

In this lesson, children will learn to add two 3-digit numbers where no exchange is necessary. They will use a written column method and begin with the 1s, then the 10s and then the 100s.

Small steps

→ Previous step: Addition and subtraction patterns
→ **This step: Adding two 3-digit numbers (1)**
→ Next step: Adding two 3-digit numbers (2)

NATIONAL CURRICULUM LINKS

Year 3 Number – Addition and Subtraction

- Add and subtract numbers with up to three digits, using formal written methods of columnar addition and subtraction.
- Add and subtract numbers mentally, including: a three-digit number and ones, a three-digit number and tens, a three-digit number and hundreds.

ASSESSING MASTERY

Children can explain how the column written method relates to the place value of the addition. They can write the addition accurately and use the columns to add two numbers accurately and efficiently.

COMMON MISCONCEPTIONS

Children may resort to the written method, even when a mental method is more appropriate. Ask:
- *When you look at this addition, what method do you think is most useful? Why?*

No exchange is required in this lesson, which may lead children to assume that the method works in any order, perhaps by adding the 100s first. Ask:
- *Can you think of any reason why we might always start by adding the ones column?*

STRENGTHENING UNDERSTANDING

Ensure children have access to place value equipment to support their understanding of place value. Encourage children to relate the numbers to the concrete elements of problems based around real-life contexts before building their confidence in abstract calculations.

GOING DEEPER

There are opportunities at various points through the lesson to explore additions which produce the same answers. Challenge children to explain how these are achieved through manipulating the place value of the additions and to describe how to find all the solutions systematically.

KEY LANGUAGE

In lesson: digit, ones (1s), tens (10s), hundreds (100s)

Other language to be used by the teacher: column

STRUCTURES AND REPRESENTATIONS

Place value grids, column addition, base 10 equipment

RESOURCES

Mandatory: base 10 or other place value equipment

Optional: digit cards

 In the eTextbook of this lesson, you will find interactive links to a selection of teaching tools.

Before you teach 

- Can children add two 2-digit numbers?
- Do you have the necessary place value equipment ready to support learning in this lesson?
- Can children confidently use place value equipment in a place value grid?

Discover

 Pair work

- Question **1** a): *How might Richard write his addition? Are there different ways?*
- Question **1** a): *Could Richard use any pictures or equipment to help him?*
- Question **1** a): *What does each digit in 321 and 541 represent?*
- Question **1** b): *Is there more than one answer to this question?*

 The key to question **1** a) is encouraging children to explore how to represent the additions in a way that allows them to find the total accurately. Some children may choose to use equipment, others may try to represent the addition on a number line. Allow children to explore and discuss the various representations and to justify their decisions. Question **1** b) encourages children to think carefully about the place value in the numbers, although some may approach this with a trial and error method to begin with.

 Children will find it enriching to have their own digit cards to move and manipulate throughout the lesson. If these are not available, they could make their own by cutting out paper rectangles. Having place value equipment available should allow children to make a solid connection and build upon their learning from the previous lesson.

Question **1** a): 326 + 541 = 867

Question **1** b): Richard will get the same total if he swaps the 2 hundreds digits, the 2 tens digits or the 2 ones digits.

Share

 Whole class teacher led

- Question **1** a): *How has each digit been represented?*
- Question **1** a): *Do you need to count to find the total for each column?*
- Question **1** b): *How could you use the place value grid to explain how to find other possible solutions?*

 The focus for this part of the lesson is on highlighting how the place value grid, combined with the base 10 equipment, is a very clear way of organising the addition. It helps clarify how to find the total by adding the digits in order. Children should use their knowledge of number bonds (rather than counting strategies) to find the total for each column. Question **1** b) effectively demonstrates how the total of each column remains constant even when the 2 hundreds digits are swapped. Encourage children to explore how the same is true if the 2 tens or the 2 ones digits are swapped.

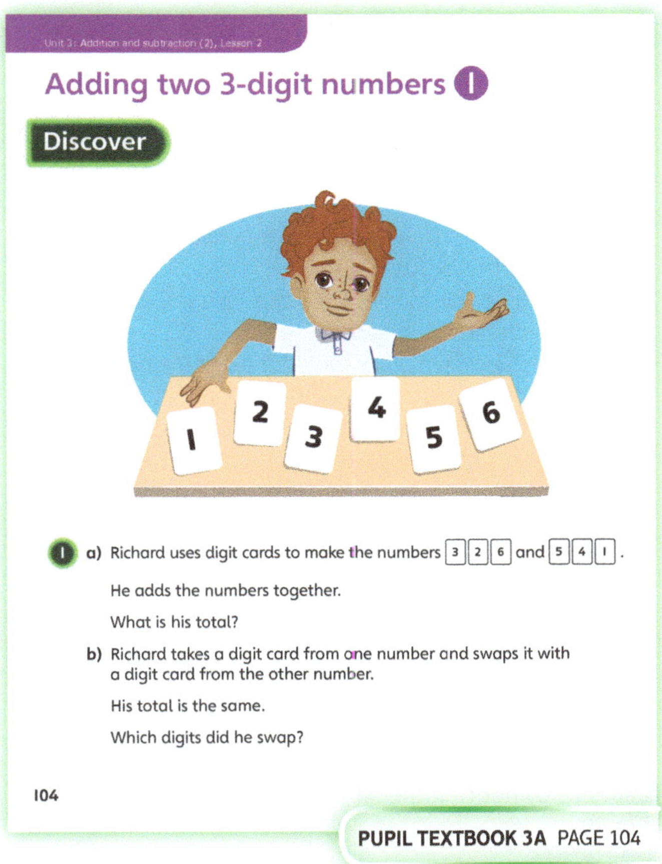

PUPIL TEXTBOOK 3A PAGE 104

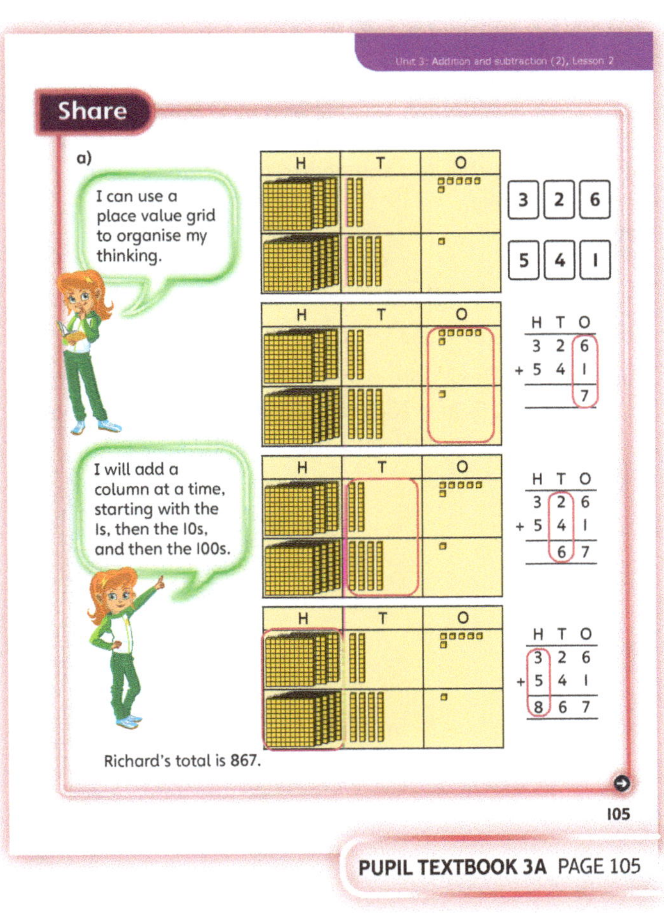

PUPIL TEXTBOOK 3A PAGE 105

Think together

 Whole class teacher led (I do, We do, You do)

ASK

- Question ❶ : *Which known number bonds will help you find the totals of the 1s, 10s and 100s?*
- Question ❷ : *How many 1s, 10s and 100s will Jamilla need to show with the equipment?*
- Question ❷ : *Which order should she add the digits in? Does it matter?*
- Question ❸ : *Where is the best place to start? Are there different ways of making a 9 or a 3 using the digit cards 1, 2, 3, 4, 5 and 6?*

IN FOCUS Develop children's confidence and fluency around the link between the place value representations and the column method. By the time children have completed question ❸, they may be confident enough to solve the additions without the support of place value equipment but they should still be encouraged to explain and justify their calculations using equipment and the language of 100s, 10s and 1s. In question ❸, children should explore different solutions by noticing the different pairs of digits that total 3 or 9.

STRENGTHEN Children should use digit cards alongside place value equipment to explore how moving the digits from one column to another changes the place value of the number. This should also change the way the number is represented using equipment. Remind them that they are working with the digit cards 1, 2, 3, 4, 5 and 6.

DEEPEN Question ❸ has multiple solutions. Children should try to justify whether they have found all the possible options and to describe the method they have used to achieve this.

ASSESSMENT CHECKPOINT Can children use known number bonds to find the totals for each column in question ❷ efficiently? Are children able to use the written column layout accurately?

ANSWERS

Question ❶ : 142 + 356 = 498

Question ❷ : 413 + 562 = 975

Question ❸ : The 2 ones digits must be 1 and 2. Then the 10s and 100s digits will be any combination of the pairs 6 and 3, and 4 and 5.

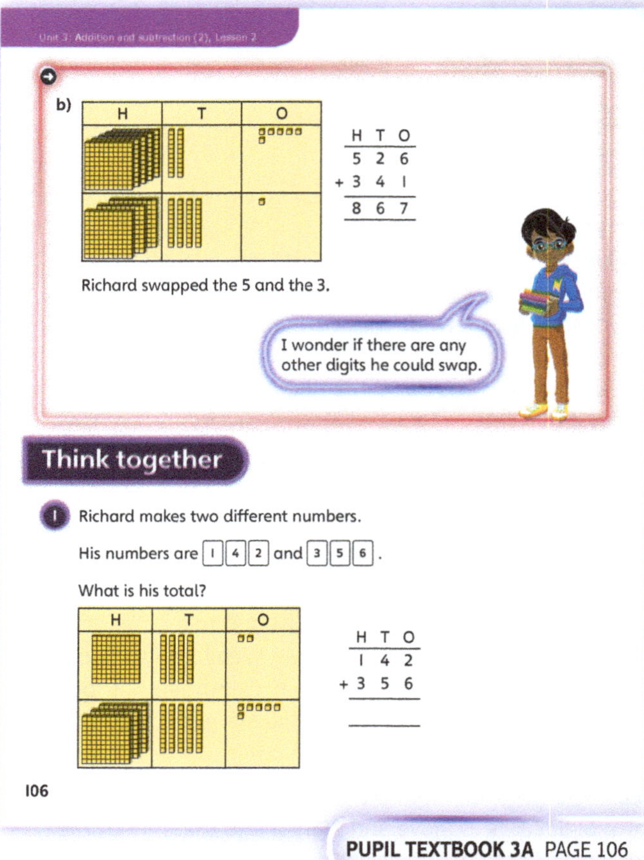

PUPIL TEXTBOOK 3A PAGE 106

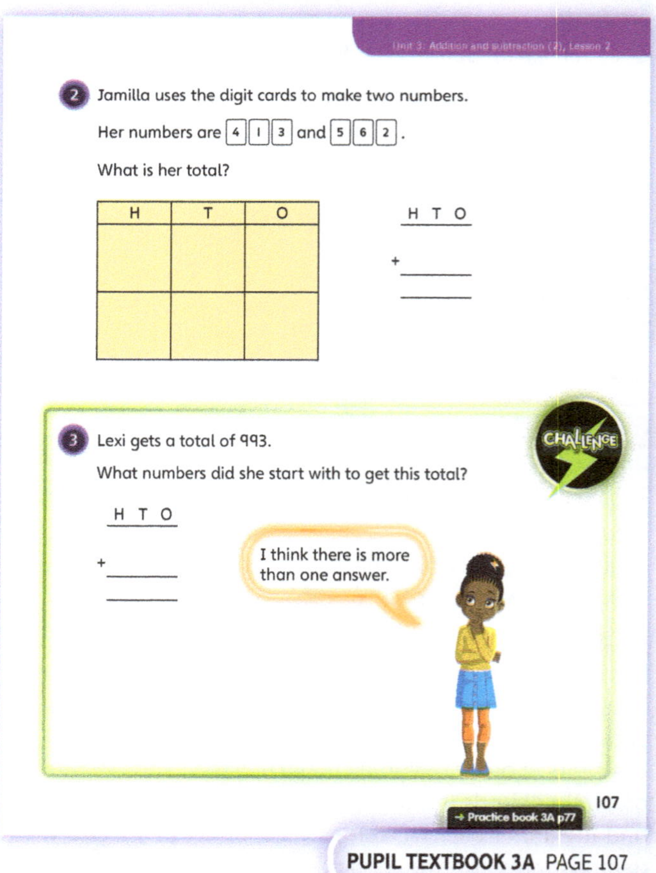

PUPIL TEXTBOOK 3A PAGE 107

Practice

WAYS OF WORKING   Independent thinking

IN FOCUS

In question **1** b) children should recognise that the lack of equipment in the tens column represents a '0'. Question **1** c) is only part filled in, requiring children to realise that the total is given but some of the place value equipment is missing. They need to complete the missing place value equipment in the grid to find the addition.

Question **2** prompts children to solve calculations in a more abstract fashion, including using a mental method where appropriate rather than always reverting to the written column method. Question **4** encourages children to reason about the missing digits based on their knowledge of parts and wholes. Question **6** challenges children to use their understanding of variation to use the initial calculation to deduce the totals for related additions.

STRENGTHEN   Although only question **1** directly prompts the use of equipment, children should have access to this to support their thinking. They should, however, use known bonds to find the totals and not rely on counting the equipment. Children may also find it helpful to have a column addition scaffold to support their layout.

DEEPEN   Questions **5** and **6** are solved most effectively through deduction and a deep analysis of the numbers to be added. Encourage children to justify their answers through reasoning although they may find it helps to begin question **5** with a 'trial and improve' method. Similarly, some children may need to explore the additions in question **6** by using a column method for the first few examples, before pausing to look for a pattern to support their reasoning.

ASSESSMENT CHECKPOINT   Solved accurately, question **4** demonstrates a sound understanding of how the method relates to place value. Question **2** demonstrates fluency and accuracy and question **3** shows competency in applying the method to real contexts.

ANSWERS   Answers for the **Practice** part of the lesson appear in the separate **Practice and Reflect answer guide**.

Reflect

WAYS OF WORKING   Pair work

IN FOCUS   The important thing for children to notice is how the transcription error (writing 143 instead of 134) leads to an error. This highlights how children should check their working out against what the question is actually asking. The other mistake – incorrect addition of the hundreds column – prompts them to focus on accuracy.

ASSESSMENT CHECKPOINT   Can children explain the errors that have been made in terms of place value?

ANSWERS   Answers for the **Reflect** part of the lesson appear in the separate **Practice and Reflect answer guide**.

After the lesson 

- Are children confident in representing the additions using place value equipment on a place value grid?
- Do children use known number bonds to find column totals?
- Can children deduce missing digits and related answers, rather than simply relying on following a procedure without any deep understanding?

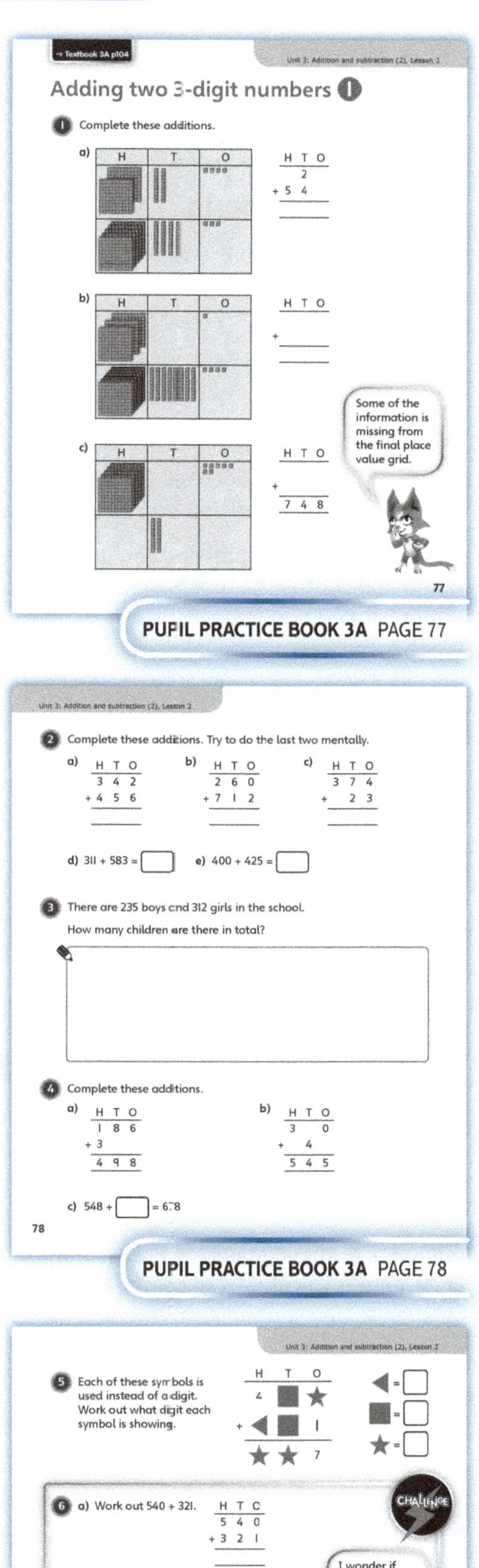

PUPIL PRACTICE BOOK 3A PAGE 77

PUPIL PRACTICE BOOK 3A PAGE 78

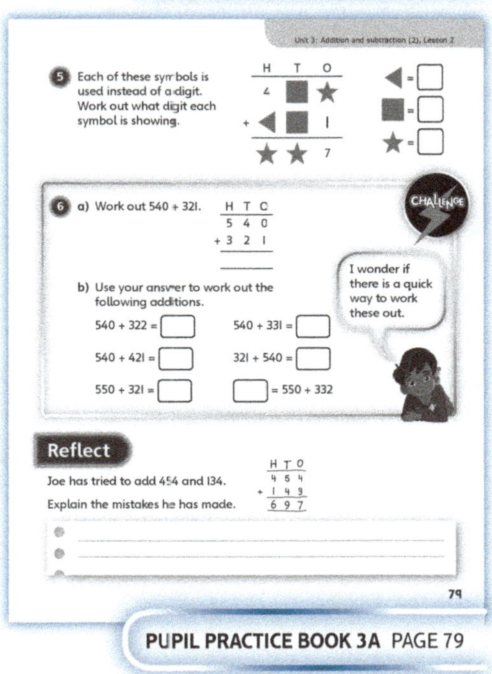

PUPIL PRACTICE BOOK 3A PAGE 79

Adding two 3-digit numbers ②

Learning focus

In this lesson, children will learn to add two 3-digit numbers where exchange may be necessary, and to recognise when it is or is not necessary.

Small steps

→ Previous step: Adding two 3-digit numbers (1)
→ **This step: Adding two 3-digit numbers (2)**
→ Next step: Subtracting a 3-digit number from a 3-digit number (1)

NATIONAL CURRICULUM LINKS

Year 3 Number – Addition and Subtraction

- Add and subtract numbers with up to three digits, using formal written methods of columnar addition and subtraction.
- Add and subtract numbers mentally, including: a three-digit number and ones, a three-digit number and tens, a three-digit number and hundreds.
- Solve problems, including missing number problems, using number facts, place value, and more complex addition and subtraction.

ASSESSING MASTERY

Children can add two 3-digit numbers and decide whether there is an exchange in none, one or two of the columns. They can explain how the exchange is presented in columnar written methods and demonstrate this using place value equipment.

COMMON MISCONCEPTIONS

Some children may add mentally, starting with adding the 100s, then the 10s and then the 1s. If children do not begin by adding the 1s, then the 10s and then the 100s, adding where exchange is required can be difficult and prone to inaccuracy. Ask:

- *Which part should we add first? Why do we add this part first?*

STRENGTHENING UNDERSTANDING

Remind children of similar work in Unit 2, where they had to exchange when adding some 1- and 2-digit numbers. Make sure children experience the concept of exchange through manipulating base 10 equipment on a place value grid.

GOING DEEPER

Completing missing digits in given column subtractions is a very good way to prompt a deeper understanding of the concept of exchange. Question 4 of Lesson 4 in this unit of the **Practice Book** contains examples of this type of problem. Challenge children to develop their own missing digit calculations. Can they create examples where there are multiple solutions?

KEY LANGUAGE

In lesson: exchange, ones (1s), tens (10s), hundreds (100s), addition

Other language to be used by the teacher: column method

STRUCTURES AND REPRESENTATIONS

Column addition, place value grids

RESOURCES

Mandatory: base 10 equipment, place value grids

 In the eTextbook of this lesson, you will find interactive links to a selection of teaching tools.

Before you teach 

- Have children used exchange before?
- Do they know what it means to exchange 10 ones or 10 tens?

Discover

 Pair work

- Question **1** a): *What number bonds can help us find the total of 1s, 10s and 100s?*
- Question **1** a): *Does your understanding of these number bonds suggest that you will need to exchange to answer this problem?*
- Question **1** a): *Is there anything we have learnt before that will help us to solve this accurately?*
- Question **1** b): *Can you think of different ways to solve this question?*

 In question **1** a) it is important that children recognise there is a need for exchange, as the sum of the 1s digits is greater than 9. Children should recognise this from similar learning in Unit 2.

 Children should represent the numbers in the questions with place value equipment such as base 10, arranging the equipment in a way that demonstrates an efficient and accurate model of the addition.

Question **1** a): 126 + 217 = 343

Question **1** b): 343 + 50 = 393 (or 343 + 20 = 363 and
363 + 30 = 393)

PUPIL TEXTBOOK 3A PAGE 108

Share

 Whole class teacher led

- Question **1** a): *Can you see how the exchange is demonstrated using the base 10 equipment?*
- Question **1** a): *How is the exchange shown in the written column method?*
- Question **1** a): *Why do we not also exchange 10 tens?*
- Question **1** b): *What is the same and what is different about the methods used to solve this question?*

 Learning should focus on how the exchange is represented by the written method and understand how this links to the way it is represented using the place value equipment. Children should discuss why we add the ones column first and then continue from right to left, in terms of the effect on the exchange. Children should also discuss the two different ways of solving question **1** b), which is a two-step problem. Children could decide which they think is the most efficient method, or which they feel best represents the problem. Ask: *Which method do you think might lead to mistakes? Why?*

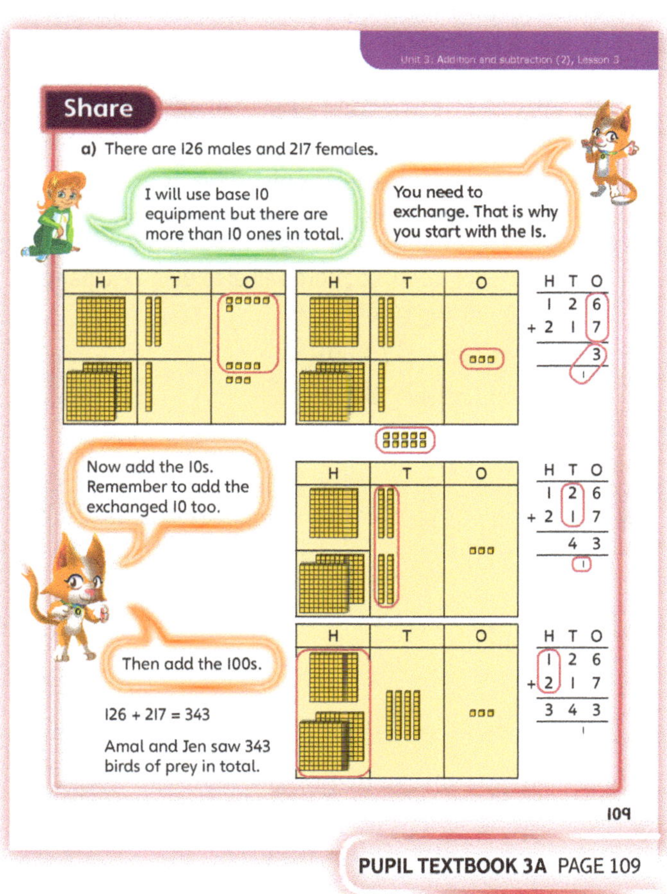

PUPIL TEXTBOOK 3A PAGE 109

Think together

WAYS OF WORKING Whole class teacher led (I do, We do, You do)

ASK

- Questions ❶ and ❷: *Do you think you will have to do an exchange of 1s?*
- Question ❶: *What number bonds do you need to know in order to add up the different digits?*
- Question ❷: *Where shall we write the exchange?*

IN FOCUS The focal point for this part of the lesson is on developing confidence and accuracy when exchanging either 1s, 10s or both. Children should build the habit of working from right to left, and discuss how this accommodates any exchange as the calculation builds.

STRENGTHEN Represent the exchanges using place value equipment. It may help some children to visualise or use ten frames to explore the exchange but they should always be encouraged to use known bonds rather than a counting-equipment strategy to total the digits.

DEEPEN Question ❸ of this section presents a slight misconception. It may not be a mistake children make themselves, but they should discuss and try to understand the reasoning behind Max's mistake. Challenge children to think of a number of calculations with the same property. What is the common element?

ASSESSMENT CHECKPOINT Can children justify their answer to Question ❷ in terms of the exchange of the ones, tens and hundreds columns?

ANSWERS

Question ❶: 262 + 251 = 513

Question ❷: 157 + 166 = 323

Question ❸ a): Max's idea seems plausible, as only the 1s digits total more than 9. However, there will be a 'knock-on' effect of exchanging 10 ones for 1 ten, so there will also be an exchange of 10 tens.

184 + 217 = 401

Question ❸ b): Answers will vary.

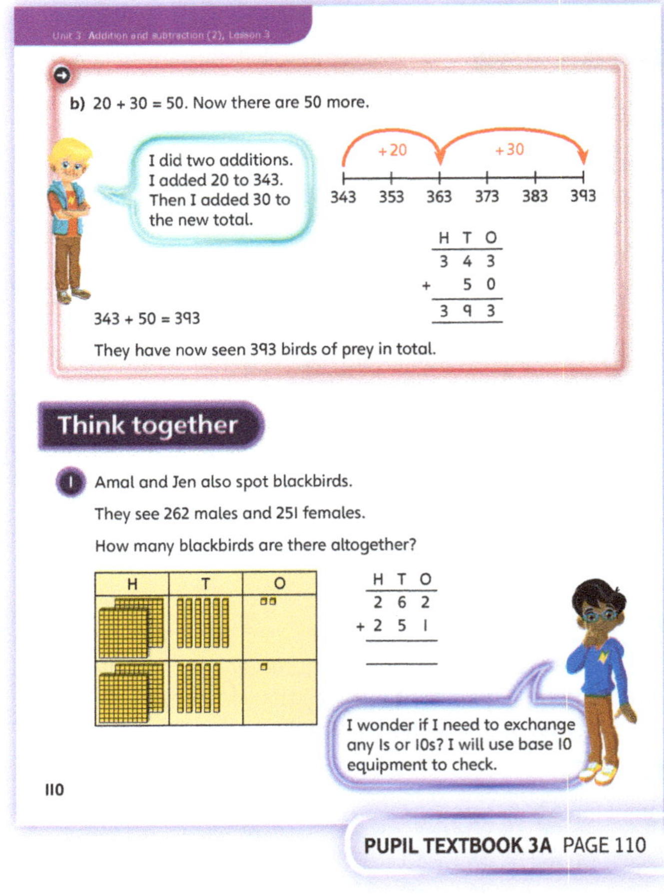

PUPIL TEXTBOOK 3A PAGE 110

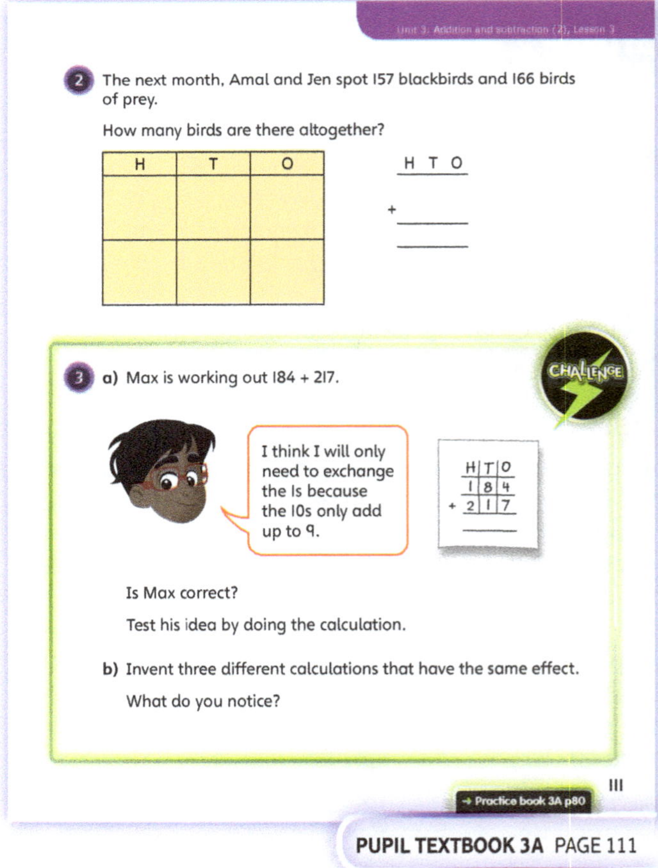

PUPIL TEXTBOOK 3A PAGE 111

Practice

WAYS OF WORKING Independent thinking

IN FOCUS These questions are designed to help children develop from using concrete or pictorial representations of place value, to become more confident and fluent in solving abstract calculations presented in columns. Children will also develop their reasoning around when an exchange is or is not necessary. This decision-making process is made distinct from the process of performing the addition itself, so that children always maintain mathematical thinking alongside procedural fluency.

STRENGTHEN Prompt children to pause before looking at a calculation and look for tell-tale signs of where an exchange might be necessary. Can they spot any bonds that will total more than 9?

DEEPEN Question 4 has multiple solutions and children can be prompted to explain why they are confident they have found all the possible combinations.

THINK DIFFERENTLY Question 3 may lead to misconceptions as solving these requires taking into account the effect of exchange. If needed, children could use base 10 equipment for support in modelling the first couple of examples in question 3 a) to build confidence.

ASSESSMENT CHECKPOINT Question 2 exhibits a deep understanding of the concept of exchange. Can children explain why they are linking each calculation to the correct bubble?

ANSWERS Answers for the **Practice** part of the lesson appear in the separate **Practice and Reflect answer guide**.

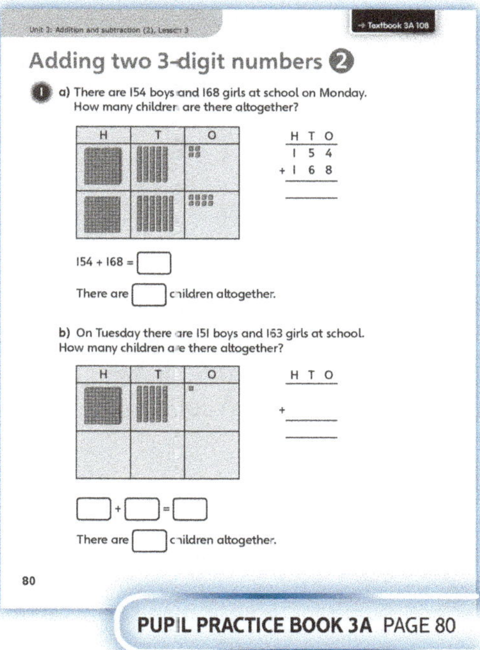

PUPIL PRACTICE BOOK 3A PAGE 80

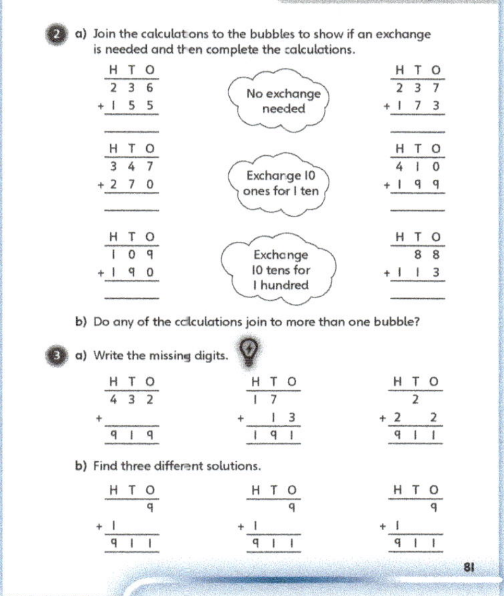

PUPIL PRACTICE BOOK 3A PAGE 81

Reflect

WAYS OF WORKING Pair work

IN FOCUS This question should allow children to use the language of place value and exchange in order to explain the mistake. They could discuss and rehearse their response in pairs before deciding how to write their answer as clearly and accurately as possible.

ASSESSMENT CHECKPOINT Can children explain the mistake using the language of place value rather than simply performing the calculation to find the 'right answer'?

ANSWERS Answers for the **Reflect** part of the lesson appear in the separate **Practice and Reflect answer guide**.

After the lesson 

- Are children confident representing the exchange in column addition?
- Can children spot when an exchange is or is not necessary?
- Do children look out for common mistakes like forgetting to add the exchanged digit?

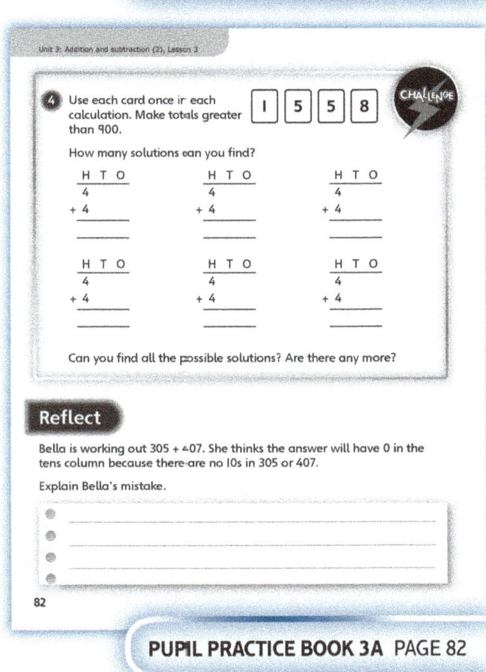

PUPIL PRACTICE BOOK 3A PAGE 82

Subtracting a 3-digit number from a 3-digit number ❶

Learning focus

In this lesson, children will learn to subtract a 3-digit number from another 3-digit number where no exchange is necessary. They represent the subtraction as a written column subtraction.

Small steps

→ Previous step: Adding two 3-digit numbers (2)

→ **This step: Subtracting a 3-digit number from a 3-digit number (1)**

→ Next step: Subtracting a 3-digit number from a 3-digit number (2)

NATIONAL CURRICULUM LINKS

Year 3 Number – Addition and Subtraction

- Add and subtract numbers with up to three digits, using formal written methods of columnar addition and subtraction.
- Add and subtract numbers mentally, including: a three-digit number and ones, a three-digit number and tens, a three-digit number and hundreds.

ASSESSING MASTERY

Children can subtract a 3-digit number from a 3-digit number using a written column method where no exchange is required. Children are increasingly confident in attempting to subtract a 3-digit number from a 3-digit number using a mental method, when appropriate.

COMMON MISCONCEPTIONS

Children may continue to find it confusing when subtraction is sometimes in the context of 'take away' and sometimes in the context of 'find the difference'. Ask:
- *What number do we need to subtract? What is the whole we are subtracting from?*

STRENGTHENING UNDERSTANDING

It may support some children to represent the numbers using base 10 equipment in order to explore and visually see subtraction as 'take away'.

GOING DEEPER

Explore the link between 'take away' and 'find the difference'. Encourage children to represent the whole number and then take away a part from the whole, by representing this with place value equipment. Then model the same subtraction as a find the difference. Discuss how the parts and the wholes are related in each of these different ways of thinking about subtraction.

KEY LANGUAGE

In lesson: subtract, subtraction, digit, mentally, **mental method**, logically, **multiple**

Other language to be used by the teacher: hundreds (100s), whole, part

STRUCTURES AND REPRESENTATIONS

Base 10 equipment, place value grids

RESOURCES

Mandatory: place value equipment (base 10, place value counters, cards and grids)

Optional: spinners, dice, paper clips, number lines

 In the eTextbook of this lesson, you will find interactive links to a selection of teaching tools.

Before you teach 

- Can children successfully identify the wholes and parts in a calculation (for example, in 35 – 23 = 12)?
- Are they able to use different methods to represent an addition or subtraction?

Discover

WAYS OF WORKING Pair work

ASK

- Question **1** a): *What is the value of each of the digits in the number Luis creates with the spinner?*
- Question **1** a): *How could Luis represent his subtraction?*
- Question **1** a): *What is the whole that Luis is subtracting from?*
- Question **1** b): *How many different numbers could Isla make using the digits 1, 6 and 6?*

IN FOCUS These questions are set in a mathematical game context. In question **1** a), children may notice that they could make different numbers to subtract from 999, if they choose to reorder the digits for the 1s, 10s and 100s. For example, Luis chooses 352, but he could choose 523, 532, 253, 235 or 325. The important thing to notice is that they are subtracting from 999, which is the whole in each subtraction.

PRACTICAL TIPS Children could play this game themselves, using spinners or dice. If they do not have a spinner, they could make one using a paper clip as the spinning 'arrow'.

ANSWERS

Question **1** a): 999 – 352 = 647

Question **1** b): 999 – 661 = 338, 999 – 616 = 383,
999 – 166 = 833

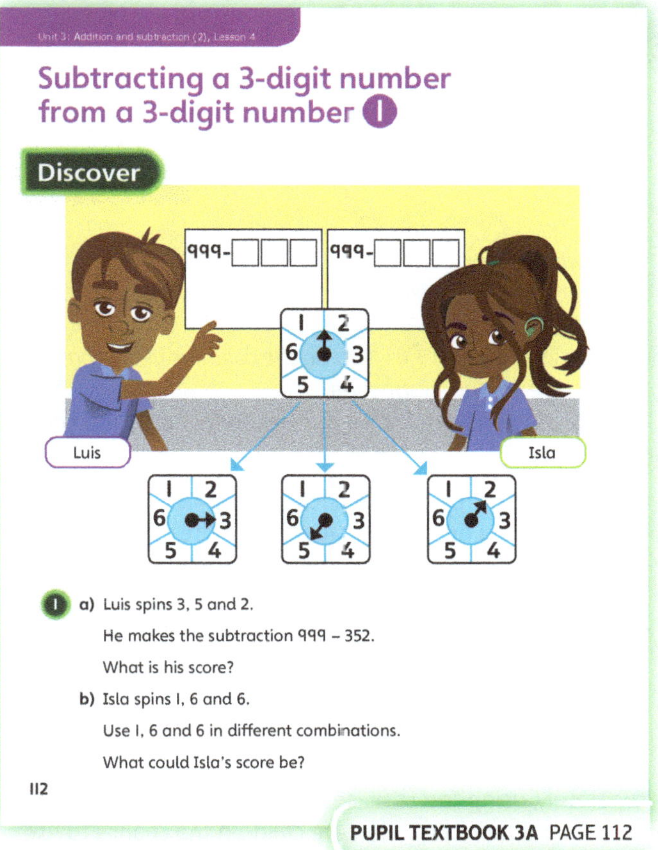

PUPIL TEXTBOOK 3A PAGE 112

Share

WAYS OF WORKING Whole class teacher led

ASK

- Question **1** a): *Which number bonds help solve the subtraction for each digit?*
- Question **1** a): *How has the whole been represented?*
- Question **1** a): *Which part has been subtracted?*
- Question **1** a): *Can you see the link between the column method and how it is shown on the number line?*
- Question **1** b): *Why does the order in which you use the digits make a difference to the answer?*

IN FOCUS In question **1** a) children should notice that the place value equipment is being used in a different way from when we model addition. In the additions, both numbers to be added are made, and then the totals combined. However, in the subtraction, only the whole number is represented, and then the digits are subtracted from this (by crossing out or physically removing them). The number line is also used to represent another way of visualising the subtraction. Children should discuss what is the same and what is different about the two approaches. This is especially appropriate when thinking of subtraction as 'taking away'. Children may want to explore how this is related to finding the difference at a later stage, or at a relevant point in the lesson. To explore this, children may make the whole and then, to compare the whole with a part, you look at the part that is left once the other part has been taken away.

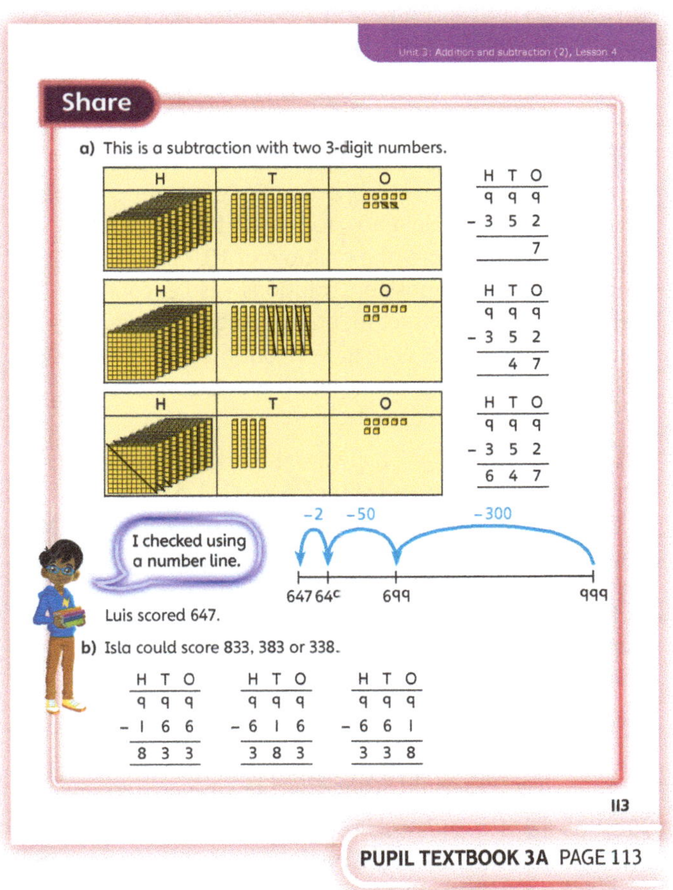

PUPIL TEXTBOOK 3A PAGE 113

Think together

WAYS OF WORKING Whole class teacher led (I do, We do, You do)

ASK

- Questions ❶, ❷ and ❸ : *How many 1s, 10s or 100s are in the whole?*
- Questions ❶ and ❷ : *How many 1s, 10s or 100s are being subtracted?*
- Questions ❶ and ❷ : *How do the columns help represent the stages of the subtraction?*
- Question ❶ : *What known number bonds help solve each step?*

IN FOCUS The important aspect of this phase is for children to gain fluency and accuracy in the use of column subtraction, alongside understanding how it relates to place value. Children should continue to recognise why and how the subtractions can be written as columns, and use the language of 1s, 10s and 100s to explain how known number bonds can be used to solve the stages of the calculation efficiently and accurately.

STRENGTHEN Represent each part of the subtraction separately, using the 1s place value equipment first, then the 10s and then the 100s. Give children the chance to use different types of equipment to develop a sense of the place value itself, rather than simply following a process without generating the deep understanding required.

For question ❸, some children may find a trial and error approach is a useful way to begin the problem, rather than trying to solve it all in one step. Encourage children to try some different combinations and see if the resulting subtractions match any of the conditions as stated by the four children.

DEEPEN Question ❸ encourages children to make decisions about number properties. For example, challenge some children to explain how to choose the 1s digit to generate an even score or a multiple of 10, rather than simply listing all the possible solutions. Also encourage them to begin to subtract mentally and apply logic. For example, they may be able to work out mentally that 999 – 964 or 946 will produce an answer that is less than 100, meeting Reena's condition.

ASSESSMENT CHECKPOINT Can children find three different subtractions to satisfy question ❷ and solve them accurately? This includes one result where the answer is less than 100.

ANSWERS

Question ❶ : 999 – 435 = 564
Jamilla's score is 564.

Question ❷ : 678 – 446 = 232, 678 – 464 = 214, 678 – 644 = 34
Ebo could score 232, 214 or 34.

Question ❸ : Mo and Ambika need to subtract a number with a 9 in the ones column. Reena needs to subtract a number with the 9 in the hundreds column. Andy needs to subtract 496.

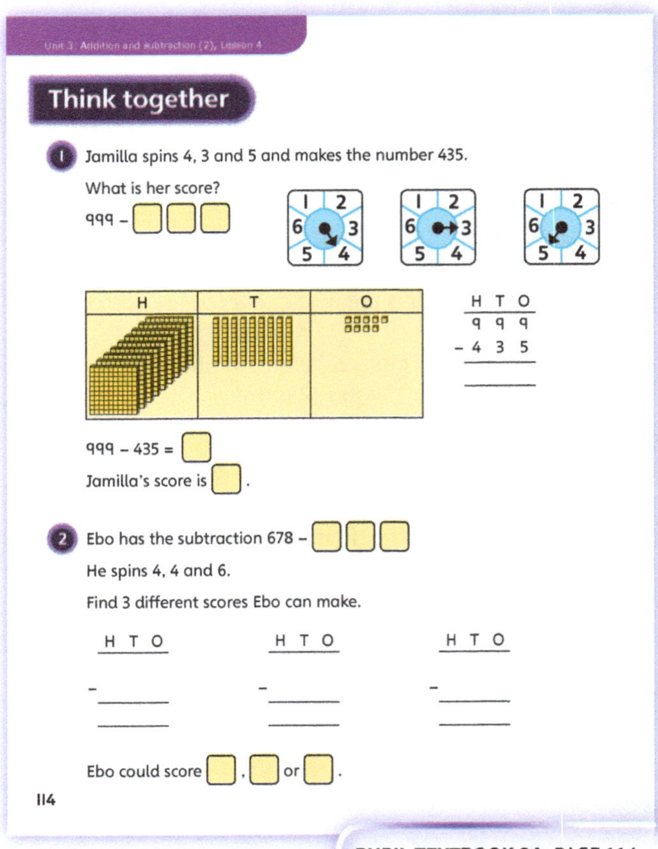

PUPIL TEXTBOOK 3A PAGE 114

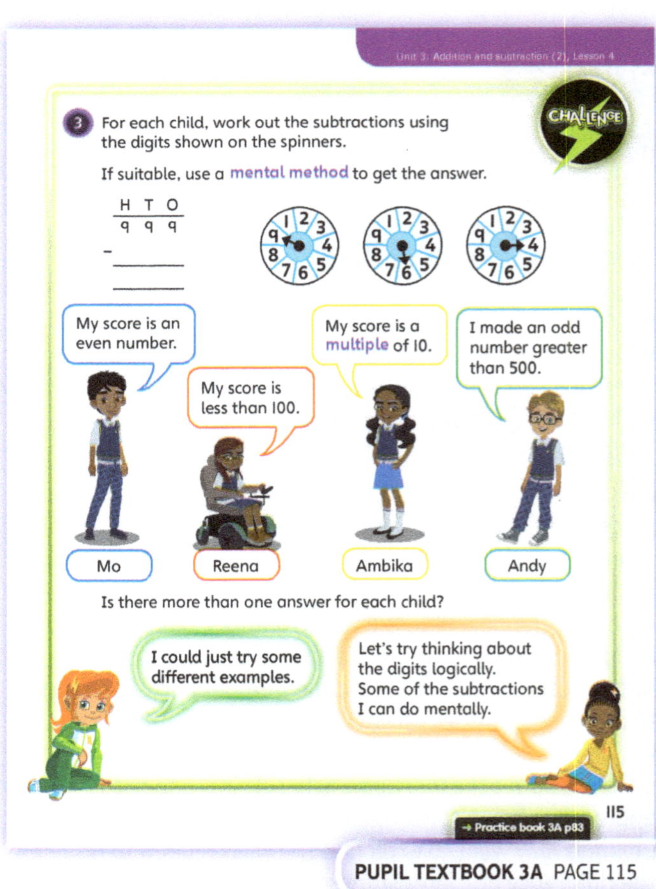

PUPIL TEXTBOOK 3A PAGE 115

Practice

WAYS OF WORKING Independent thinking

IN FOCUS In question **1**, children generate understanding and fluency with continued scaffolding of place value equipment. In questions **2** and **3**, children develop fluency in the abstract calculations and see how find the difference and take away are both related as subtractions, even attempting to mentally work out one or more of the answers – if suitable. The important aspect is that children increase in fluency and accuracy as they move through the exercises. Questions **4** and **5** prompt them to apply mathematical thinking to probe the concept more deeply, through missing digits and by exploring the properties of the numbers generated through subtraction.

STRENGTHEN Make sure children have access to base 10 equipment, place value cards and place value grids to help scaffold their thinking. However, support children to become more fluent with the abstract calculations by asking them to visualise what the equipment would look like. Encourage them to attempt to solve the calculations before checking using the real equipment.

DEEPEN Question **4** a) prompts children to think of a story problem to match their calculation. Challenge children to generate subtraction stories that match both 'take away' and 'find the difference' scenarios.

ASSESSMENT CHECKPOINT If children can complete the subtractions in question **3** accurately and efficiently, then that should indicate sound fluency alongside an understanding of place value. This is especially the case for 688 – 34 and for the last two calculations where they may attempt to use a mental method or create their own column addition.

ANSWERS Answers for the **Practice** part of the lesson appear in the separate **Practice and Reflect answer guide**.

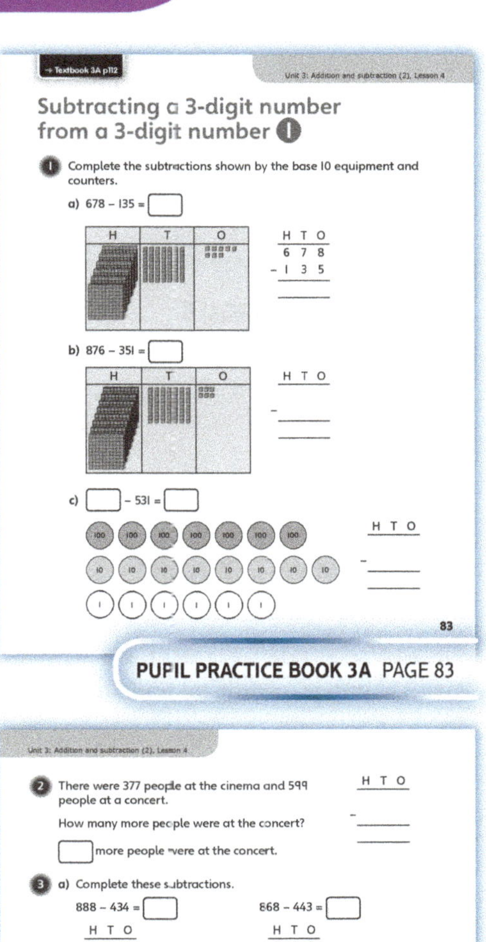

PUPIL PRACTICE BOOK 3A PAGE 83

PUPIL PRACTICE BOOK 3A PAGE 84

Reflect

WAYS OF WORKING Pair work

IN FOCUS Children should discuss the different methods of representing subtractions they have been using in the lesson, and then decide how to replicate that for the **Reflect** question.

ASSESSMENT CHECKPOINT Can children explain the place value of each part of the subtraction?

ANSWERS Answers for the **Reflect** part of the lesson appear in the separate **Practice and Reflect answer guide**.

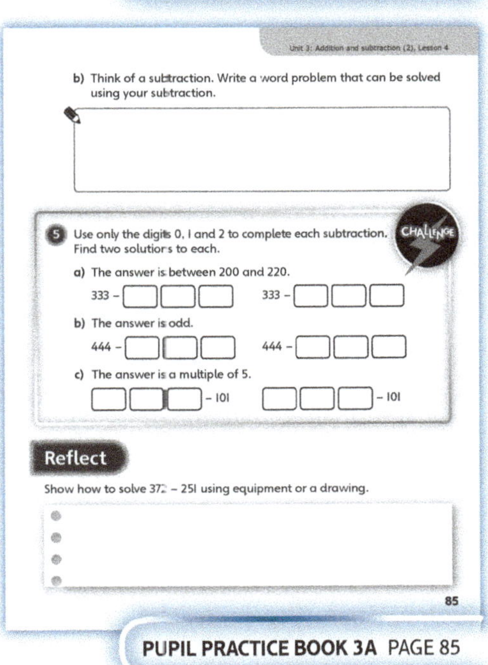

PUPIL PRACTICE BOOK 3A PAGE 85

After the lesson 

- Do children have a good understanding of why writing in columns makes the process more efficient and accurate? Did children attempt to use a mental method to answer any of the questions?
- Do children use their knowledge of known number bonds to solve the steps of the calculations?
- Do children have a deep enough understanding of the approach to be able to solve missing number or missing digit problems?

Subtracting a 3-digit number from a 3-digit number ②

Learning focus

In this lesson, children will develop their fluency with column subtraction of 3-digit numbers, to include calculations where exchange is necessary across one or two columns.

Small steps

→ Previous step: Subtracting a 3-digit number from a 3-digit number (1)
→ **This step: Subtracting a 3-digit number from a 3-digit number (2)**
→ Next step: Estimating answers to additions and subtractions

NATIONAL CURRICULUM LINKS

Year 3 Number – Addition and Subtraction

- Add and subtract numbers with up to three digits, using formal written methods of columnar addition and subtraction.
- Add and subtract numbers mentally, including: a three-digit number and ones, a three-digit number and tens, a three-digit number and hundreds.
- Solve problems, including missing number problems, using number facts, place value, and more complex addition and subtraction.

ASSESSING MASTERY

Children can represent column subtractions involving exchange across one or two columns, and can explain when and why it is necessary.

COMMON MISCONCEPTIONS

There is a danger that children assume that you always subtract the smaller digit from the larger digit, which can lead to a misconception where the subtractions require exchange. In this lesson it is important that children understand they are subtracting from a whole. Ask:
- *What is the number to subtract? What is the whole we are subtracting from?*

STRENGTHENING UNDERSTANDING

Use a part-whole model to represent the subtractions so that children can clearly identify the whole from which the part is subtracted.

GOING DEEPER

There are numerous opportunities to deepen fluency through solving missing number and missing digit problems. Further depth may also be gained through challenging children to create their own subtraction problems. Where suitable, they could also try to work without the aid of equipment or apply mental methods to try to solve some subtractions mentally.

KEY LANGUAGE

In lesson: digit, exchange, ones (1s), tens (10s), hundreds (100s), subtract, mentally

Other language to be used by the teacher: column subtraction

STRUCTURES AND REPRESENTATIONS

Place value grid, number line

RESOURCES

Mandatory: place value equipment (such as base 10 and place value counters)

Optional: number lines, part-whole models

 In the eTextbook of this lesson, you will find interactive links to a selection of teaching tools.

Before you teach 

- Do children know what is different between the kinds of exchange in addition and in subtraction? Do they understand this?
- Can they recognise which of these requires an exchange: 157 – 72 or 175 – 72?

Discover

 Pair work

ASK

- Question **1** a): *What is the whole number?*
- Question **1** a): *What is the part that is being subtracted?*
- Question **1** a): *Can we use the same approach as in the previous lesson?*

IN FOCUS In both questions the vital point is that children recognise the context requires a subtraction, and then that the numbers involved will require an exchange.

PRACTICAL TIPS It may help children to sketch their own version of the staircase, and draw a stick figure on to the sketch. Alternatively, they could role-play a dramatic sketch. They may find it lends some sense of reality to the context if they discuss times they may have had to walk a long way or climb a large number of steps. The sense of 'are we there yet?' is familiar to many children, and may help them understand the context of the problem.

ANSWERS

Question **1** a): 361 − 147 = 214;
      Aki has 214 steps left to climb.

Question **1** b): 361 − 157 = 204 or 214 − 10 = 204;
      Olivia has 204 steps left to climb.

PUPIL TEXTBOOK 3A PAGE 116

Share

 Whole class teacher led

ASK

- Question **1** a): *Why do we need to make an exchange?*
- Question **1** a): *Which part of the written column method shows that we have exchanged 1 ten for 10 ones?*
- Question **1** a): *Why has the 6 been crossed out and a 5 been written instead? What do the 6 and the 5 represent?*
- Question **1** a): *Why do we need to start by subtracting the 1s?*
- Question **1** b): *Can you work out 214 − 10 mentally? What mental strategies or methods would you use to help you?*

IN FOCUS The focus of question **1** a) is to recognise the need for an exchange and choose the appropriate exchange. Children need to represent the exchange in the written method and perform the subtraction beginning with the 1s and working through the 10s and then the 100s. Children could also discuss alternative approaches to solving question **1** b), such as calculating 147 + 10 = 157 and then performing 361 − 157. They could compare this method with the one shown and discuss which method they think best represents how to think about the problem.

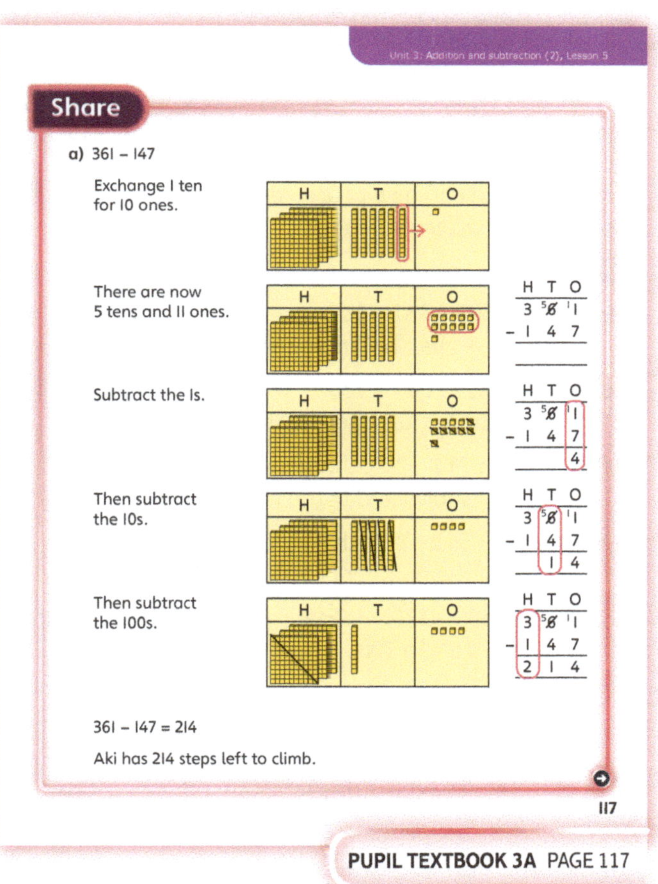

PUPIL TEXTBOOK 3A PAGE 117

Think together

WAYS OF WORKING Whole class teacher led (I do, We do, You do)

ASK

- Question **1** : *What is the whole? What part is being subtracted?*
- Question **2** : *Is an exchange necessary?*
- Question **2** : *Why do we need to begin by subtracting the 1s?*
- Question **4** : *What can be done if there is a zero in the column you need to exchange from?*

IN FOCUS Questions **1** and **2** look at different subtractions with different known parts. The important aspect of questions **1** and **2** (and indeed with all the questions in this phase of the lesson) is for children to recognise where the exchange is needed, and how to represent it in written methods. Children should build confidence and fluency, alongside understanding, by using the language of place value and manipulatives to represent the process.

STRENGTHEN Children should explore the exchange necessary using a range of place value equipment such as place value counters, base 10 or any other suitable equipment.

DEEPEN Question **3** requires children to recognise and explain some common mistakes. Challenge children to think of some word problems that would match the two subtractions.

ASSESSMENT CHECKPOINT Can children correctly complete the subtractions for question **3** a) and b)? This will indicate whether they have sufficient understanding of both the need to identify and subtract from the whole and also of the exchange process: the order that the exchange needs to be completed in and how to accurately show the exchange when using a written method.

ANSWERS

Question **1** : 525 − 361 = 164
Lexi has 164 more steps left to climb.

Question **2** : 525 − 387 = 138
Emma has already climbed 138 steps.

Question **3** a): 314 − 253; the 10s digits have been subtracted in the wrong order.

Question **3** b): 553 − 255; the exchange has been made but the number of 10s has not been reduced.

Question **4** : 506 − 328 = 178; you need to exchange both a 10 and a 100.

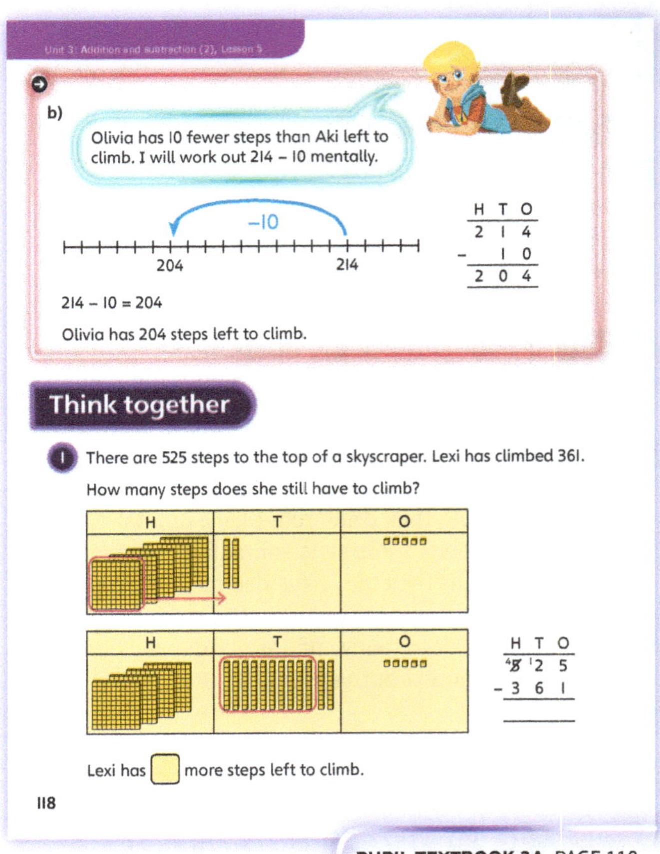

PUPIL TEXTBOOK 3A PAGE 118

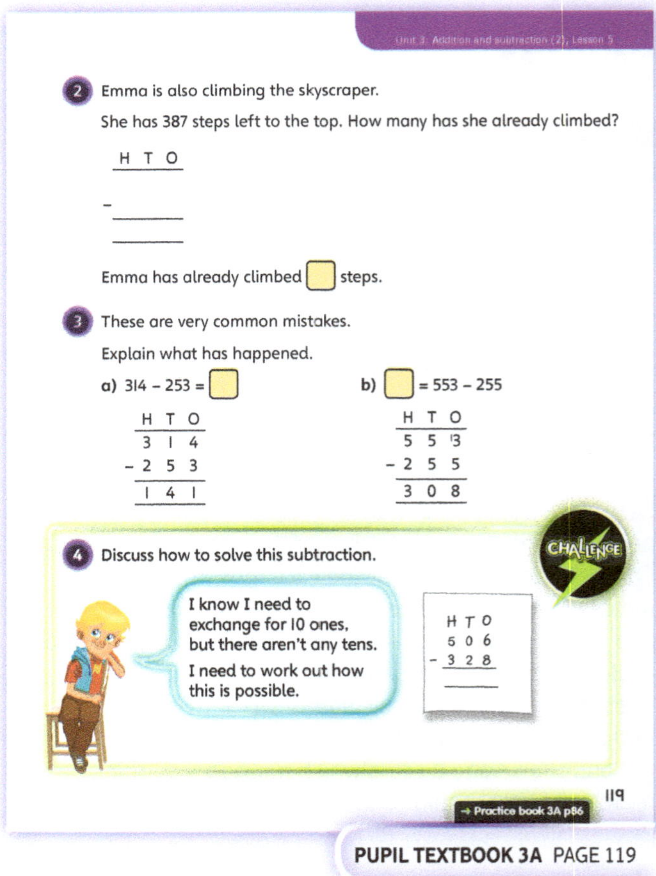

PUPIL TEXTBOOK 3A PAGE 119

Practice

WAYS OF WORKING Independent thinking

IN FOCUS The key is for children to develop fluency and confidence by working on questions with concrete materials, progressing to calculations which are presented as abstract subtractions.

Question ❶ has story problems supported by representations of place value equipment. Question ❷ develops children's ability to perform the written method, including one part where they have to set out the calculation themselves.

STRENGTHEN Make sure that children have access to place value equipment to support their thinking and to help model the required exchange. Children should all be encouraged to use known number bonds to solve the steps of the subtractions.

DEEPEN Question ❻ challenges children to examine and reason about mathematical generalisations surrounding subtraction. Most children will find it beneficial to begin with a few different trials and discuss their observations of the results they collect.

THINK DIFFERENTLY Question ❺ requires children to recognise the original calculation from the workings and equipment presented. This prompts children to reverse engineer the calculation from the mechanics of the method.

ASSESSMENT CHECKPOINT If children are able to explain how they found the missing digits in question ❹, then they have a good level of understanding.

ANSWERS Answers for the **Practice** part of the lesson appear in the separate **Practice and Reflect answer guide**.

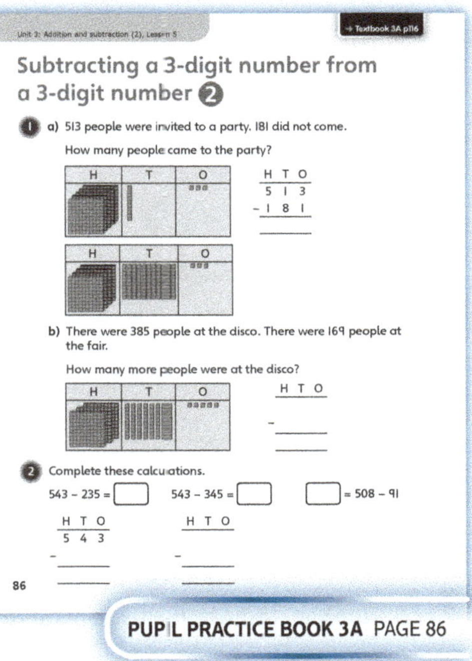

PUPIL PRACTICE BOOK 3A PAGE 86

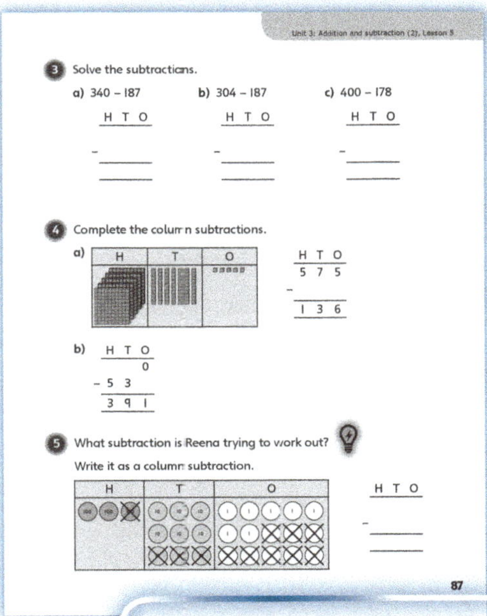

PUPIL PRACTICE BOOK 3A PAGE 87

Reflect

WAYS OF WORKING Independent thinking

IN FOCUS Children should be able to invent a number of different subtractions that fulfil this requirement. After they have decided on their calculation, they can discuss them in pairs. They should justify their answer using the language of place value and exchange.

ASSESSMENT CHECKPOINT Can children explain how they could find multiple subtractions that meet the requirement?

ANSWERS Answers for the **Reflect** part of the lesson appear in the separate **Practice and Reflect answer guide**.

After the lesson ⏸

- Can children explain how to represent an exchange in written column subtraction?
- Are children able to use bonds within 20 to solve the subtractions in each place value position?
- Can children exchange across more than one column when necessary?

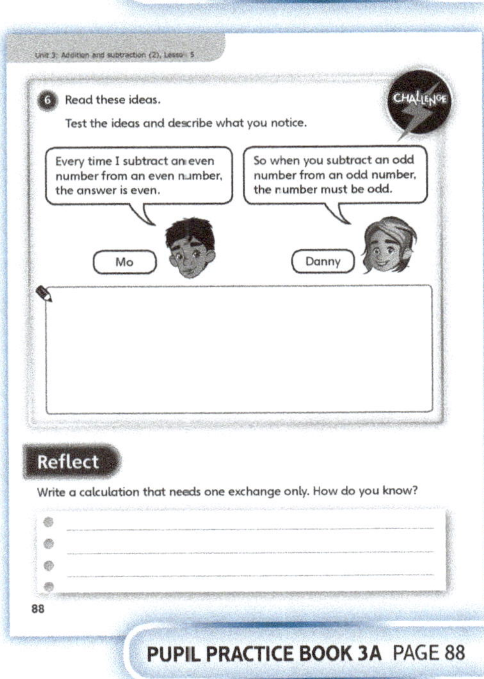

PUPIL PRACTICE BOOK 3A PAGE 88

Estimating answers to additions and subtractions

Learning focus

In this lesson, children will develop skills of estimation and approximation to allow them to make simple checks of additions and subtractions. They have not yet learnt rounding, so the approach builds on number sense and approximate position on a number line.

Small steps

→ Previous step: Subtracting a 3-digit number from a 3-digit number (2)
→ **This step: Estimating answers to additions and subtractions**
→ Next step: Checking strategies

NATIONAL CURRICULUM LINKS

Year 3 Number – Addition and Subtraction

Estimate the answer to a calculation and use inverse operations to check answers.

ASSESSING MASTERY

Children can use a rough approximation to estimate answers to calculations by adding 100s mentally.

COMMON MISCONCEPTIONS

There may be some confusion over the difference in meaning between estimation (making a justified or educated 'guess' based on the information available) and approximation (choosing a rough value to stand in where a precise value is not known). Ask:

• *Is 101 + 201 approximately 200 or 300?*
• *Estimate how many children are in our classroom today.*

STRENGTHENING UNDERSTANDING

Many children find it confusing to estimate or approximate, thinking that a precise answer is always necessary or preferable. They may find it difficult to accept that a rough answer can sometimes be more useful or appropriate.

Use a washing line with 100s pegged at regular intervals. Give children the opportunity to discuss which numbers would be placed near to the 100s and to explore the relative locations of numbers to either side of a 100 mark.

GOING DEEPER

There are some contexts where either an over-estimate or an under-estimate is always preferable. For example, if you need at least 22 water bottles for a school trip, then having 30 is better than having 20, even though 20 is the closer approximation. Discuss this with children and ask them if they can see where this consideration might come into some of the problems in the lesson.

KEY LANGUAGE

In lesson: estimate, estimation, approximate, **approximation**, **approx.**, **approximately**, about

Other language to be used by the teacher: roughly

STRUCTURES AND REPRESENTATIONS

Number line, column method, table

RESOURCES

Mandatory: number lines

Optional: washing line, pegs, 100s numbers to 1,000, paper clips, cubes

 In the eTextbook of this lesson, you will find interactive links to a selection of teaching tools.

Before you teach 

• Can children confidently suggest numbers near 100? How about far from 100?
• Do they understand the meaning of 'about' and 'roughly' in the context of estimating?

Discover

WAYS OF WORKING Pair work

ASK

- Question **1** a): *What do you think 'approx.' means? Have you seen it before?*
- Question **1** a): *Can you give examples of numbers that are close to 200, 300 or 400?*
- Question **1**: *How could you show the different approximations on a number line?*

IN FOCUS The focus of the context of question **1** a) is to discuss the meaning of the words 'approx.' and 'approximately' and to understand how the number 290 relates to the approximate number of 200, 300 or 400 matchsticks.

PRACTICAL TIPS One strategy for introducing children to the concepts of estimation and approximation is to ask them to count a large number of small objects, such as a pot full of paper clips or a tray full of cubes. They could estimate the total and then check by counting, perhaps by grouping in 10s.

ANSWERS

Question **1** a): Amelia needs the 'Approx. 300' bag.

Question **1** b): Ebo counted the 'Approx. 200' bag.

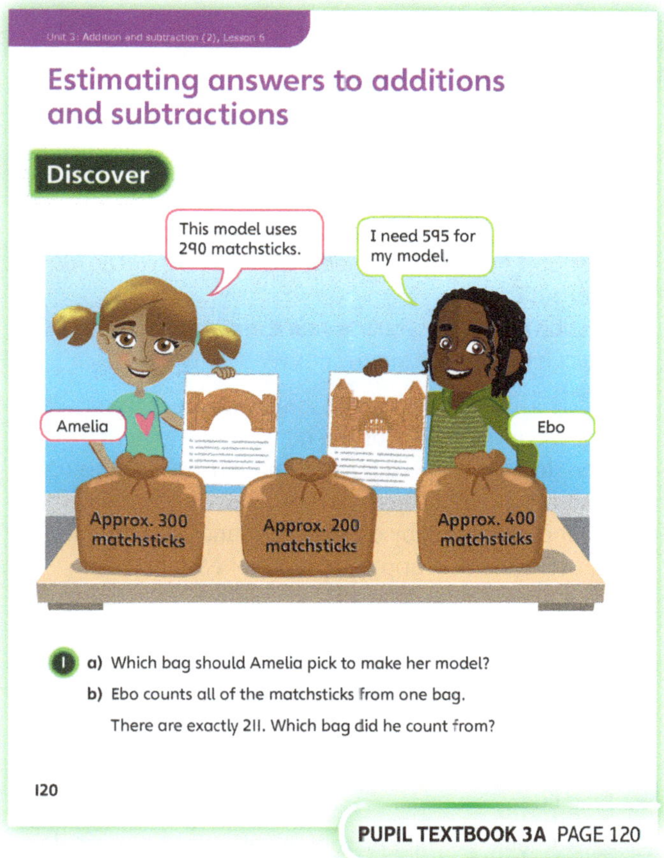

PUPIL TEXTBOOK 3A PAGE 120

Share

WAYS OF WORKING Whole class teacher led

ASK

- Question **1** a): *Can you point to where you think 299 would be on the number line?*
- Question **1** a): *How do you know that 290 is closer to 300 than 200?*
- Question **1** a): *Can you suggest any numbers that would be even closer to 300 than 290? How do you know?*
- Question **1** b): *Why is 211 from the 'Approx. 200' bag when 211 is more than 200?*

IN FOCUS This phase of the lesson uses positions on a number line in both parts of question **1** to build a 'number sense' for approximations to the nearest 100. Note that pupils are not being taught to round at this stage, but rather to use their visual intuition and their understanding of place value to make decisions about approximations.

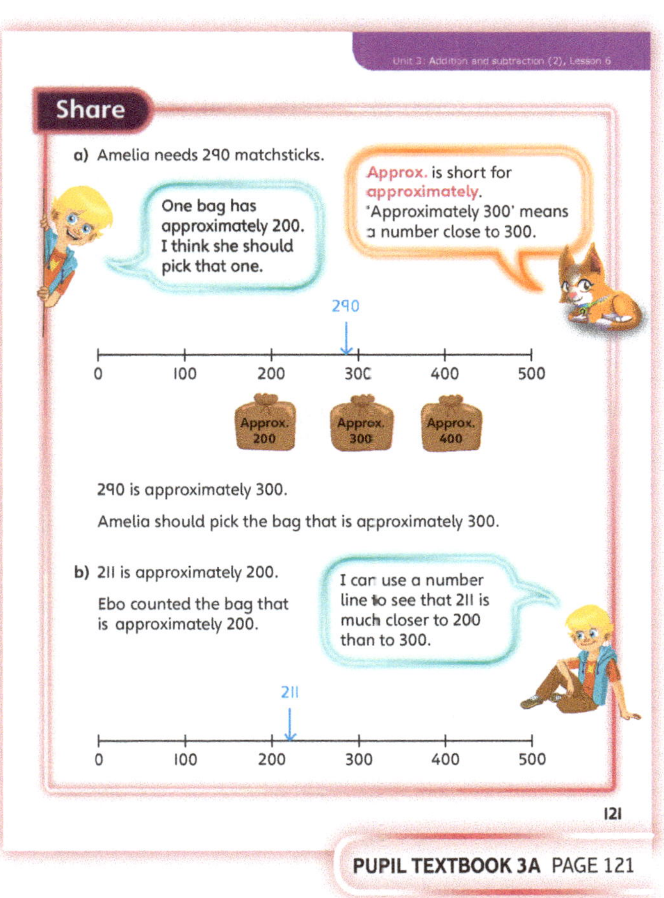

PUPIL TEXTBOOK 3A PAGE 121

Think together

WAYS OF WORKING Whole class teacher led (I do, We do, You do)

ASK

- Question **1**: *What number is each letter close to?*
- Question **2**: *How can we tell if it is nearer to 200 or 300; 400 or 500?*
- Question **3** b): *Can you use your estimate in question **1** b) to calculate the approximate answer mentally?*

IN FOCUS This phase of the lesson introduces the concepts in stages. Question **1** uses the number line to help children develop their skills of estimation, by recognising proximity to a round 100s number. Question **2** builds the skill of approximation by choosing a round 100s number to represent the best approximation for each number. Question **3** a) then suggests how children could use these two skills to find an approximate answer to calculations, with question **3** b) prompting children to consider how this might be applied as a checking strategy.

STRENGTHEN Help children to develop a sense of approximation by pegging numbers in position on a washing line.

DEEPEN Challenge children to think of contexts or problems in which over-estimates or under-estimates are preferable.

ASSESSMENT CHECKPOINT Can children use estimation to predict reasonable answers to the calculations in question **3**?

ANSWERS

Question **1**: Estimates should be within these ranges:

A) 80–90, B) 205–210, C) 310–320, D) 380–395, E) 410–420, F) 485–495

Question **2**: Approximations should be marked near these numbers:

701→700, 590→600, 199→200, 275→300, 109→100, 612→600, 999→1,000, 707→700, 11→0

Question **3** a): 381 + 398 is approximately 400 + 400 = 800.

Question **3** b): A reasonable approximation is 500 – 300, so the answer should be roughly 200. She should try the calculation again.

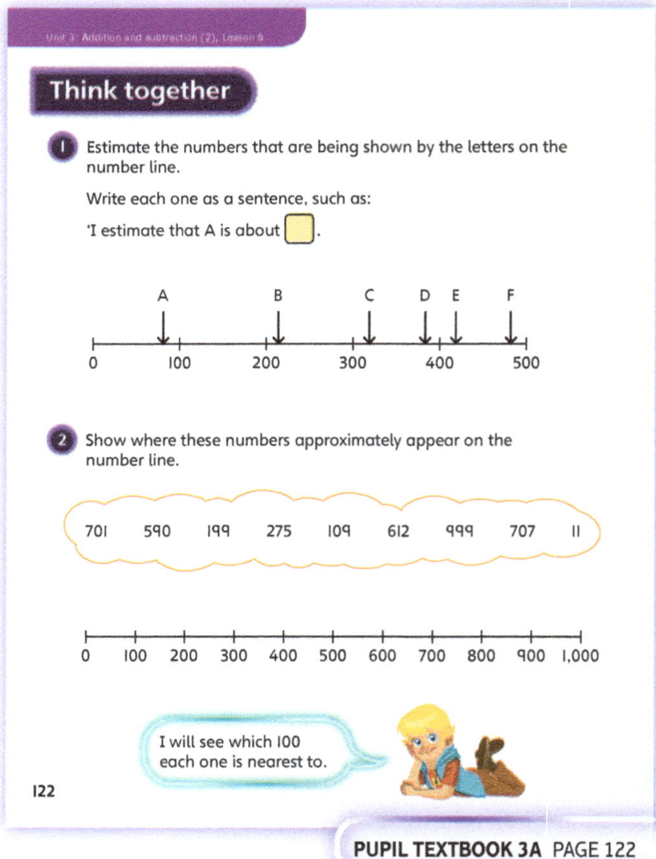

PUPIL TEXTBOOK 3A PAGE 122

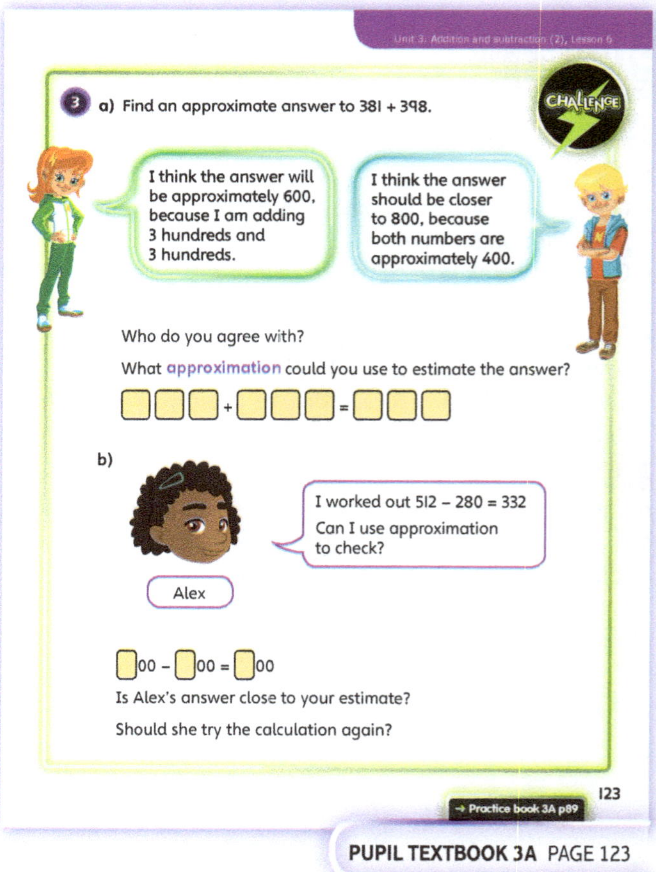

PUPIL TEXTBOOK 3A PAGE 123

Practice

WAYS OF WORKING Independent thinking

IN FOCUS Questions **1** and **2** practise the key skills of estimation and approximation. Questions **3** and **4** prompt children to use these skills to find approximate answers to additions and subtractions. Question **5** requires children to use the estimations as a checking strategy.

STRENGTHEN Ask children to represent the numbers on 0–1,000 number lines and use the representations to support their approximations.

DEEPEN Question **6** starts to challenge children to think about under- and over-estimates. Although it seems that Andy's idea (that it might be a bit more or a bit less than 500) is reasonable, given that the numbers are less than 200 and 300, the answer has to be less than 500.

THINK DIFFERENTLY Question **5** prompts children to find the likely mistake by using estimation as a checking strategy.

ASSESSMENT CHECKPOINT Answers to question **5** will demonstrate if children have understood how the concept can be used in relation to calculations.

ANSWERS Answers for the **Practice** part of the lesson appear in the separate **Practice and Reflect answer guide**.

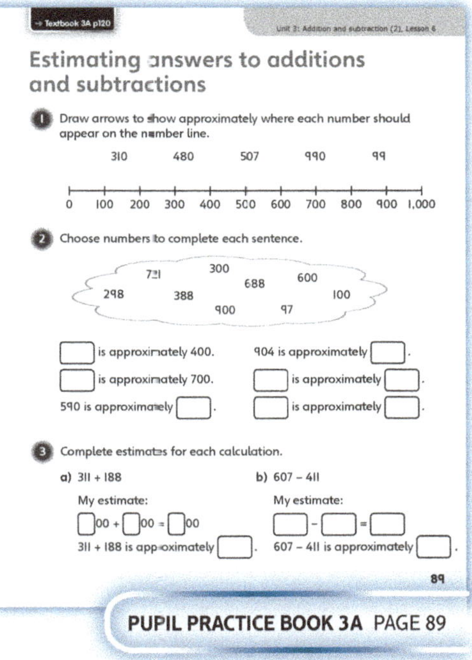

PUPIL PRACTICE BOOK 3A PAGE 89

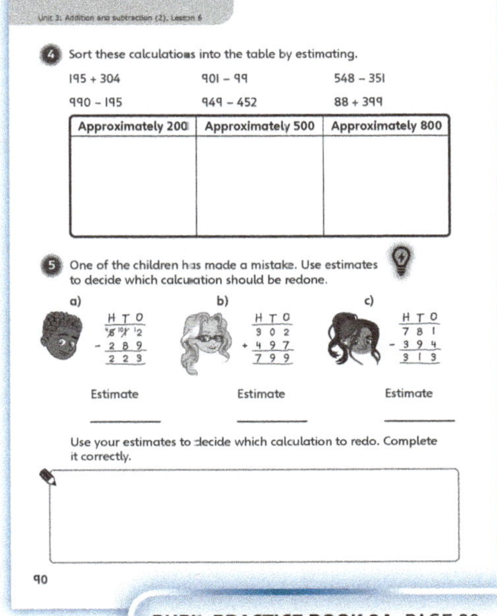

PUPIL PRACTICE BOOK 3A PAGE 90

Reflect

WAYS OF WORKING Pair work

IN FOCUS Children should discuss and agree reasonable estimates for each calculation and then discuss mental approaches to the calculations.

ASSESSMENT CHECKPOINT Can children make estimates without resorting to the process of column methods immediately?

ANSWERS Answers for the **Reflect** part of the lesson appear in the separate **Practice and Reflect answer guide**.

After the lesson ⏸

- Can children justify their approximations and estimates?
- Are children able to estimate answers to calculations?
- Can children use estimates to decide if a reasonable answer has been given for a calculation?

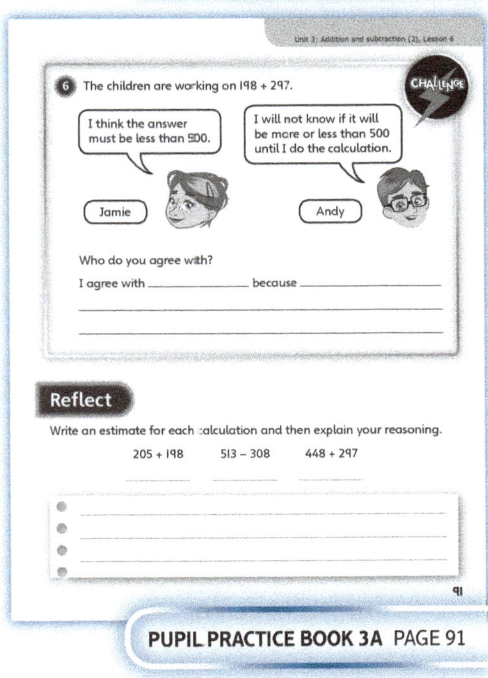

PUPIL PRACTICE BOOK 3A PAGE 91

Checking strategies

Learning focus

In this lesson, children will learn to use inverse operations and fact families as checking strategies. They will also use them to help make appropriate calculations more efficient as mental strategies.

Small steps

→ Previous step: Estimating answers to additions and subtractions
→ **This step: Checking strategies**
→ Next step: Problem solving – addition and subtraction (1)

NATIONAL CURRICULUM LINKS

Year 3 Number – Addition and Subtraction

Estimate the answer to a calculation and use inverse operations to check answers.

ASSESSING MASTERY

Children can use the part-whole concept to find an inverse operation to check an addition or a subtraction. Also, they are able to decide when it is more efficient to use mental strategies over written methods.

COMMON MISCONCEPTIONS

Once children learn written methods for calculations, they may revert to the algorithm rather than first thinking about the most efficient or accurate way to solve a calculation. The use of fact families in this lesson prompts children to make decisions about when a mental method may be more appropriate than the written algorithm. Ask:

- *What fact family could we use to work out 198 – 10 = ___? What if we try 198 – ___ = 10; or 10 + ___ = 198; or 198 = ___ + 10? Can we answer this mentally now, with the aid of the fact family?*

STRENGTHENING UNDERSTANDING

The part-whole model is used throughout the lesson, to help children to visualise the families of facts. Ask: *Can you think of the way to solve this that is least likely to produce a mistake?*

GOING DEEPER

Challenge children to make the link with the learning in the previous lesson about estimates and approximations. Children could employ and adjust approximations in calculations such as 399 + ___ = 601, by working out 400 + ___ = 600 mentally, and then adjusting to find the precise answer.

KEY LANGUAGE

In lesson: subtraction, part-whole, approximately, fact family

Other language to be used by the teacher: addition, approximate, estimate, inverse operation

STRUCTURES AND REPRESENTATIONS

Part-whole model, number lines

RESOURCES

Optional: part-whole model, number line, scrap paper

 In the eTextbook of this lesson, you will find interactive links to a selection of teaching tools.

Before you teach 

- Do children know what a fact family is? Can they find other facts in the family of this calculation: 3 + 5 = 8?
- What methods can they apply to check if they have the correct answer to a subtraction?

Discover

WAYS OF WORKING Pair work

ASK

- Question ❶ a): *What is the whole and what are the parts represented in each calculation?*
- Question ❶ a): *Can you think of related additions based on these parts and wholes?*
- Question ❶ a): *How can you decide if a calculation is correct?*

IN FOCUS For question ❶ a), children should discuss whether the calculations seem to have reasonable answers. They may employ a strategy from the previous lesson, which used estimation and approximation. They should also identify the parts and the whole from each calculation and then represent them in a part-whole model. They can then use that structure to find an addition to check the calculations.

PRACTICAL TIPS Children should use part-whole models to replicate the calculations, perhaps by writing the numbers on scraps of paper so that they do not appear permanent as this is a lesson about checking strategies where calculations may or may not be correct.

ANSWERS

Question ❶ a): The parts of Ambika's calculation do not match the whole, so this calculation should be checked again.

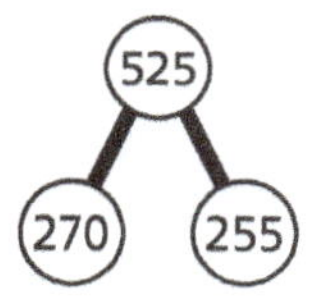

270 + 255 = 525   427 + 332 = 759

Question ❶ b): 755 − 427 = 328

Share

WAYS OF WORKING Whole class teacher led

ASK

- Question ❶ a): *Can you spot how the additions match the parts of the original subtraction?*
- Question ❶ a): *Which calculation should be checked carefully?*
- Question ❶ a): *Can you see what the likely mistake was?*
- Question ❶ b): *Can you draw a new part-whole model to check that 755 − 427 = 328 is correct?*

IN FOCUS Question ❶ a) shows how to use the part-whole model to identify an addition that can be used to check the subtraction. Children should discuss how one of the additions produces a different whole from the parts in the original answer. This means that it should be checked again. Where the parts add to produce the same whole as in the original calculation, this should give confidence that the original was correct. Children should note that this is not conclusive proof of being correct or incorrect but it is a helpful way of spotting likely errors.

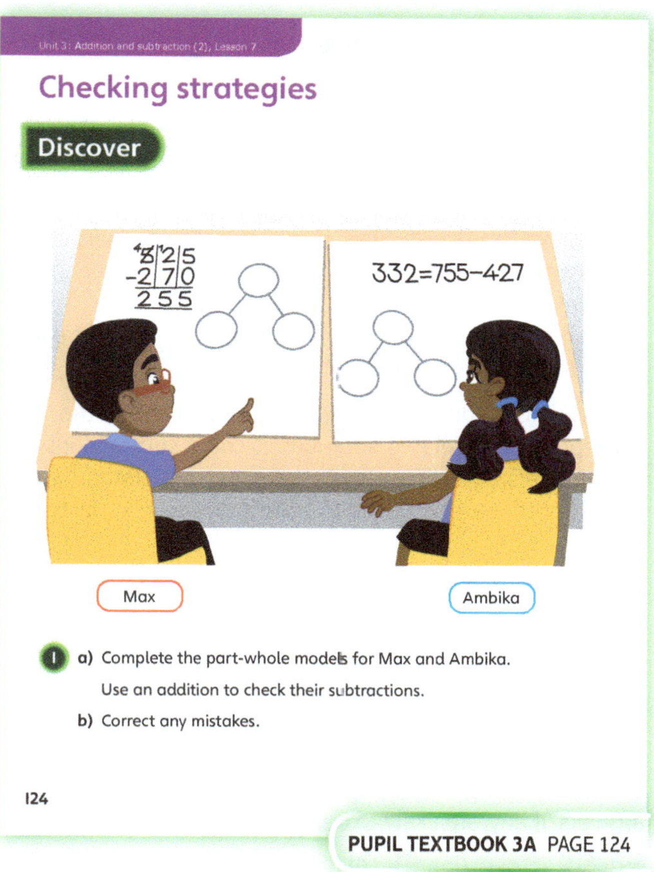

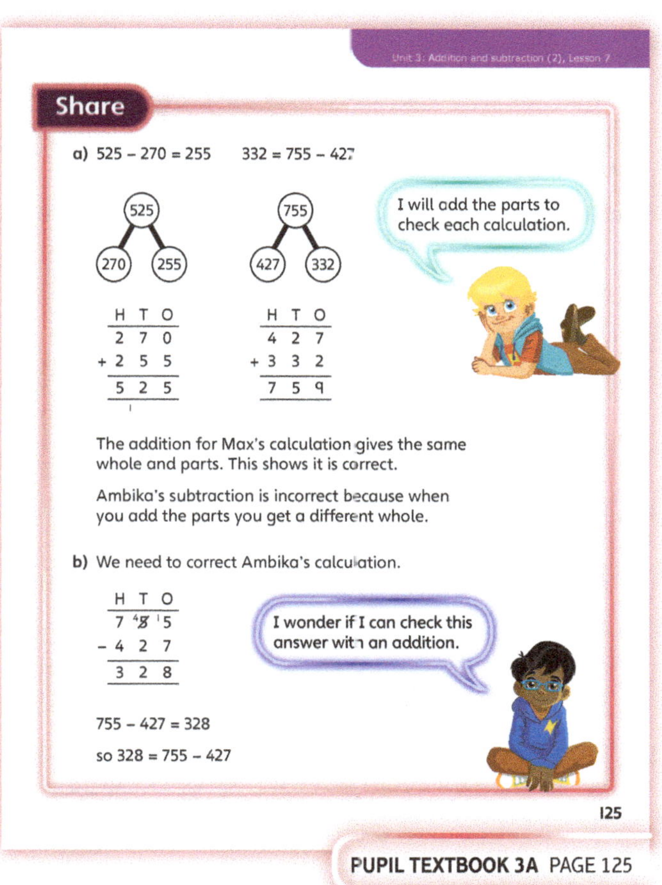

Think together

WAYS OF WORKING Whole class teacher led (I do, We do, You do)

ASK

- Question **1** : *Can you identify the parts and the wholes?*
- Question **1** : *Which operation could you use to check a subtraction?*
- Question **3** : *Can you identify an easier or more efficient way to solve this calculation?*

IN FOCUS Questions **1** and **2** continue to model the concept of using related additions to check subtractions, and vice versa. The important aspect is to understand why this is appropriate, in terms of parts and wholes or fact families.

Question **3** develops thinking further to make decisions about which methods are efficient or appropriate, including when to use mental methods.

STRENGTHEN Encourage the use of the part-whole model, including writing the elements of the calculation on scraps of paper so that they do not seem permanent. This supports the idea of using a checking strategy to determine which answers, if any, are likely to need checking or changing.

DEEPEN In question **3**, children may find that the fact families prompt clear mental methods. For example, when solving 501 – 499 = ___, the column method requires multiple exchanges and is prone to error as well as being inefficient. When written as 499 + ___ = 501, children may see that there is a simple mental addition that can find the answer efficiently.

ASSESSMENT CHECKPOINT Can children identify a mental method for solving 710 – ___ = 690, based on fact families?

ANSWERS

Question **1** a): Subtraction 612 – 371 = 341 needs checking again.

Question **1** b): 141 = 712 – 571 is correct.

Question **2** a): 812 – 477 = 335, so the addition is incorrect. 334 + 477 = 811

Question **2** b): 812 - 521 = 291, so the addition is incorrect. 912 = 521 + 391

Question **3** a): Mistake 1: Emma has exchanged incorrectly. Mistake 2: 0 – 9 is not 9.

Question **3** b): 501 = 499 + 2, so 501 – 499 = 2 Both methods help.

Question **3** c): 690 + 20 = 710, so 710 – 20 = 690

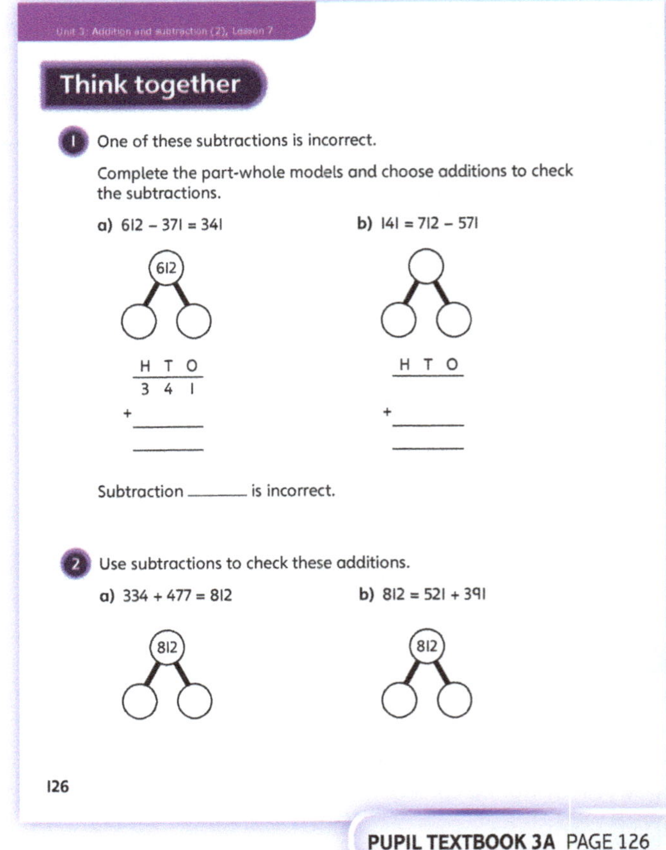

PUPIL TEXTBOOK 3A PAGE 126

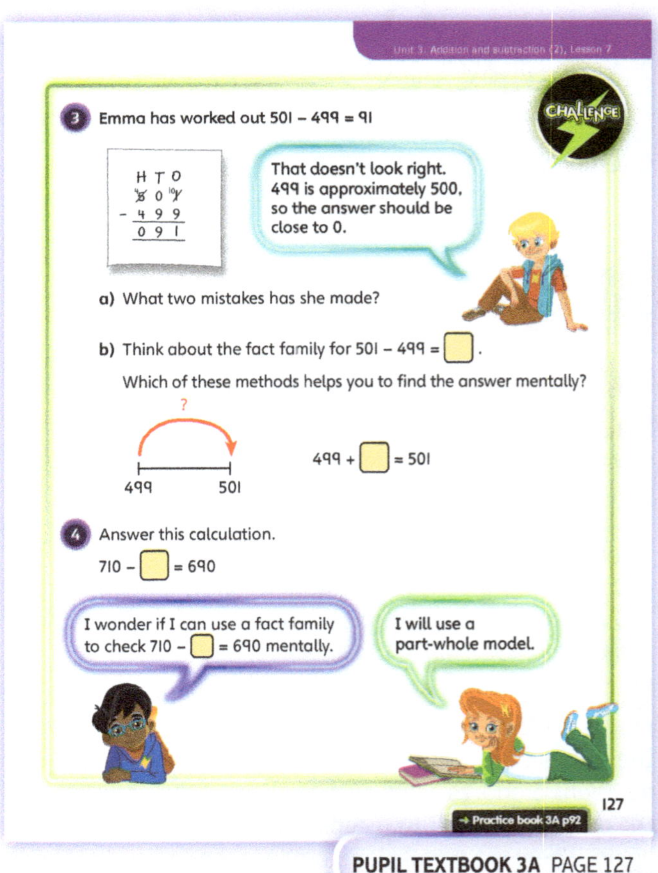

PUPIL TEXTBOOK 3A PAGE 127

Practice

WAYS OF WORKING Independent thinking

IN FOCUS Questions **1** to **3** encourage children to use inverse operations and make decisions about which calculations are likely to be correct. Questions **4** and **5** prompt children to use fact families to make decisions about how to use mental calculation methods to solve the calculations efficiently.

STRENGTHEN The part-whole model is the ideal way to represent the structure needed, and to inform which operation to use as a checking strategy. Build children's confidence first with some simpler calculations within 100.

DEEPEN Question **6** develops the concept that fact families (in conjunction with adjustments to calculations) can be used to widen the related facts that can be found. When one fact is known, other related facts can also be found. This should give children a deeper understanding of how the parts of the fact family relate to each other and thus increase their confidence in using fact families to extrapolate other related answers.

THINK DIFFERENTLY Question **5** asks children to notice the link between addition and subtraction in a calculation that might be inefficiently solved using the column method. The number line allows a much simpler visual model of the problem, which is less likely to result in a mistake.

ASSESSMENT CHECKPOINT Answers to questions **2** and **3** will demonstrate a good level of understanding of the main objective for the lesson. It encourages children to recognise the parts and whole in a calculation and then successfully employ an inverse operation to check their answer.

ANSWERS Answers for the **Practice** part of the lesson appear in the separate **Practice and Reflect answer guide**.

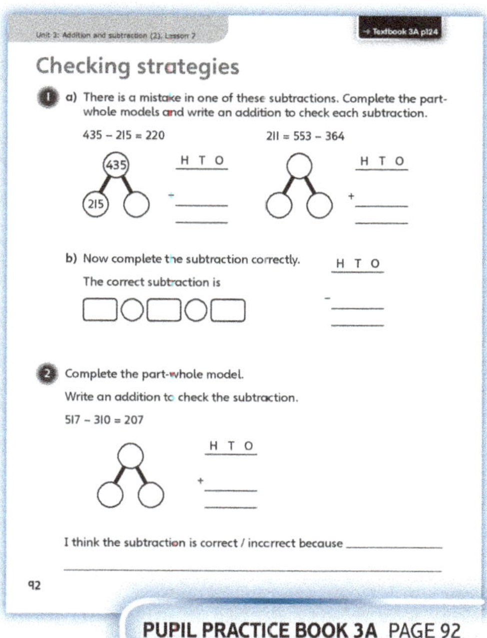

PUPIL PRACTICE BOOK 3A PAGE 92

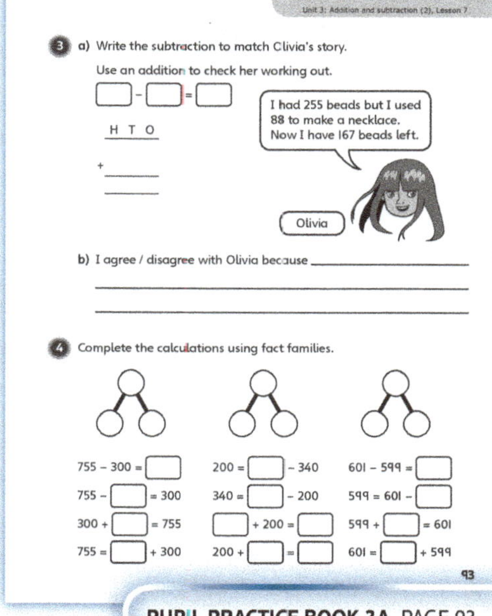

PUPIL PRACTICE BOOK 3A PAGE 93

Reflect

WAYS OF WORKING Pair work

IN FOCUS Children should compare the methods from the previous lesson with the current lesson. The prompt is open-ended, and this will allow children plenty of freedom to demonstrate the depth of their understanding of the different checking strategies.

ASSESSMENT CHECKPOINT Can children justify their reasoning using the language of parts and wholes?

ANSWERS Answers for the **Reflect** part of the lesson appear in the separate **Practice and Reflect answer guide**.

After the lesson ⏸

- Do children understand how to use the fact families to find related calculations?
- Can children decide whether a checking calculation confirms or casts doubt on a calculation?
- Can children use fact families, where appropriate, to inform mental calculation strategies?

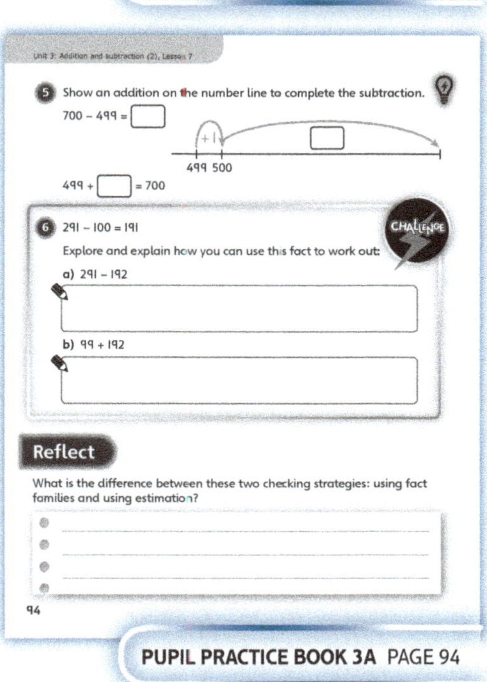

PUPIL PRACTICE BOOK 3A PAGE 94

Problem solving – addition and subtraction ❶

Learning focus

In this lesson, children will use a single bar model to represent word problems that require addition or subtraction.

Small steps

→ Previous step: Checking strategies
→ **This step: Problem solving – addition and subtraction (1)**
→ Next step: Problem solving – addition and subtraction (2)

NATIONAL CURRICULUM LINKS

Year 3 Number – Addition and Subtraction

Solve problems, including missing number problems, using number facts, place value, and more complex addition and subtraction.

ASSESSING MASTERY

Children can use a bar model to represent the context of a story problem and to identify the operation required based on whether the missing information is the whole or a part.

COMMON MISCONCEPTIONS

Children can struggle to identify the missing information in story problems and so find it difficult to know which operation to use. The bar model can represent the structure of a problem to support decisions about which calculation to use. However, the model itself does not yield the numerical answer and so children will need to know that they are required to complete the calculation using a method they have learnt previously. Try posing a story problem, such as: *Farmer Beth has 310 animals on her farm. 200 are sheep and the rest are cows. How many cows does she have?* Then ask:
• *What information are you told in this word problem? What information is missing?*
• *If we model this in a bar diagram, is it the whole or a part that we are missing? Do we need to add or subtract in order to find the missing information?*

STRENGTHENING UNDERSTANDING

Children can find it difficult to draw the bars and the braces to their own satisfaction. Using strips of paper which children can fold, tear or cut to an appropriate length can support children in creating their own bar models. The process of creating the models can help to cement the idea of how the model works.

GOING DEEPER

More than one bar model can be used to represent a problem successfully. Ask children to identify which problems could be represented in different ways and make decisions about which models they think best represent a given context.

KEY LANGUAGE

In lesson: subtraction, approximation, bar model

Other language to be used by the teacher: operation, approximate, raise, raised

STRUCTURES AND REPRESENTATIONS

Bar model

RESOURCES

Optional: strips of paper

 In the eTextbook of this lesson, you will find interactive links to a selection of teaching tools.

Before you teach 

• Are children able to represent a simple addition as a bar model?
• Are they able to represent a simple subtraction as a bar model?
• Can they confidently identify whether it is the whole or a part that is missing in a word problem?

Discover

 Pair work

ASK

- Question **1** a): *Can you identify the prices you need to use?*
- Question **1** a): *How can you prove that an addition is the appropriate operation to use?*
- Question **1** a): *What sort of diagram could you use?*
- Question **1** b): *What does 'approximation' or 'approximate' mean?*

IN FOCUS In question **1** a) it is important that children recognise what information they need to find in the picture in order to find the solution. Children may recognise that the operation required is addition, but some children may not be able to justify why this requires addition rather than subtraction.

PRACTICAL TIPS Spend some time discussing the items in the shop context. Children could create price labels for the items identified in each problem. When representing the bars, children could cut strips of paper to represent the different parts of the bar.

ANSWERS

Question **1** a): 275 + 99 = 374

Question **1** b): 275 + 100 = 375

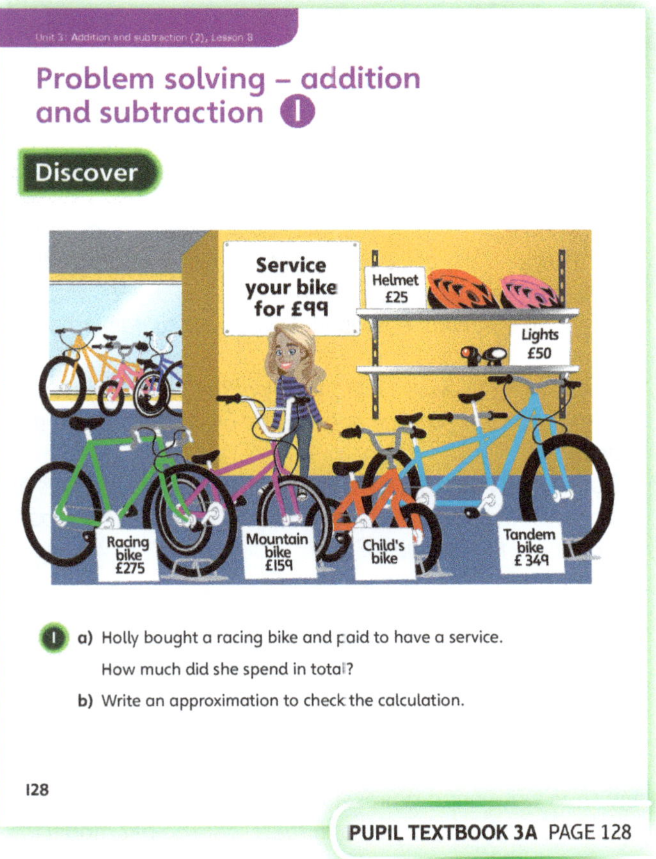

PUPIL TEXTBOOK 3A PAGE 128

Share

 Whole class teacher led

ASK

- Question **1** a): *What does the '?' represent on the bar model?*
- Question **1** a): *How does the bar model show we need to use the operation of addition?*

IN FOCUS The bar model very clearly represents the concept of the parts and wholes. The brace with the question mark identifies that in question **1** a) it is the whole that needs to be calculated. The most important aspect of this phase of the lesson is recognising that the bar model supports the relationship between the different aspects of a story problem, and can justify which calculation needs to be used to find the solution.

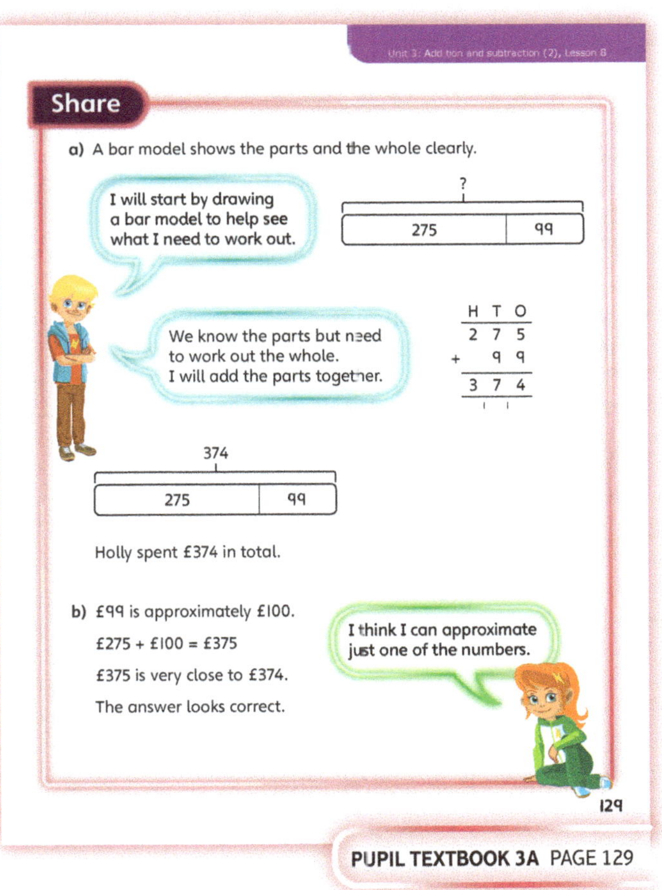

PUPIL TEXTBOOK 3A PAGE 129

Think together

WAYS OF WORKING Whole class teacher led (I do, We do, You do)

ASK

- Question ❶ : *Which information do you know – the parts or the whole? Where do you need to look to find the missing part information?*
- Question ❶ : *What operation do you need to do to find the whole?*
- Question ❷ : *Can you identify any of the missing parts from the picture? Do you need to add or subtract to find the answer?*
- Question ❸ a): *Whose idea do you think best represents the story problem? Do both methods work? Can you draw a bar diagram with three parts?*

IN FOCUS Question ❷ is important because it prompts children to recognise that the missing information is one of the parts, rather than the whole. Therefore, they need to identify that the appropriate operation to solve this question is a subtraction.

STRENGTHEN Use strips of paper for children to fold or cut and then write on to represent the different parts of a bar model. Using concrete items to represent the bar model may help children to visualise more clearly the relationship between the parts and the whole and understand which operation they need to perform in order to solve a given problem.

DEEPEN Question ❸ requires the addition of three numbers. Some children may use one bar to represent the addition of the first two parts and then draw a second bar for the addition of the first total with the remaining part. Other children may combine all three parts into one diagram, and represent the steps of the calculation by showing a brace for two parts, and then adding the third part to the total shown by that brace.

ASSESSMENT CHECKPOINT Can children confidently explain how the different elements of the bar models represent aspects from the story problems?

ANSWERS

Question ❶ : 159 + 25 = 184: Dad spent £184.

Question ❷ : 468 − 349 = 119: The child's bike cost £119.

Question ❸ a): 275 + 25 + 50 = 350: Toshi spent £350.

Question ❸ b): ⬜ + 25 + 25 = 399, 399 − 25 − 25 = 349: The bike cost £349, so it must have been a tandem.

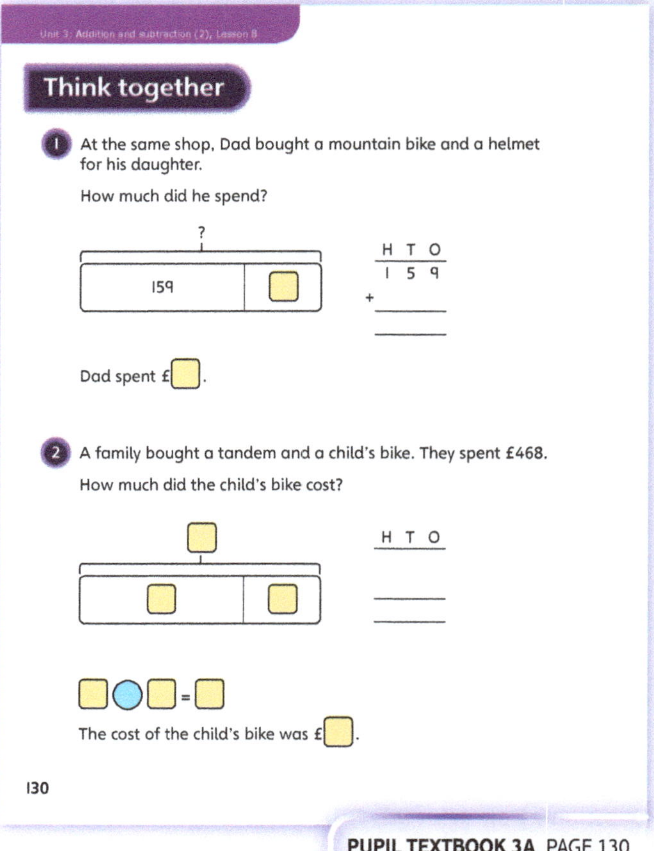

PUPIL TEXTBOOK 3A PAGE 130

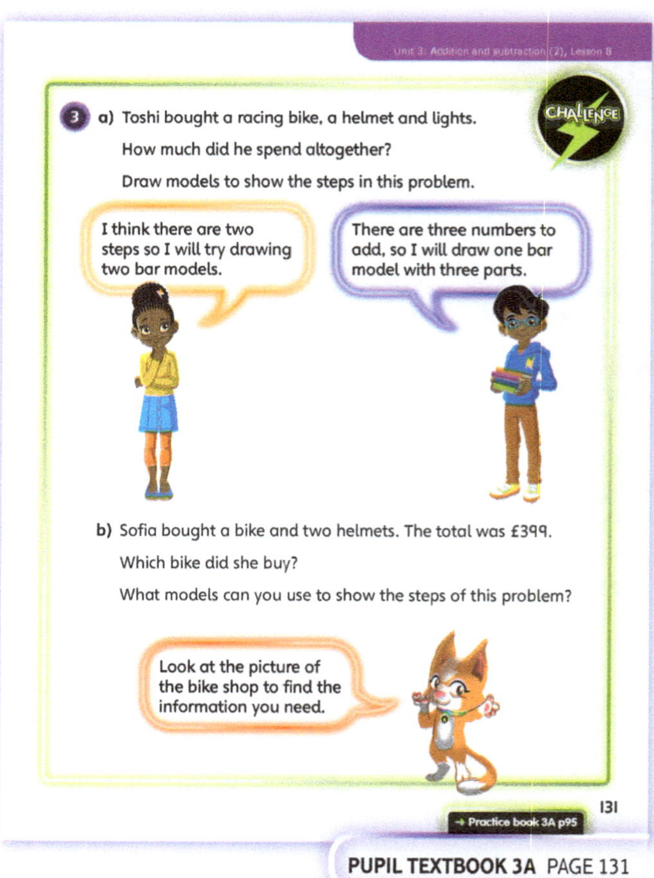

PUPIL TEXTBOOK 3A PAGE 131

Practice

WAYS OF WORKING Independent thinking

IN FOCUS Question **2** prompts children to focus on which bars are useful and accurate representations of a problem, as opposed to which merely appear to be relevant at first glance.

Question **3** challenges children to create their own bar model. This focuses their attention on which elements of the model they already know, which information they need to find and therefore which operation is needed to solve the question.

STRENGTHEN Children may find it boosts their confidence to draw their diagrams on a whiteboard first. They can then adapt it to fit once they have started to explore the problems through the diagram itself.

DEEPEN Question **5** requires children to notice the relationship of the parts to one another. Children may well make the initial mistake of performing 125 + 55 then subtracting the 180 from 500. Prompt them to recognise that they need to find the three numbers that total 500 and not just two.

THINK DIFFERENTLY Question **4** prompts children to consider and recognise where the incorrect operation has been used. Alex has jumped to a conclusion because she is trying to find the answer too quickly.

ASSESSMENT CHECKPOINT If children can explain and justify their decisions for their answer to question **2**, then they should be demonstrating a good grasp of how the bar models represent problems.

ANSWERS Answers for the **Practice** part of the lesson appear in the separate **Practice and Reflect answer guide**.

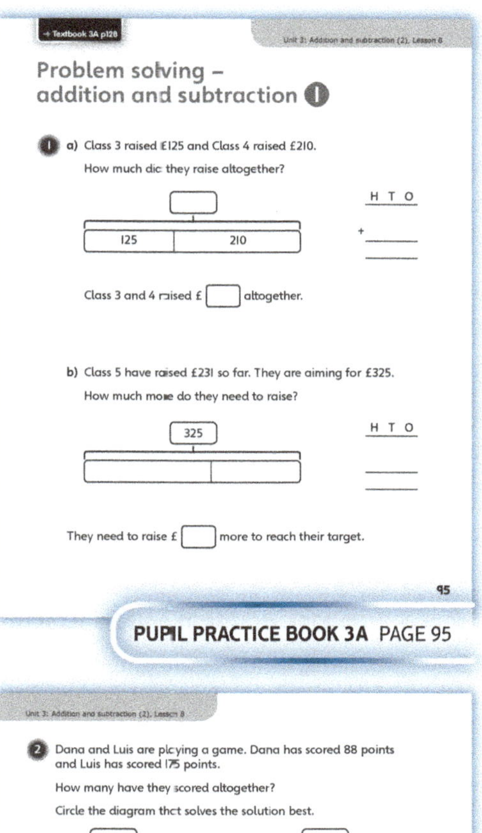

PUPIL PRACTICE BOOK 3A PAGE 95

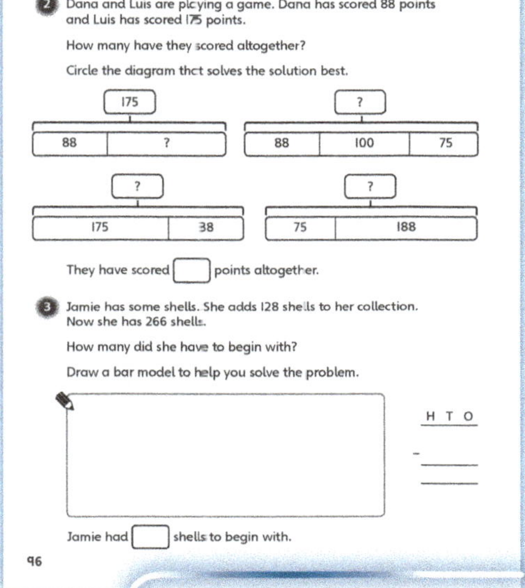

PUPIL PRACTICE BOOK 3A PAGE 96

Reflect

WAYS OF WORKING Independent thinking

IN FOCUS Rather than focusing on children's ability to draw bar diagrams, this challenges children to think about how the different elements of the bar model really relate to the context of the problem.

ASSESSMENT CHECKPOINT Have children recognised that the bar represents a problem where a whole is known and a part is known, and so written a story problem that requires a subtraction?

ANSWERS Answers for the **Reflect** part of the lesson appear in the separate **Practice and Reflect answer guide**.

After the lesson ⏸

- Can children explain how the bar model relates to the particular context of a problem?
- Can children identify the missing information required from a problem, and whether it is a whole or a part?
- Have children developed some confidence in drawing their own bar models to solve problems?

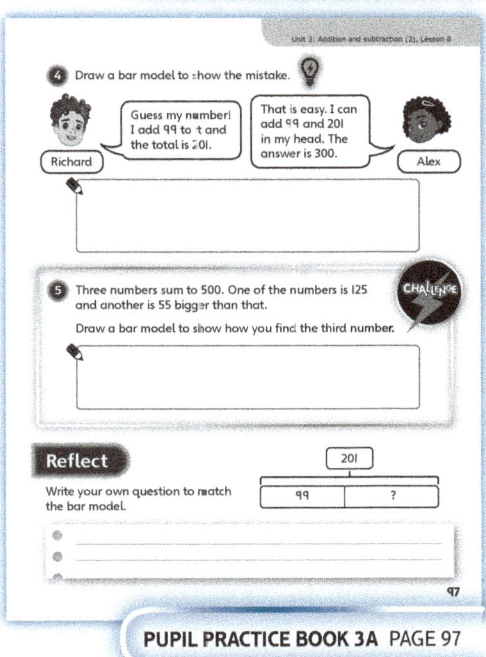

PUPIL PRACTICE BOOK 3A PAGE 97

Problem solving – addition and subtraction ②

Learning focus

In this lesson, children will develop their use of the bar model to include two bars to represent comparison and to tackle problems with two or more steps.

Small steps

→ Previous step: Problem solving – addition and subtraction (1)

→ **This step: Problem solving – addition and subtraction (2)**

→ Next step: Multiplication – equal grouping

NATIONAL CURRICULUM LINKS

Year 3 Number – Addition and Subtraction

Solve problems, including missing number problems, using number facts, place value, and more complex addition and subtraction.

ASSESSING MASTERY

Children can represent and justify problem-solving decisions through the use of comparison bar models.

COMMON MISCONCEPTIONS

Misconceptions can arise when children encounter comparison bar models, as opposed to single bars. Children may find it difficult to know how to represent the whole, and how to reason between the two bars. Ask:

- *Which part of this model represents the whole?*
- *What are the parts that make up the whole?*

STRENGTHENING UNDERSTANDING

Allow children to represent the bars using strips of paper, or resources such as coloured rods. The use of concrete materials will aid visualisation and strengthen their understanding of the relationship between the parts and the whole.

GOING DEEPER

Challenge children to create their own story problems or alternative story problems to match the bars.

KEY LANGUAGE

In lesson: subtract, bar model, part

Other language to be used by the teacher: runs (in a cricket context), operation, column addition, column subtraction, whole, missing information

STRUCTURES AND REPRESENTATIONS

Bar model, column addition and subtraction

RESOURCES

Mandatory: bar models

Optional: strips of paper, coloured rods, lengths of string

 In the eTextbook of this lesson, you will find interactive links to a selection of teaching tools.

Before you teach 

- Do children understand what it means to 'find the difference'?
- Can they confidently create a bar showing 18 and another showing 5 more than that?

Discover

WAYS OF WORKING Pair work

ASK

- Question ❶ : *How can you show which operation is needed?*
- Question ❶ : *What numbers are relevant in solving this problem?*

IN FOCUS The important aspect of question ❶ a) is to recognise that it is a comparison problem, and so understand that it can be solved using a find the difference subtraction. Are children able to represent this using a bar model?

PRACTICAL TIPS Some children may want to represent the amounts using place value equipment or strips of paper to create a bar model. In question ❶ a) they should arrange the equipment in a way that allows for the direct comparison this problem requires.

ANSWERS

Question ❶ a): 454 − 128 = 326

Question ❶ b): 128 + 105 + 83 = 316

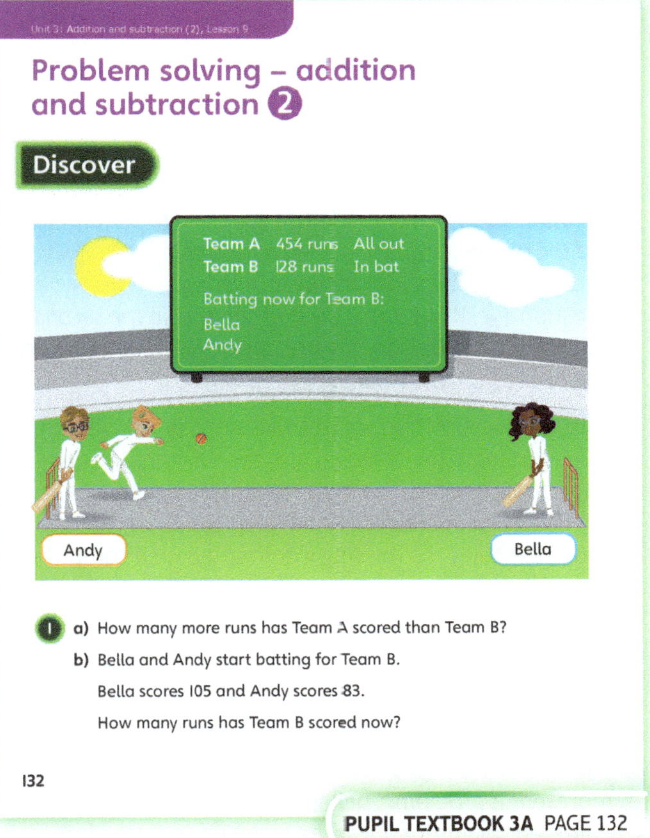

PUPIL TEXTBOOK 3A PAGE 132

Share

WAYS OF WORKING Whole class teacher led

ASK

- Question ❶ a): *Can you explain how the bar model represents the comparison clearly?*
- Question ❶ a): *Which part of the bar model shows the information we need to work out?*
- Question ❶ b): *Can you think of an alternative way to solve this?*

IN FOCUS Question ❶ a) demonstrates the power of a comparison model for problems which involve finding the difference. The important point is that children notice how to represent the missing information on the diagram.

For question ❶ b), children may discuss how to find the solution in a different order. For instance, they may find a way of adding the two new scores first, then adding that onto the 128 runs team B had already.

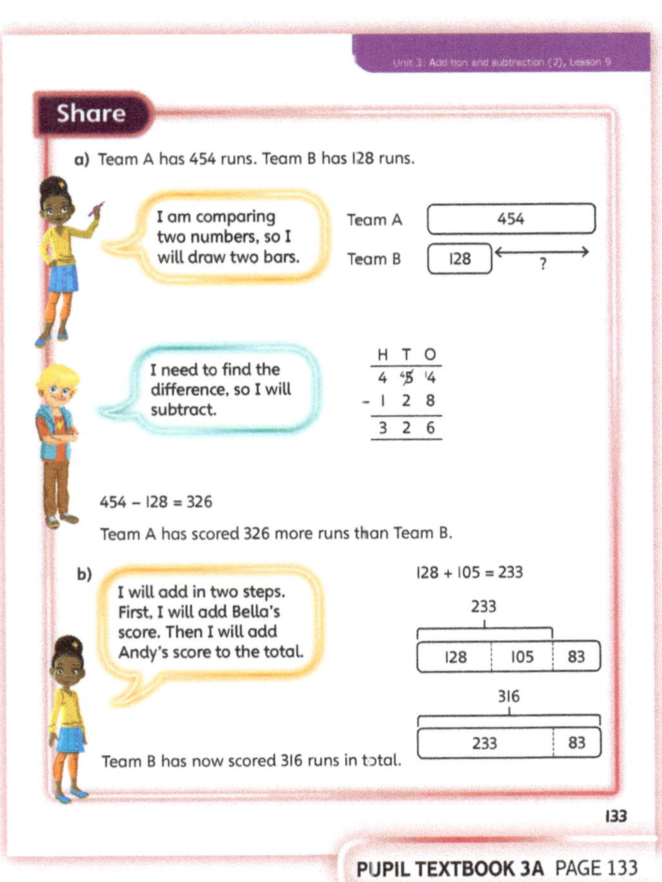

PUPIL TEXTBOOK 3A PAGE 133

Think together

 Whole class teacher led (I do, We do, You do)

- Question **1** : *What operation do you need to perform to find the difference between the two scores?*
- Question **2** : *What information is missing?*
- Question **2** : *What do we need to subtract in the second column subtraction? What is the whole and what do we need to subtract from the whole?*

 Question **2** requires two steps of calculation, including a single bar addition and then a comparison bar, to find the difference. The important aspect of this phase of the lesson is for children to develop some confidence with representing the missing information on a comparison bar model.

 Encourage children to use strips of paper, coloured rods or lengths of string to represent the two bars required. This will help them to visualise the parts and whole of the bar model.

 Question **3** presents two different bar models to represent a problem. Challenge children to explore how the different diagrams show the same information but in different ways. Ask them to justify which approach they think better represents the story context in terms of how clear it makes the path to the solution.

 Can children explain how the two steps of the calculation in question **2** are represented in different aspects of the bar model?

Question **1** : 451 – 317 = 134; Isla's team scored 134 more runs.

Question **2** : 165 + 56 = 221 and then 320 – 221 = 99; Jamilla and Emma need 99 more runs to get the same score as Mo and Lexi.

Question **3** : 188 + 188 + 56 = 432

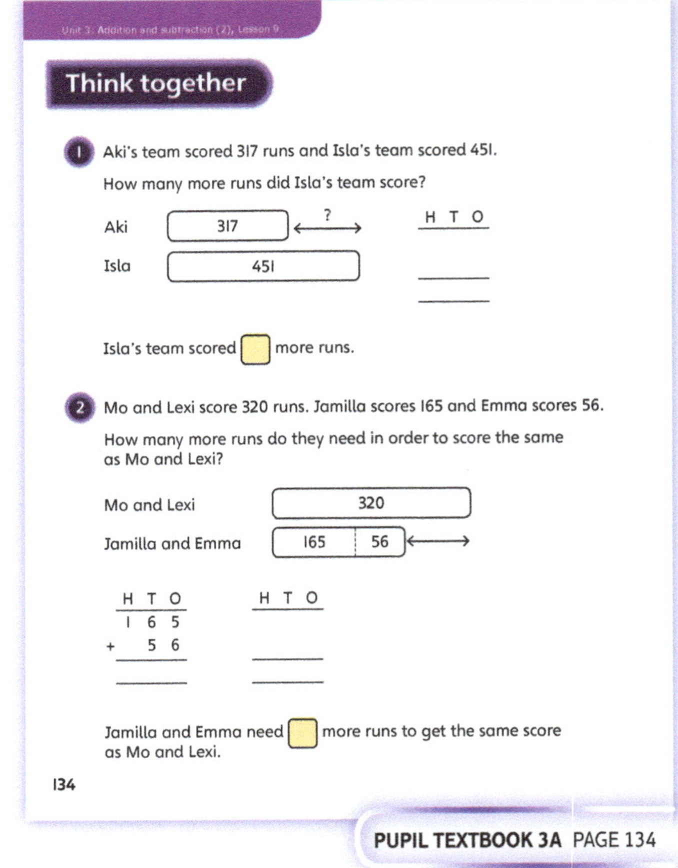

PUPIL TEXTBOOK 3A PAGE 134

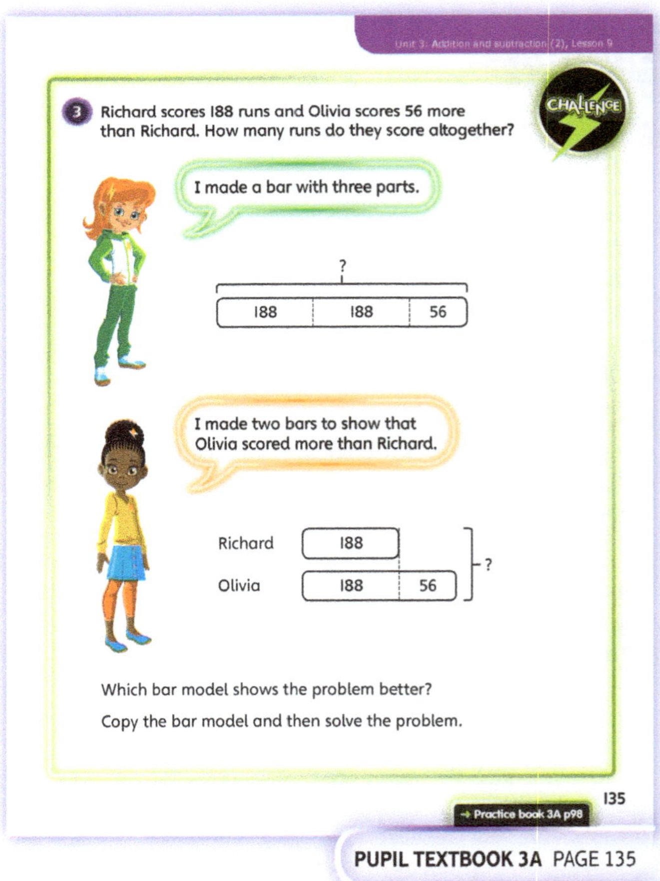

PUPIL TEXTBOOK 3A PAGE 135

Practice

WAYS OF WORKING Independent thinking

IN FOCUS Questions ❶ and ❷ provide children with the bar models; they have to add the numbers and identify the calculation required. Questions ❸ and ❹ extend this by challenging children to draw and label their own bar models to represent the problem.

STRENGTHEN Many of the problems require more than one step and children may struggle to identify how to begin. Encourage them to use the bar model concept to sketch the elements of the problem without trying to solve it at all. The first stage is simply to represent the given information and then the second stage is to identify and organise the calculation by looking carefully at the diagram.

DEEPEN Question ❺ requires children to recognise how to represent the whole. They need to find the difference and start with Ebo's number in order to find Zac's number and accurately complete the bar model. Ensure children understand that it is only the shaded parts of the bar model that add up to 801.

THINK DIFFERENTLY Not every question is best represented using a comparison bar. For example, question ❹ b) is best represented as a single bar. Question ❸ uses a comparison, but the difference is known, and they have to use this information to find one of the parts.

ASSESSMENT CHECKPOINT Confident solutions to question ❹ will represent a good level of understanding.

ANSWERS Answers for the **Practice** part of the lesson appear in the separate **Practice and Reflect answer guide**.

Reflect

WAYS OF WORKING Group work

IN FOCUS This focuses not on the use of a bar to get an answer, but on the decisions about how to represent different problems using different bar models.

ASSESSMENT CHECKPOINT Can children explain when the comparison bar is appropriate and when it is not?

ANSWERS Answers for the **Reflect** part of the lesson appear in the separate **Practice and Reflect answer guide**.

After the lesson ⏸

- Can children represent a 'find the difference' problem as a comparison model?
- Do children know how to decide which bar models to draw?
- Can children identify the required missing information?

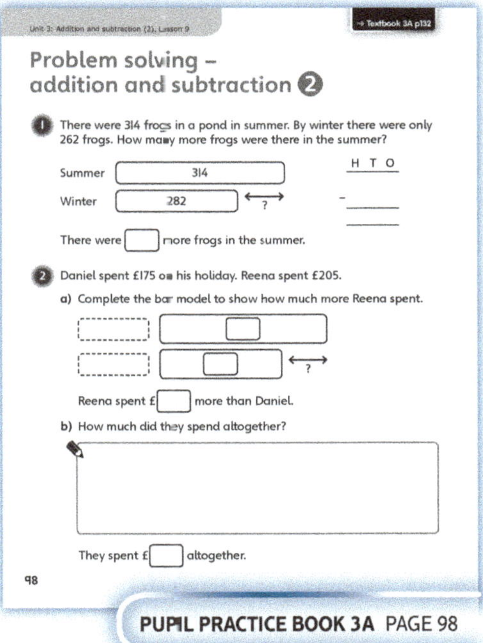

PUPIL PRACTICE BOOK 3A PAGE 98

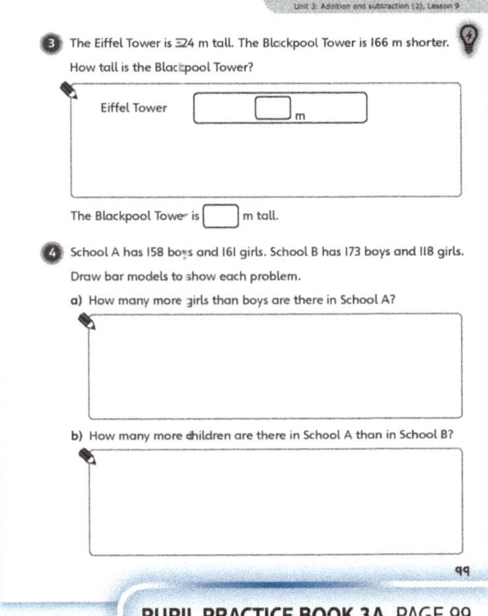

PUPIL PRACTICE BOOK 3A PAGE 99

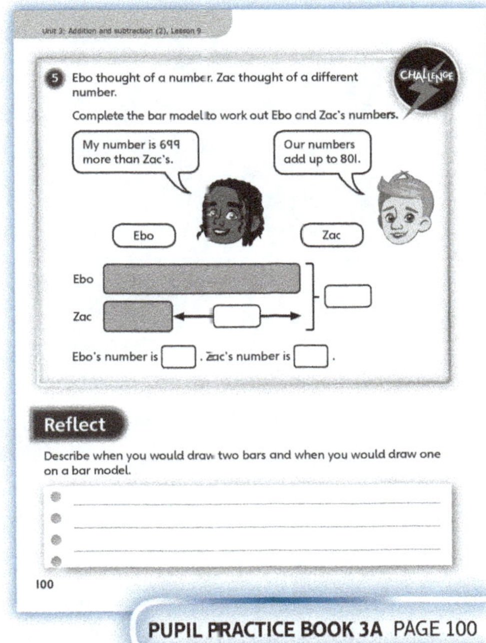

PUPIL PRACTICE BOOK 3A PAGE 100

End of unit check

Don't forget the *Power Maths* unit assessment grid on p26.

WAYS OF WORKING Group work adult led

IN FOCUS

- Questions ❶ to ❸ all focus on accuracy and efficiency using the column method and exchange.
- Questions ❹ and ❺ focus on the checking strategies of approximation and using an inverse operation (in this case, an addition to check a subtraction).
- Question ❻ focuses on problem solving. This is a SATs-style question.

ANSWERS AND COMMENTARY

Children who have mastered this unit will be able to add or subtract numbers with up to 3 digits within 1,000. They will be able to justify whether or not an exchange is necessary and be able to explain the effect of doing an exchange in terms of place value. Children will also be able to justify an answer through checking strategies of approximation, estimation and the use of inverse operations.

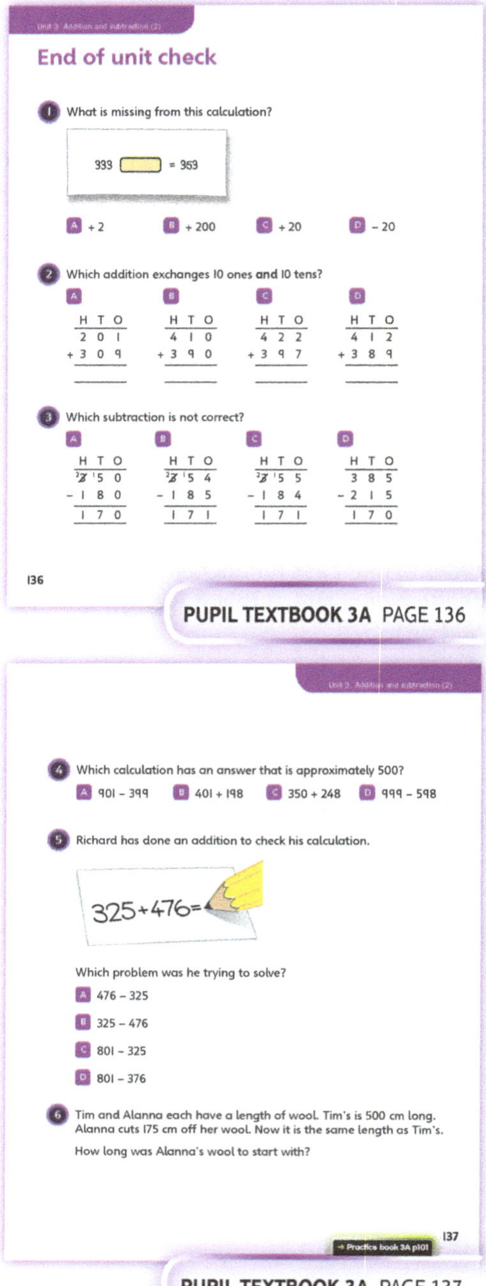

PUPIL TEXTBOOK 3A PAGE 136

PUPIL TEXTBOOK 3A PAGE 137

Q	A	WRONG ANSWERS AND MISCONCEPTIONS	STRENGTHENING UNDERSTANDING
1	C	A and B suggest a lack of understanding of place value. D indicates children misinterpreted the missing number problem as a subtraction.	Place value: allow children to model their reasoning and justify decisions using place value equipment.
2	D	A may suggest they are confused when zero is a place holder.	Use the equipment to discuss the decisions and the calculations, but not as a counting strategy to explain the concepts.
3	B	A, C or D suggest children are not confident using exchange.	
4	A	B and C suggest children just added the 100s. D indicates children may find it difficult to approximate to 1,000.	Children should also be encouraged to use any other models or images that support their reasoning, such as part-whole models or number lines.
5	C	A, B or D suggest they are not confident with the relationship between wholes and parts in additions and subtractions.	
6	675 cm	An answer of 325 cm suggests children have not understood parts and wholes in this context.	

My journal

WAYS OF WORKING Independent thinking

ANSWERS AND COMMENTARY Answers will vary. Children do not have to write the answers, though they may choose to explore the best methods to find the answers as a way of deciding which are more complex. Instead, children should make and justify decisions about the different levels of complexity or difficulty in solving the different calculations.

If children are struggling to form an opinion or find it difficult to move beyond simply finding the answers, ask:
- *Which question would you choose to attempt first?*
- *Are there any you would solve in a different way from the others?*
- *What is the same and what is different about the calculations?*
- *Which do you think would be most likely to cause a mistake?*

If children are struggling to explain their reasoning, ask:
- *Could you explain how many steps each option takes?*
- *What do people tend to get wrong in class?*
- *Have you made any mistakes in this unit that you are reminded of here?*

Power check

WAYS OF WORKING Independent thinking

ASK
- *What numbers could you add and subtract before this unit?*
- *Do you know any new skills?*
- *Have you built more confidence in any areas?*
- *Do you feel confident knowing when and how to use exchange?*
- *Can you check your own answers now?*

Power play

WAYS OF WORKING Pair work

IN FOCUS Use this Power play to give children the opportunity to explore calculations in the context of a strategic game. The game can be adapted, as per Sparks' suggestion, to deepen the challenge by attempting to create an answer close to a specified number. Allow children to use the numbers generated by their spins in whatever position or order they like in their addition. This will encourage strategic thinking and challenge their understanding of place value.

ANSWERS AND COMMENTARY Supply pairs of children with a 0–5 spinner or a dice. Children will be choosing their digits to try to block their opponent, although the spinner provides a random element to the game. This random element will mean that children will aim for and 'hope' for certain numbers to come up on the spinner (or dice), meaning they will be using the mathematical skill of predicting, with a particular focus on place value in the context of addition.

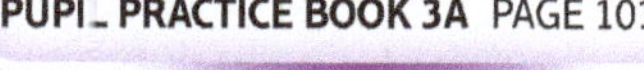

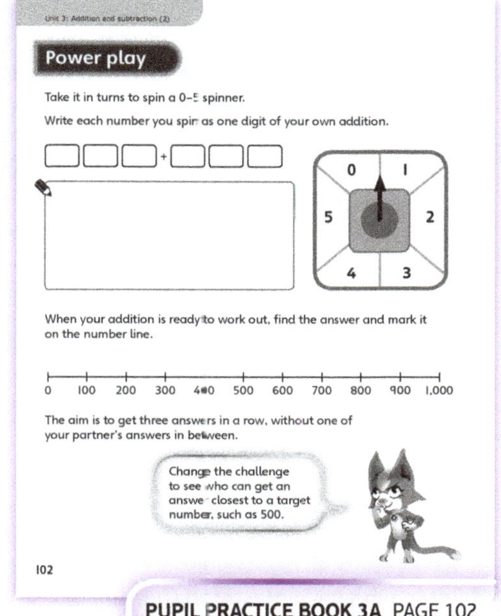

PUPIL PRACTICE BOOK 3A PAGE 101

PUPIL PRACTICE BOOK 3A PAGE 102

After the unit ⏸

- Can children explain why exchange is required for some calculations, and how it relates to place value?
- Can children justify and explain their answers using checking strategies and sound understanding of different methods?

Strengthen and **Deepen** activities for this unit can be found in the *Power Maths* online subscription.

Unit 4
Multiplication and division ①

Mastery Expert tip! "Rapid recall and knowledge of times-tables is vital to free up valuable working memory. I make time for children to regularly practice these, so that they can improve their speed of recall. We use arrays to help children visualise 2 × 3 as an array of 6 counters, and then they're more likely to remember the fact."

WHY THIS UNIT IS IMPORTANT

This unit builds on recognising equal groups. Three lessons are spent exploring in depth each of the times-tables that children need to know in Year 3, encouraging rapid recall. Children are reminded of the difference between equal sharing and equal grouping and then move on to look at when division problems may have a remainder of sorts. Although a full understanding of remainder is not essential in Year 3, children do need to have a basic understanding of it. There are two lessons that focus on problem solving, and using the bar model to represent simple one-step multiplication and division problems. This reinforces multiplication as repeated addition. Children then move on to solve simple two-step problems that involve all of the four operations.

WHERE THIS UNIT FITS

→ Unit 3: Addition and subtraction (2)

→ **Unit 4: Multiplication and division (1)**

This unit builds on children's work in Year 2, where multiplication and division are introduced and equal and unequal groups are explored. It also builds on equal sharing and equal grouping. This unit provides essential preparation for beginning to multiply and divide 2-digit numbers by 1-digit numbers in the spring term, and also for working with fractions. Knowledge of times-table facts is also essential.

Before they start this unit, it is expected that children:

- know what it means when groups are equal and not equal
- know that multiplication can be seen as repeated addition and division as repeated subtraction
- know that an array shows two multiplications, such as $5 \times 4 = 4 \times 5$.

ASSESSING MASTERY

Children who have mastered this unit will start to know their 3, 4, and 8 times-tables off by heart. They will know when to multiply and will understand the difference between equal grouping and sharing. They will know that some divisions do not always give a whole answer and can have a remainder. They will be able to represent multiplication and division problems using a bar model.

COMMON MISCONCEPTIONS	STRENGTHENING UNDERSTANDING	GOING DEEPER
Children may think that $0 \times a = a$. This is a common mistake that is made with multiplication facts.	Reinforce the importance of knowing times-tables. 0 groups of anything is 0 and any number of groups of 0 is 0.	Children use their knowledge of times-tables to work out problems such as 13×4, or 24×4.
Children may not always recognise if they need to do a multiplication or division to find the solution.	Use counters to represent the objects in groups and then use a bar model to represent the situation. You can use counters initially to represent the bars.	Ask children to write simple one-step and multi-step multiplication and division problems and to explain when a problem requires more than one step.

Unit 4: Multiplication and division 

WAYS OF WORKING

Introduce the unit using teacher-led discussion. Ask the children to think back to their work in Year 2 and identify why the bananas are in unequal groups. Question children to see which of Flo's words they recognise.

STRUCTURES AND REPRESENTATIONS

Arrays: This model shows the total of a multiplication and reinforces commutativity. It can also be used to demonstrate sharing and grouping.

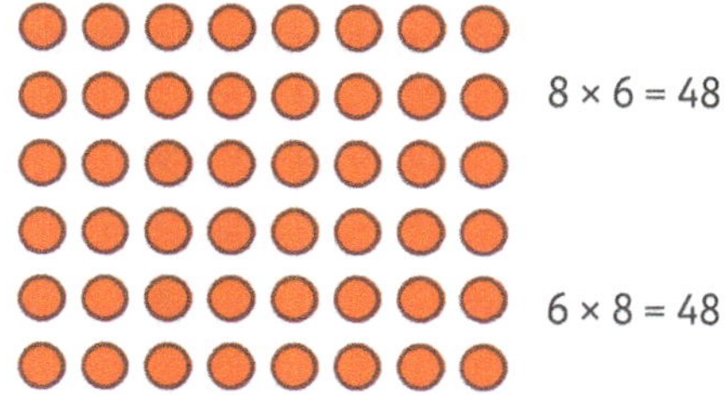

$8 \times 6 = 48$

$6 \times 8 = 48$

Bar model: This model represents the situation in multiplication and division word problems and shows the link between multiplication and repeated addition.

$6 \times 5 = 30$

| 5 | 5 | 5 | 5 | 5 | 5 |

30

| 6 | 6 | |

$5 \times 6 = 30$

30

| 6 | 6 | 6 | 6 | 6 |

KEY LANGUAGE

There is some key language that children will need to know as part of the learning in this unit.

➜ equal groups, unequal groups, shared equally

➜ multiply (×), multiplication statement, multiplication fact, multiplication sentence, divide (÷), division statement, division fact

➜ times-table

➜ group, share

➜ whole, left over, remainder

➜ one-step, two-step, multi-step

➜ array, bar model, number line

➜ pattern

➜ count up, total, double, method

➜ repeated addition

PUPIL TEXTBOOK 3A PAGE 138

PUPIL TEXTBOOK 3A PAGE 139

Multiplication – equal grouping

Learning focus

In this lesson, children will recap their knowledge from Year 2 and be able to recognise equal groups. For any equal groups, children should be able write down a multiplication sentence, know how it links to repeated addition and know how to find the answer.

Small steps

→ Previous step: Problem solving – addition and subtraction (2)
→ **This step: Multiplication – equal grouping**
→ Next step: Multiplying by 3

NATIONAL CURRICULUM LINKS

Year 3 Number – Multiplication and Division

- Write and calculate mathematical statements for multiplication and division using the multiplication tables that they know, including for two-digit numbers times one-digit numbers, using mental and progressing to formal written methods.
- Recall and use multiplication and division facts for the 3, 4 and 8 multiplication tables.
- Solve problems, including missing number problems, involving multiplication and division, including positive integer scaling problems and correspondence problems in which *n* objects are connected to *m* objects.

ASSESSING MASTERY

Children can recognise equal groups and write down the correct multiplication sentence for each group. Children can recognise groups that are not equal and say why they are not equal.

COMMON MISCONCEPTIONS

Children may think that groups that do not look the same are not equal. For example, if jam tarts on a plate have a different arrangement, children may think the groups are not equal, even though there are four on each plate. Ask:
- *How many are in each group? Are there the same amount in each group? Does this mean they are equal?*

Children over-rely on repeated addition to work out multiplication sentences. Encourage children to start to know their times-table facts to help instant recall of the answers. Ask:
- *How else can you say 3 + 3 + 3? Is there a multiplication that matches that addition? Which times-table does it remind you of?*

STRENGTHENING UNDERSTANDING

Children who are struggling with the concepts in this lesson should use cubes or counters to represent the objects. For example, the number of towers of cubes will represent the number of groups. Children should then be able to compare whether there are the same number in each group by comparing the heights or lengths of the towers. Ask: *Are the heights the same? Does this mean the groups are equal? Why? What would happen if they were not equal?*

GOING DEEPER

Ask children to show their own equal groups and unequal groups. For example, say: *Make me 4 groups that are equal. Can you write a multiplication sentence for each one? Draw me a story to represent 3 × 5. Explain your thinking.*

KEY LANGUAGE

In lesson: multiplication (×), equal groups, **multiplication sentence**, grouping, groups, equal, total, **repeated addition**

Other language to be used by the teacher: not equal, times-table, multiply, multiplication statement

STRUCTURES AND REPRESENTATIONS

arrays

RESOURCES

Mandatory: cubes and counters

Optional: number lines

 In the eTextbook of this lesson, you will find interactive links to a selection of teaching tools.

Before you teach 

- Do children know some 2, 5, and 10 times-table facts?
- Can they recognise equal groups and unequal groups?
- Can they write a multiplication statement for any equal groups?

Discover

 Pair work

- Question **1** a): *Which groups are equal? How do you know? How can you use counters to show the groups are equal or not equal? Can you make D so that is showing equal groups?*
- Question **1** b): *What multiplication statements can you see? Can you work out the answers? How did you work out the answers? Did you just know some?*

 Question **1** a) is used to recap learning from Year 2. Children should discuss in pairs whether the groups are equal or unequal. They may use cubes and counters to compare or may be able to see straight away. It is important for children to be able to reason why the groups are equal or not and not just say that they are. Question **1** b) determines whether children can write down the corresponding multiplication sentence for this and work out the corresponding answer. Children may need to work out C using repeated addition.

 Counters and cubes should be available for children to represent the objects in the different groups.

Question **1** a): A, B, C are equal groups.

Question **1** b): A: $4 \times 2 = 8$, B: $3 \times 5 = 15$, C: $3 \times 4 = 12$

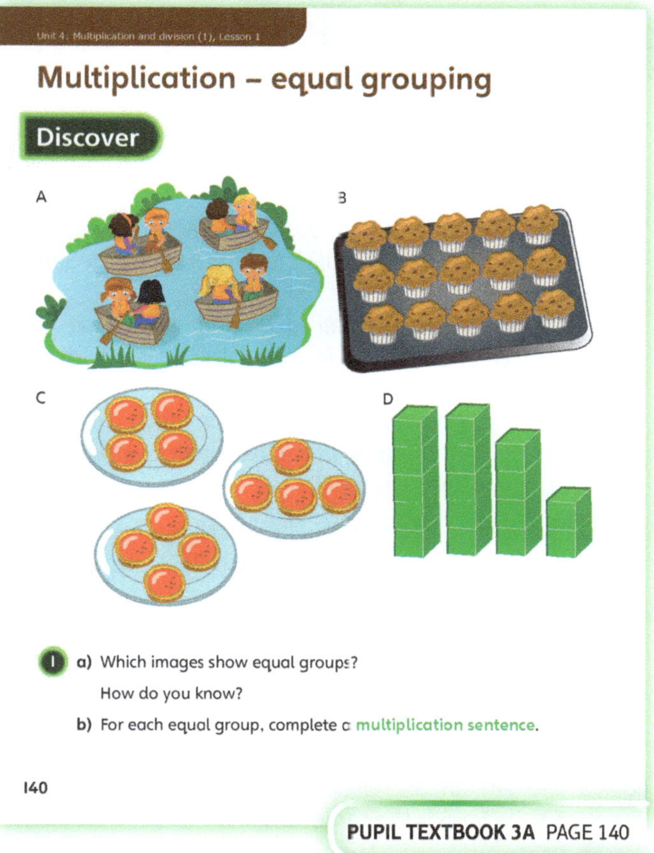

PUPIL TEXTBOOK 3A PAGE 140

Share

 Whole class teacher led

- Question **1** a): *How many groups can you see? How many are in each group? Can you see why they are equal groups? Is there a different way of saying 3 groups of 5 muffins?*
- Question **1** b): *Do you know any of these off by heart? What other ones do you know? How did you work out 3×4, as we have not done the 4 times-table yet?*

 In question **1** a) children provide responses to what they have just done. Together, count the number of groups for each diagram. Count the number in each group and show each time whether or not they are equal. For B, discuss how the array shows that 3×5 is the same as 5×3. Show the groups circled. For C, discuss with children that although the groups look different there are the same number in each group, so the groups are equal. When moving to question **1** b), remind children of the need to know their multiplication facts off by heart. Give children a method (such as repeated addition, linked to counting up by the same amount each time) to help them, in case they cannot remember.

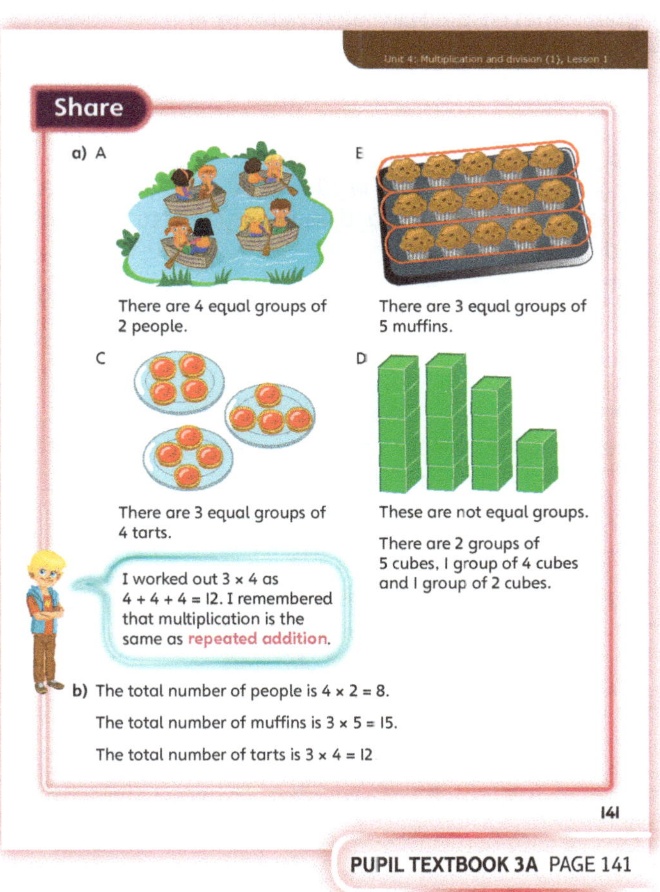

PUPIL TEXTBOOK 3A PAGE 141

Think together

WAYS OF WORKING Whole class teacher led (I do, We do, You do)

ASK

- Question ❶ : *How many groups are there? How many are in each group? Are they equal? If they do not look the same, how can they be equal?*
- Question ❷ : *What is this called? What multiplication statement can you write to describe this? Can you think of another statement?*
- Question ❸ : *Why can they each see something different? Can you see what they see? Circle them with your finger. Can you see any more?*

IN FOCUS Question ❶ checks whether children know that groups are equal even if they do not look equal. Question ❷ reinforces the fact that an array can show two multiplications. Ensure that children see both. Ask children to circle the rows first and then the columns. Question ❸ looks at how an arrangement of stars can lead to different people giving different answers. Three characters discuss different equal groups that they see. Ask children to look at the diagram and circle the groups. Discuss that 5 + 5 + 5 + 5 + 5 + 5 is the same as 6 × 5.

STRENGTHEN For question ❶, get children to put the counters out on their desk as they are in the book and ask them to show that they are equal by putting them in 3 lines, each the same length. This may help them see that they are equal. Also count aloud with children. First count how many groups, with children circling the groups with their finger as they count. Then count how many there are in each group. Did they get the same amount? Does this mean the groups are equal?

DEEPEN For question ❸, ask: *Can you see any other groups? Can you see 10 groups of 3 or 10 × 3?* You could ask children to make their own equal and unequal groups to help them understand.

ASSESSMENT CHECKPOINT Can children recognise equal and unequal groups and write a multiplication sentence for equal groups?

ANSWERS

Question ❶ : There are 3 equal groups of 5 counters, so there are 15 counters in total;
5 + 5 + 5 = 15; 3 × 5 = 15.

Question ❷ : 10 × 5 = 50; 5 × 10 = 50.

Question ❸ : Circle 6 groups of 5 stars.
Circle 3 groups of 10 stars.
Circle 15 groups of 2 stars.

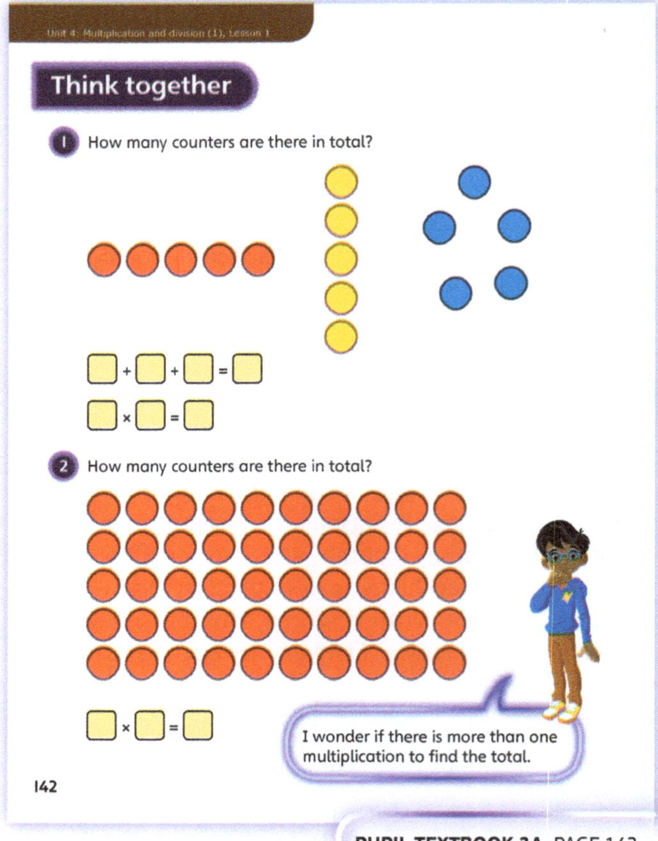

PUPIL TEXTBOOK 3A PAGE 142

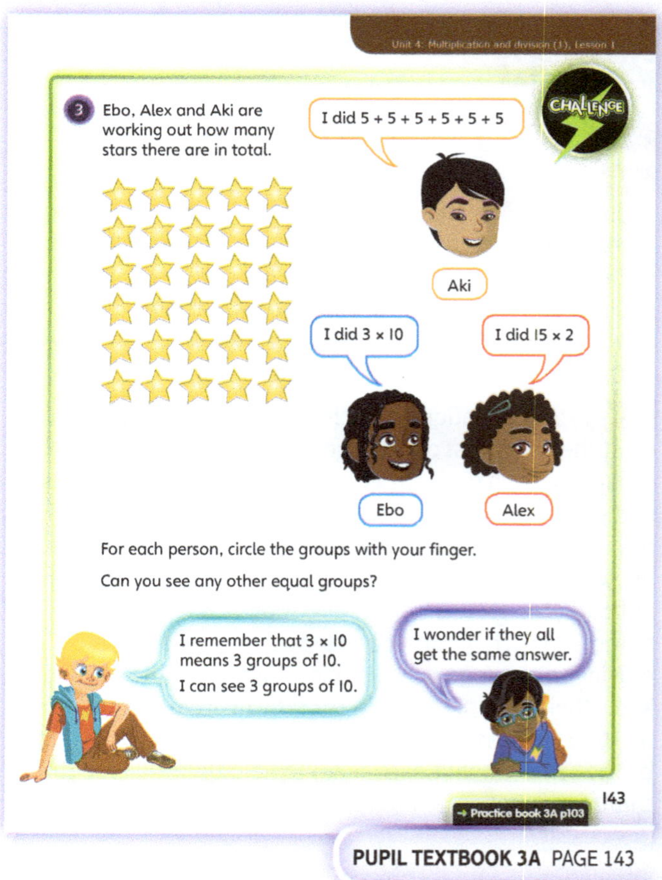

PUPIL TEXTBOOK 3A PAGE 143

Practice

WAYS OF WORKING   Independent thinking

IN FOCUS   Question ❶ requires children to work out whether equal or unequal groups are shown. Questions ❷ and ❸ require children to work out how many objects there are in total, using multiplication. Throughout these questions, the clear link with repeated addition is shown. Children may use a number line to count up in order to work out the total. Question ❻ asks children to show how arrays of symbols can show different groups.

STRENGTHEN   To support understanding of whether or not groups are equal, ask children to use cubes to represent the objects. The number of towers of cubes will represent the number of groups. Children should then be able to compare if there are the same number in each group by comparing the heights or lengths of the towers. Count aloud with children. To support knowledge of multiplication facts, encourage children to count up using a number line. Each time link the group you are counting on, to the jump on the number line. Avoid children counting up in 1s.

DEEPEN   For question ❶ c), ask: *If the first child had 3 × 10p and 1 × 20p, the second child had 2 × 20p and 1 × 10p and the third child had 1 × 50p, would the groups be equal then?* Ask children if they can see any other groups in the arrays in question ❻.

ASSESSMENT CHECKPOINT   Can children recognise equal groups and write a correct multiplication sentence for equal groups? Can children use repeated addition to help them work out multiplication sentences?

ANSWERS   Answers for the **Practice** part of the lesson appear in the separate **Practice and Reflect answer guide**.

Reflect

WAYS OF WORKING   Independent thinking

IN FOCUS   This question looks to see whether children can link equal groups to multiplication statements. Children should be able to come up with several different answers. Ask children to see how many different answers they can find. Do they all give the same answer? Why?

ASSESSMENT CHECKPOINT   Check if children can circle equal groups and then check if they write the correct multiplication sentence for those equal groups. Can they write more than one multiplication sentence?

ANSWERS   Answers for the **Reflect** part of the lesson appear in the separate **Practice and Reflect answer guide**.

After the lesson 

- Can children explain why groups are equal or not?
- Can children write a correct multiplication sentence for equal groups, with the correct answer?
- Do children know the connection between repeated addition and multiplication?

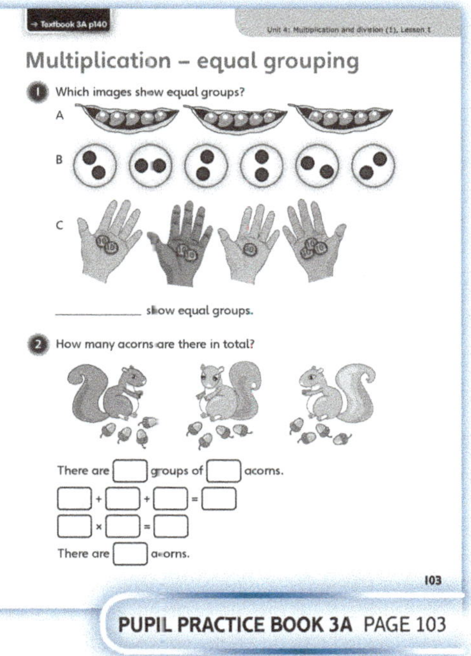

PUPIL PRACTICE BOOK 3A PAGE 103

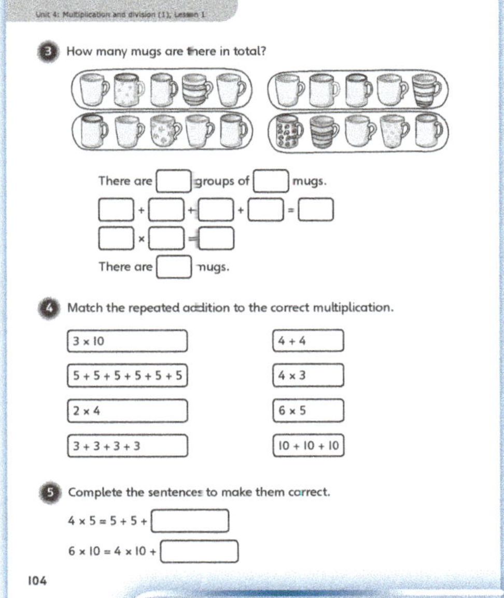

PUPIL PRACTICE BOOK 3A PAGE 104

PUPIL PRACTICE BOOK 3A PAGE 105

Multiplying by 3

Learning focus

In this lesson, children will start to understand what it means to multiply by 3. Children will see the link between repeated addition, counting up in 3s and multiplying by 3.

Small steps

→ Previous step: Multiplication – equal grouping
→ **This step: Multiplying by 3**
→ Next step: Dividing by 3

NATIONAL CURRICULUM LINKS

Year 3 Number – Multiplication and Division

- Recall and use multiplication and division facts for the 3, 4 and 8 multiplication tables.
- Solve problems, including missing number problems, involving multiplication and division, including positive integer scaling problems and correspondence problems in which *n* objects are connected to *m* objects.
- Write and calculate mathematical statements for multiplication and division using the multiplication tables that they know, including for two-digit numbers times one-digit numbers, using mental and progressing to formal written methods.

ASSESSING MASTERY

Children can form multiplication sentences involving multiplying by 3. Children can work out the answer to multiplication sentences by knowing the link with repeated addition, can use a number line to count up in 3s, and can start to remember some multiplication facts.

COMMON MISCONCEPTIONS

When children are presented with 3 groups of objects as opposed to groups of 3 objects, they may not see this as multiplying by 3. For example, they may not see that 12 groups of 3 is the same as 3 groups of 12, and that they both mean 12×3. Show a 12 by 3 array and a 3 by 12 array and ask:
- *What is the same? What is different?*

Children often start the count from 0 when working out, for example, 11×3, even if they already know 10×3. Ask:
- *How many 3s do you have? How many more 3s do you need?*

STRENGTHENING UNDERSTANDING

To help children multiply by 3, ensure that you show the direct link between the objects, a repeated addition, counting up in 3s on a number line and the multiplication statement. As children count up in 3s, put counters on the number line each time, to show the three objects.

GOING DEEPER

Ask children to use the multiplication facts they know to work out other multiplication facts. Children can also start to explore how to work out $4 \times 3 + 5 \times 3$ as one multiplication, showing by adding arrays together that this is the same as 9×3.

KEY LANGUAGE

In lesson: multiply (×), multiplication statement, total, count up, number line, array

Other language to be used by the teacher: multiplication sentence, times-table, equal groups, repeated addition, *x* groups of *y*

STRUCTURES AND REPRESENTATIONS

Number lines to show repeated addition, arrays to show multiplication commutativity

RESOURCES

Mandatory: cubes, counters, number lines

Optional: balls, plastic cups

 In the eTextbook of this lesson, you will find interactive links to a selection of teaching tools.

Before you teach ⏸

- Can children count up in 3s?
- Do children know how to form a multiplication statement from repeated addition?

Discover

 Pair work

- Question **1** a): *How many groups do you have? How many are there in each group? How can we write this as a multiplication statement? How can we work out the total? Should we count up in 1s or 3s? What can we use to help us?*
- Question **1** b): *How many groups do you have this time? How many are there in each group? What has changed? How can you work out the answer? Do you need to start from 0 again? Can you start from somewhere else? How do you know?*

 Question **1** a) is about children realising that they have 7 equal groups and there are 3 balls in each group. Using their knowledge from the previous lesson, they should start to see this as a multiplication. In order to work out the total, children should be encouraged to explore different ways of working it out and compare the method. Is the best method to count up in 1s, for example? Children should start to realise that an effective way of working out the total is to count up in 3s. They have done this in Year 2, using a number line to support them. In question **1** b), look for children who start the count at 0 again and those who start the count from where they left off previously.

 You could use balls, counters or cubes under plastic cups to re-enact the picture and help children get a feel for this activity.

Question **1** a): There are 7 cups.
There are 3 balls under each cup.
7 × 3 = 21 There are 21 balls.

Question **1** b): 8 × 3 = 24

Share

 Whole class teacher led

- Question **1** a): *How many groups can you see? How many are there in each group? How can you work out the total? What did you use to help you keep track of the count? Why is counting in 1s not the best method? What multiplication can you see?*
- Question **1** b): *Do we have to start at 0 again? Why not? What would happen if we did not have the answer to the first part, would we still start at 21? What if I had 9 cups?*

 In question **1** a) discuss the method of counting up in 1s and count aloud with children. Reiterate that this is not the most effective way of counting as it takes time. Encourage children to realise there must be a quicker way. Many children are likely to count in 3s as Flo, suggests. Show the clear link between each cup and the count. Also make links to the previous lesson, where children were reminded that multiplication is repeated addition. In question **1** b), discuss the idea of starting from 0 to work out 8 × 3 and compare this with starting at 21 and going up another 3. Ask children to decide which is quicker.

PUPIL TEXTBOOK 3A PAGE 144

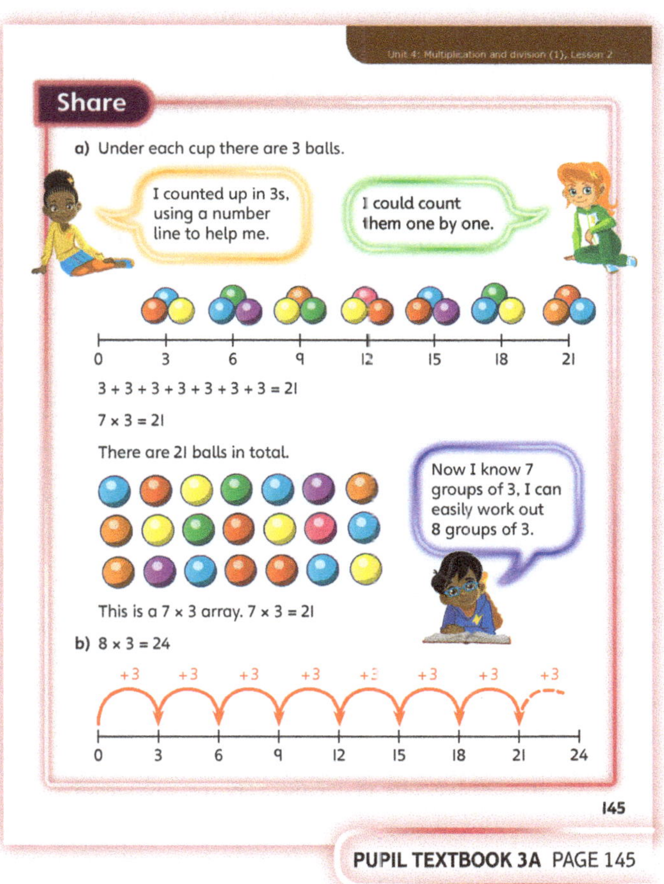

PUPIL TEXTBOOK 3A PAGE 145

Think together

WAYS OF WORKING Whole class teacher led (I do, We do, You do)

ASK

- Question **1** : *How many groups are there? How many are in each group? What has changed? If we count from 0, where do we get to? How many 3s are we adding together? What multiplication is this? What is the answer? Could we have used an answer from earlier? How?*
- Question **2** : *Do we have equal groups? How many equal groups are there? How many are in each group? What multiplication sentence is this? Do you know the answer straight away? If not, how can you work it out?*
- Question **3** : *What multiplications can you see? What is the same about each of these? What is different? How do the objects show that 3 × 5 is the same as 5 × 3?*

IN FOCUS Question **1** builds on from the **Discover** section by adding an extra cup. Discuss counting from 0 first to allow children to see the link between repeated addition and multiplication. Question **2** sets out a different context and removes the structure of repeated addition and counting, although this is what most children are likely to do. Some children may know that the answer is 18 straight away because of the previous examples. Question **3** looks at different ways of representing 5 × 3 through asking what is the same and what is different. Discuss how the total is the same each time, but the objects may be presented in many different ways. Show the link to 3 × 5 and 5 × 3.

STRENGTHEN At each stage, use concrete objects alongside a number line to help children see the link between repeated addition, counting up in 3s and multiplication. For question **1**, encourage children to put 3 counters at each point on the number line to show that this represents one cup. Children need to be secure in knowing what counting in 3s looks like (0, 3, 6, 9, and so on).

DEEPEN Encourage further thinking in question **2** by extending the calculation. Ask: *How many different ways can you work out 6 × 3? How can you work out 8 × 3 from your answer to this question?*

ASSESSMENT CHECKPOINT Can children understand what it means to multiply by 3 and use their knowledge of counting in 3s to work out the answers to the repeated addition and multiplication statements that they form?

ANSWERS

Question **1** : 3 + 3 + 3 + 3 + 3 + 3 + 3 + 3 + 3 = 27;
9 × 3 = 27 There are 27 balls.

Question **2** : 6 × 3 = 18
There are 18 hats.

Question **3** : The total is the same each time; the arrangements are different (some are in arrays, others are not and the arrays are different ways around).

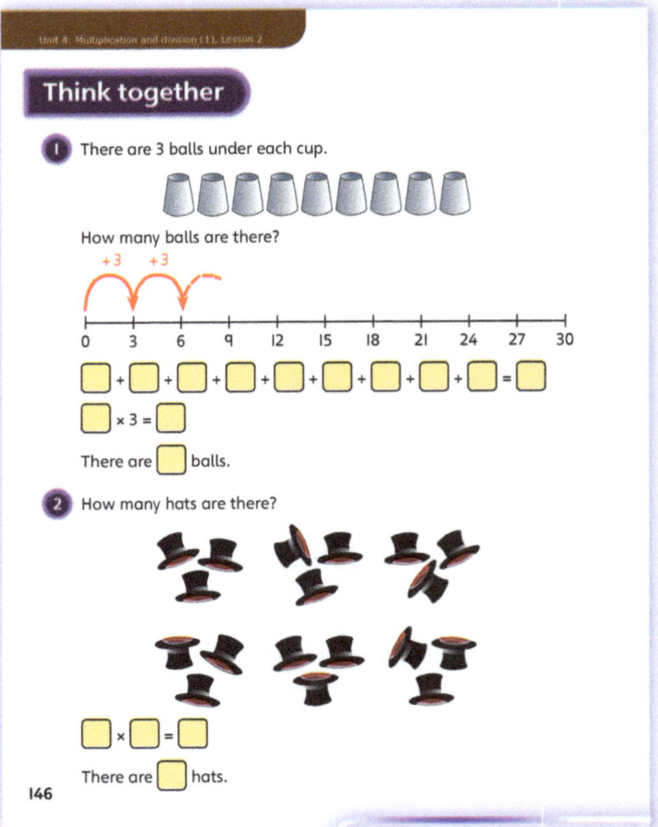

PUPIL TEXTBOOK 3A PAGE 146

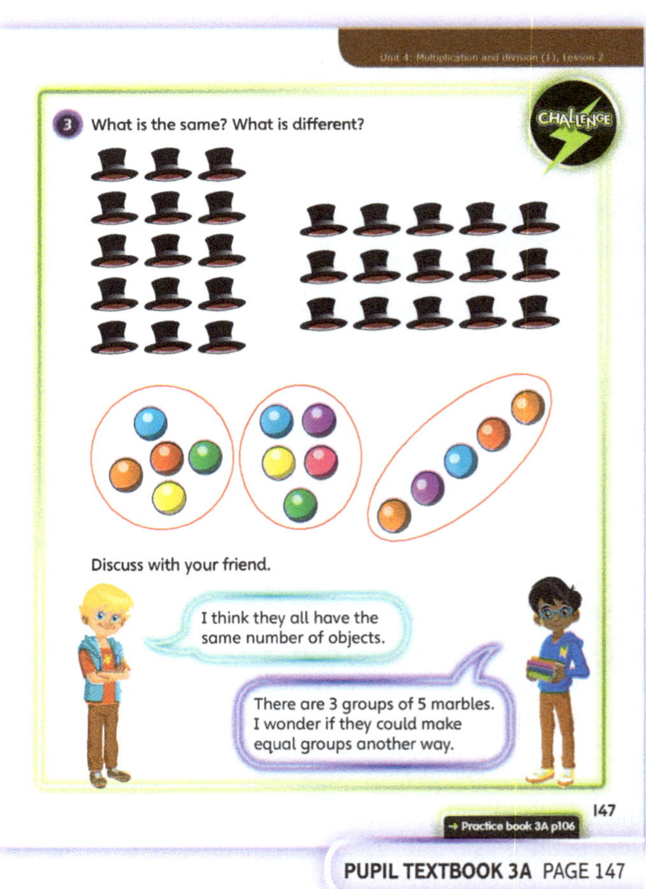

PUPIL TEXTBOOK 3A PAGE 147

Practice

WAYS OF WORKING Independent thinking

IN FOCUS Questions **1** and **2** provide practice for children to form the correct multiplication statements for situations. Encourage children to use a number line and counting up in 3s to help them work out the answers to the calculations. Some children may already start to know some 3 times-table facts off by heart and so will be able to just write the answer. Encourage checking of answers. Question **4** links multiplying by 3 with addition. Ash asks children to realise that it is just one multiplication. Question **5** reinforces the idea that you do not always have to start the count from 0.

STRENGTHEN To support multiplying by 3, use concrete objects alongside a number line. Count aloud with children in 3s from 0, each time showing the jump on the number line. For question **4**, show that this can be written as one multiplication by asking children to join the two arrays together to make one big array. Children could use counters on their desk. Encourage them to see that they need to rotate one of the arrays in order to join it.

DEEPEN In question **3**, children need to realise that 3 × 12 is the same as 12 × 3. For question **4**, ask: *How can this be a single multiplication? How can you show it using counters?* Question **6** links to children's prior understanding of odd and even numbers and asks them if an odd number multiplied by 3 always gives an odd number. Can children find a way of showing this is true for any odd number? Do even numbers multiplied by 3 ever give odd numbers?

THINK DIFFERENTLY In question **3**, children now have three groups as opposed to 3 in each group. This requires children to realise that this is still multiplying 12 by 3. Draw out the fact that 3 groups of 12 is the same as 12 groups of 3.

ASSESSMENT CHECKPOINT Can children represent a question as a multiplication sentence with ×3 and can they use counting in 3s to work out the correct answer to the multiplication?

ANSWERS Answers for the **Practice** part of the lesson appear in the separate **Practice and Reflect answer guide**.

Reflect

WAYS OF WORKING Independent thinking

IN FOCUS Children are presented with 9 × 3 = 27 and are asked to think of a story or situation that matches this. Encourage them to write or draw a story that they have not previously met. Share these as a class. Discuss what is the same and what is different. For example, some children may have three groups, others may have 3 in a group. Discuss if this is still multiplying by 3.

ASSESSMENT CHECKPOINT Check if children know that for multiplying by 3 they can either have three groups or 3 in each group.

ANSWERS Answers for the **Reflect** part of the lesson appear in the separate **Practice and Reflect answer guide**.

After the lesson 

- Can children form multiplication statements involving ×3 from word problems?
- Can children use 10 × 3 = 30 to work out 11 × 3 and 12 × 3?
- Do children know how to use repeated addition and counting up in 3s to multiply by 3?

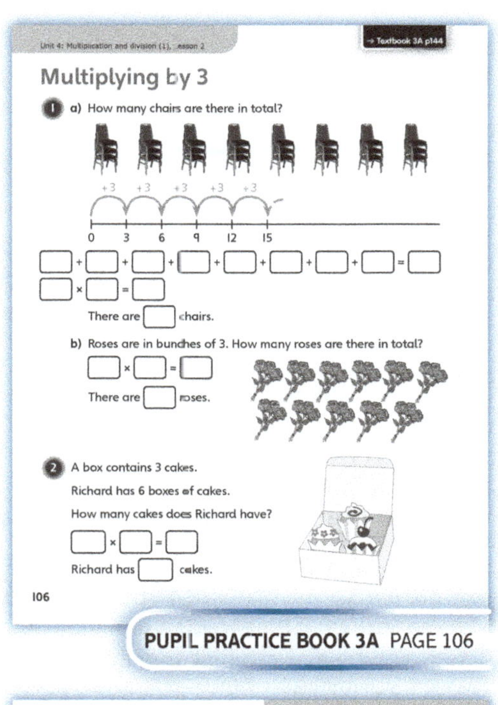

PUPIL PRACTICE BOOK 3A PAGE 106

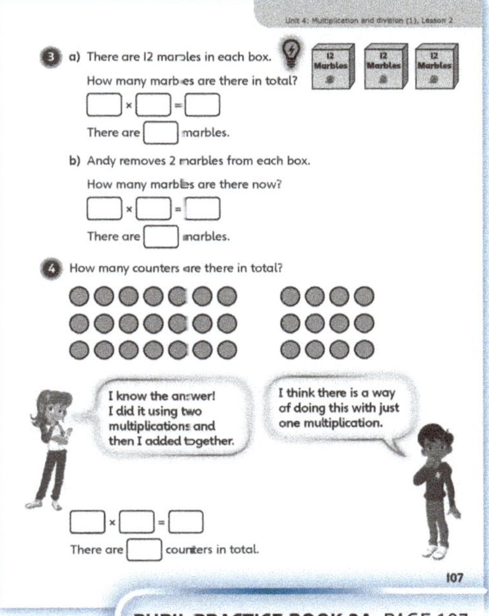

PUPIL PRACTICE BOOK 3A PAGE 107

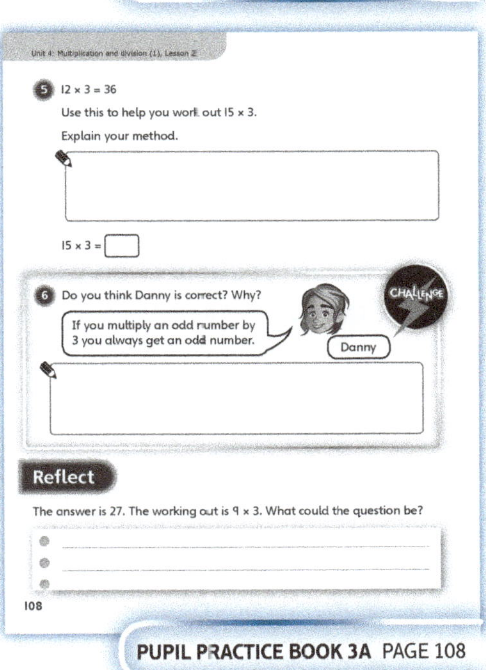

PUPIL PRACTICE BOOK 3A PAGE 108

Dividing by 3

Learning focus

In this lesson, children will start to understand what it means to divide by 3. Children will see that a division statement can be used to represent either equal grouping or sharing.

Small steps

→ Previous step: Multiplying by 3
→ **This step: Dividing by 3**
→ Next step: 3 times-table

NATIONAL CURRICULUM LINKS

Year 3 Number – Multiplication and Division

- Recall and use multiplication and division facts for the 3, 4 and 8 multiplication tables.
- Write and calculate mathematical statements for multiplication and division using the multiplication tables that they know, including for two-digit numbers times one-digit numbers, using mental and progressing to formal written methods.
- Solve problems, including missing number problems, involving multiplication and division, including positive integer scaling problems and correspondence problems in which *n* objects are connected to *m* objects.

ASSESSING MASTERY

Children can form division sentences from a grouping or a sharing situation, and they know the difference between grouping and sharing. Children can use repeated subtraction and counting back in 3s to work out the result of a division, as well as seeing how an array can help them.

COMMON MISCONCEPTIONS

Children may think that dividing always means sharing. Present children with examples to show that it can also mean grouping. Ask:

- *Can you tell me a story where you share and another where you group? What is the difference between your two stories?*

STRENGTHENING UNDERSTANDING

Work with concrete objects to show the difference between grouping and sharing. For grouping, explicitly link each group with the steps in repeated subtraction. Use a number line to show how grouping can be seen as repeated subtraction by counting back in 3s (or subtracting 3). To help understand sharing, first ask children to share the objects one-by-one and then ask children to take three at a time and allocate one to each group. Also ensure that children know what each number in each division statement represents and allow them to see that division can be either grouping or sharing.

GOING DEEPER

Ask children to show two different word problems for $12 \div 3 = 4$. What is the same about their stories? What is different? Can they show that 3 is the number of groups, but that it could also be the amount in each group? Can children use $36 \div 3$ to work out $42 \div 3$?

KEY LANGUAGE

In lesson: division statement, equally, group, method, array, multiplication fact

Other language to be used by the teacher: grouping, sharing, multiply, divide

STRUCTURES AND REPRESENTATIONS

Number lines, arrays

RESOURCES

Mandatory: cubes, counters, number lines

 In the eTextbook of this lesson, you will find interactive links to a selection of teaching tools.

Before you teach 

- Can children count back in 3s from 30 to 0?
- Can children share a set of objects between a given number of groups?
- Can children put a set of objects into groups of a given size?

Discover

 Pair work

ASK

- Question **1** a): *How many cakes do we have? How many cakes go into each box? Is this an example of grouping or sharing? What is the difference? How can you work out how many boxes are needed?*
- Question **1** b): *How many cakes do we have now? How many people are we sharing them between? How can you share them? Can they be shared equally? Can you share them 3 at time? How many does each person get?*

IN FOCUS Question **1** a) is about showing division as grouping. The division sign is given to help them remember what they need to do. Children may approach this in different ways. They are likely to put the cakes into groups of 3 and count them. Try to encourage them to see that grouping can be seen as repeated subtraction. Question **1** b) provides a contrasting example, this time looking at sharing. Children may use different methods: they might share the cakes out 1 at a time or 3 at a time. They should start to understand the two reasons for division (grouping and sharing).

PRACTICAL TIPS Children could use counters to represent the cakes.

ANSWERS

Question **1** a): $18 \div 3 = 6$
6 boxes are needed.

Question **1** b): $27 \div 3 = 9$
Each person gets 9 cakes.

Share

 Whole class teacher led

ASK

- Question **1** a): *Why has Dexter done a repeated subtraction? What is happening at each stage? Can you see why division can be seen as repeated subtraction? What does each number represent in $18 \div 3 = 6$?*
- Question **1** b): *Is this the same as the previous question? Is this sharing or grouping? How do you know? Do you have to share just one at a time? What does each number represent in $27 \div 3 = 9$? What does the division sign represent this time?*

IN FOCUS In question **1** a), children are likely to have grouped objects into 3s. Dexter talks children through this method, showing it as a repeated subtraction. At each stage, model what children are doing. For example: *I put 3 cakes into a box. I have one box now. I have $18 - 3 = 15$ cakes left.* A number line is shown to model each stage of the repeated subtraction. It is important that children see the maths that is going on when they group and not just see that there are 6 groups. Look at the division statement $18 \div 3 = 6$ and ensure that children know what each number represents. Question **1** b) shows sharing the cakes 1 a time and also 3 at a time. Discuss with children if these methods give the same answer and why. Look explicitly at the difference between grouping and sharing. Children need to understand that the first method is grouping and the second is sharing.

PUPIL TEXTBOOK 3A PAGE 148

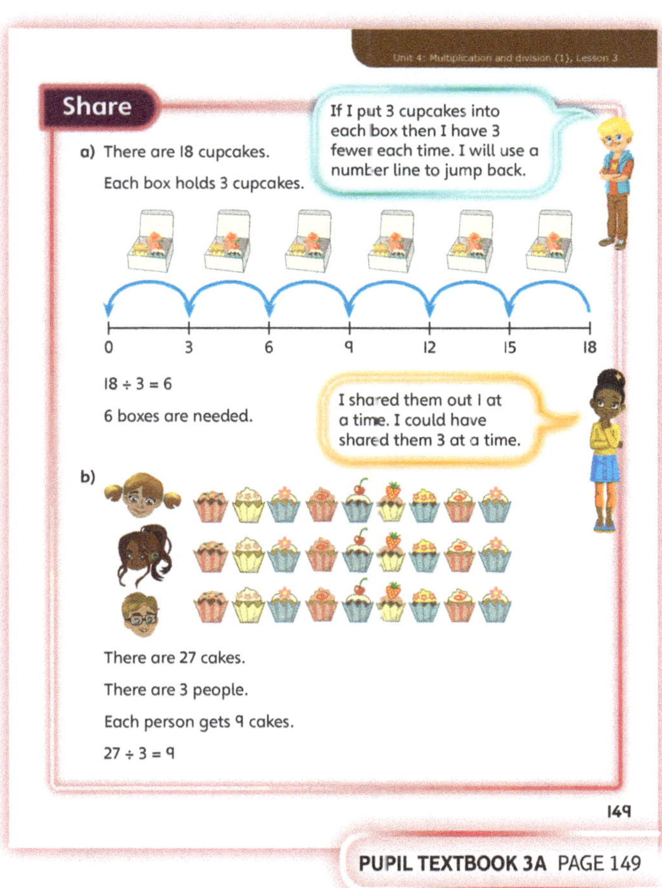

PUPIL TEXTBOOK 3A PAGE 149

Think together

WAYS OF WORKING Whole class teacher led (I do, We do, You do)

ASK

- Question **1** : *How many bread rolls are there? How many are in each pack? Is this grouping or sharing? Can you explain each step of your method?*
- Question **2** : *How many doughnuts are there? How many plates are there? Is this grouping or sharing? What division do you do? How does the array help you get the answer quicker?*
- Question **3** : *Could this be grouping or sharing? What has each person done? How can an array help? What would the array look like? How can you use multiplication to help you? Do you know a different way of getting the answer?*

IN FOCUS Question **1** is an example of grouping. Although the full structure for repeated subtraction is not shown, children may need to go through it. Encourage children to group the objects into 3s. Children may explore whether they can draw an array to help them. Question **2** is an example of sharing. Children may share one at a time or three at time. The doughnuts are presented in a 7 × 3 array. Ask children if the array can help them get the answer to the division. Explain that there are 21 doughnuts and 3 plates (number of rows), and there are 7 doughnuts on each plate (the number in each row). Link this with the division statement. Question **3** shows that a division can be seen as both equal sharing and grouping, and shows different methods that children may use. The last method links to their work on multiplication. Children can see that the answer to the division is the same as ☐ × 3 = 33.

STRENGTHEN For questions **1** and **2**, carry out the grouping and sharing with objects. Explicitly link each action to the steps in repeated subtraction or sharing. Also ensure that children know what each number in each division statement represents and allow children to see that division can be either grouping or sharing.

DEEPEN Ask children to use their division knowledge to answer a missing number multiplication problem. Ask: *How could you use division to work out ☐ × 3 = 36? Can you explain why? Can children make up their own division problems?*

ASSESSMENT CHECKPOINT Can children understand that division can either be equal grouping or sharing as well as form a division statement to answer a question? Can children work out the answer to division statements, practically, using an array or through knowledge of multiplication facts?

ANSWERS

Question **1** : There are 12 bread rolls.
There are 4 packs of bread rolls. 12 ÷ 3 = 4

Question **2** : 21 ÷ 3 = 7
7 doughnuts will go on each plate.

Question **3** : Zac used grouping.
Olivia drew a 3 × 11 array or an 11 × 3 array.
Lee used the fact 11 × 3 = 33 to help him.
You could also use sharing or repeated subtraction.

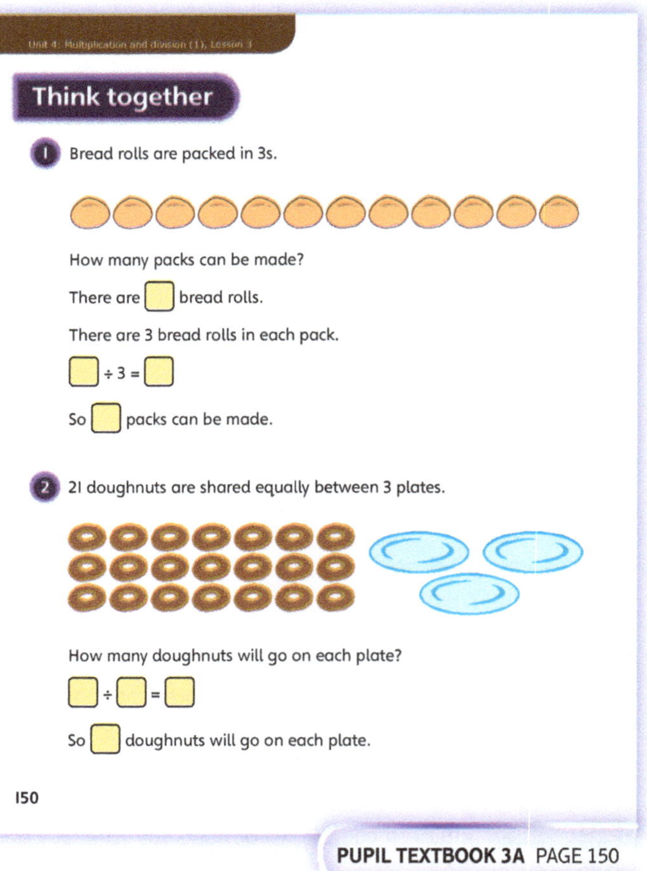

PUPIL TEXTBOOK 3A PAGE 150

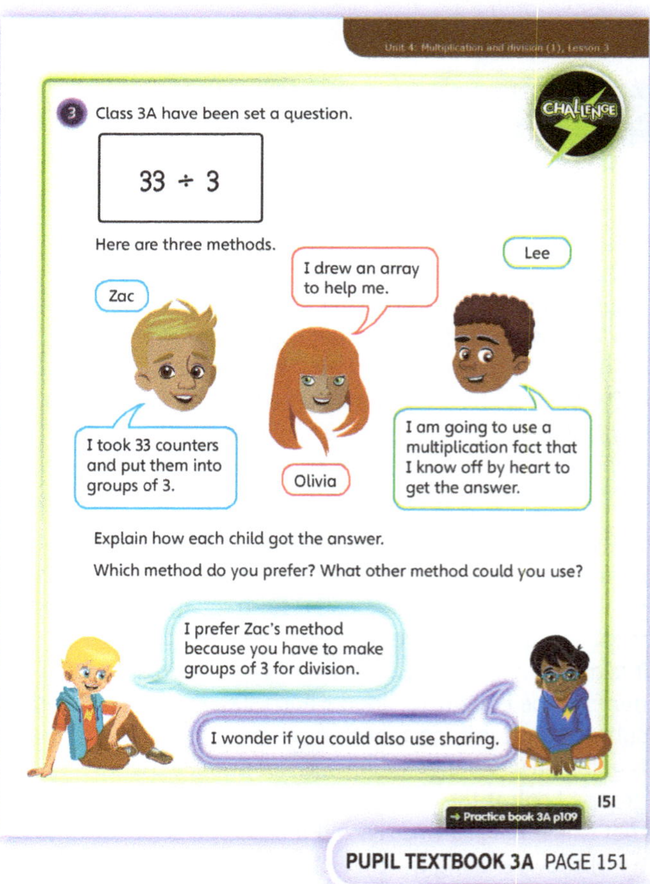

PUPIL TEXTBOOK 3A PAGE 151

Practice

WAYS OF WORKING Independent work

IN FOCUS Question ② starts to look at how arrays can be used to help solve divisions. Children should start to see that if a group of objects can be shown as either three rows or three columns then they can work out the answer to the division.

STRENGTHEN To support dividing by 3, first ensure children understand the difference between equal grouping and equal sharing. Provide simple contexts and ask them to show with counters. For example, in question ①, use 18 counters to represent the grapes and put them into 3 piles. Start by sharing 1 at a time and ask children if they could take 3 at a time instead. Link any division statements clearly with the equipment and the context. For example, in 18 ÷ 3, 18 is the number of grapes; 3 is the number of children. Division in this case means sharing. For grouping, ask children to take 3 counters at a time and show how many groups can be made. Link this to counting back on a number line and repeated subtraction. Again, link the concrete and abstract division statement.

DEEPEN In question ③ b), Ash asks how many more are needed to share the cubes equally. Ask children to use this to work out which numbers divide equally by 3. Do they notice anything about these numbers?

THINK DIFFERENTLY Question ④ provides children with a division statement and asks them to work out another. Look for children who start again from 0. Ask: *Can we start with the previous answer?* Support them in understanding what 36 ÷ 3 means. Ask: *How many more do you have now? If you are grouping, how many more groups have you got? Is this the same for sharing? How would this change the answer?*

ASSESSMENT CHECKPOINT Can children represent a question as a division sentence involving ÷ 3 and use their knowledge of counting back in 3s to work out the answer? Can children understand the difference between sharing and grouping, and know that the division sign can be used to represent both?

ANSWERS Answers for the **Practice** part of the lesson appear in the separate **Practice and Reflect answer guide**.

Reflect

WAYS OF WORKING Pair work

IN FOCUS After independently working out their explanation they should explain to their partner. Encourage children to use concrete objects to help them. Check whether they use a grouping or a sharing strategy. Ask them if they can also demonstrate the other method.

ASSESSMENT CHECKPOINT Check if children know that for dividing by 3 they can either group or share, and that they know the differences between the methods, but also know that the answer to the division will be the same.

ANSWERS Answers for the **Reflect** part of the lesson appear in the separate **Practice and Reflect answer guide**.

After the lesson 

- Can children form division statements involving grouping or sharing?
- Do children know the difference between division as grouping and division as sharing?
- Do children know how to use repeated subtraction and counting down in 3s to divide by 3? Do they know how an array can help?

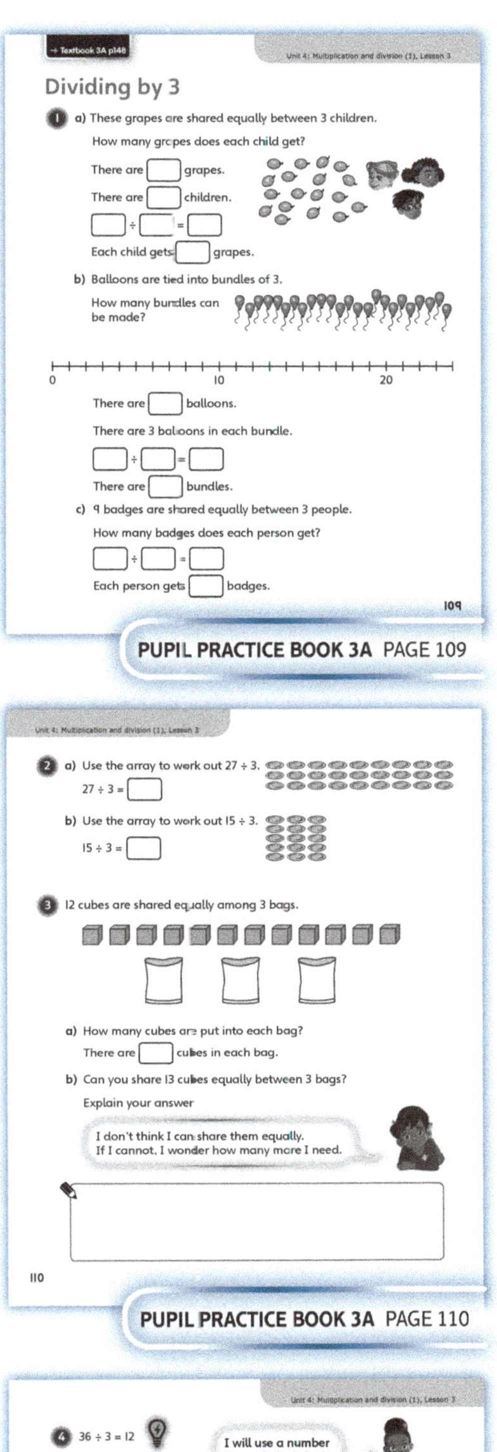

PUPIL PRACTICE BOOK 3A PAGE 109

PUPIL PRACTICE BOOK 3A PAGE 110

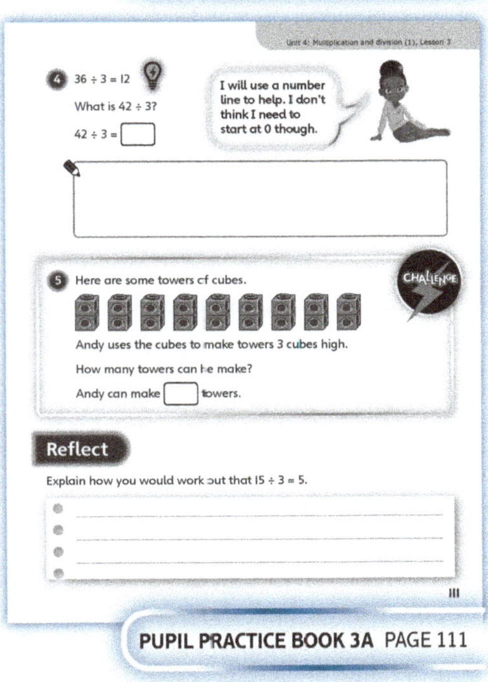

PUPIL PRACTICE BOOK 3A PAGE 111

3 times-table

Learning focus

In this lesson, children will focus on the 3 times-table. They will realise that the 3 times-table contains multiplication facts and should be able to recall associated division facts from multiplication facts.

Small steps

→ Previous step: Dividing by 3
→ **This step: 3 times-table**
→ Next step: Multiplying by 4

NATIONAL CURRICULUM LINKS

Year 3 Number – Multiplication and Division

- Recall and use multiplication and division facts for the 3, 4 and 8 multiplication tables.
- Write and calculate mathematical statements for multiplication and division using the multiplication tables that they know, including for two-digit numbers times one-digit numbers, using mental and progressing to formal written methods.
- Solve problems, including missing number problems, involving multiplication and division, including positive integer scaling problems and correspondence problems in which *n* objects are connected to *m* objects.

ASSESSING MASTERY

Children can recognise a multiplication fact from a given image and begin to develop a rapid recall of multiplication facts and associated division facts from the 3 times-table.

COMMON MISCONCEPTIONS

Children think that $0 \times 3 = 3$. This is a common mistake and it is mainly because children have not thought about the answer and instead try to recall it too quickly. Remind children that 0 × anything always gives the answer 0. Ask:

- *Biscuits are in groups of 3. There is one group of biscuits, how many biscuits? There are no groups of biscuits, how many biscuits now?*

To work out if 9×3 is greater than or equal to 7×3, children may think you have to work out each multiplication fact, but this is not the case. They could draw the groups in a straight line to compare them. Ask:

- *Can you show me why 9 groups of 3 must be greater than 7 groups of 3, without working them out?*

STRENGTHENING UNDERSTANDING

To strengthen understanding of times-table facts show them as a visual or concrete representation. For example, 1×3 could be represented by a line of 3 counters, 2×3 as two lines of 3 counters, etc. Link this to counting in 3s on a number line.

At this stage, it is fine if some children still need to use cubes or counters to represent multiplications and divisions. What is important is that children can see what the times-table facts are. Reinforce that they need to eventually develop rapid recall of the multiplication and division facts, but also need to understand what they are.

GOING DEEPER

Can children reason why $8 \times 3 > 7 \times 3$ without working out the two multiplications? Can they work out missing numbers to make statements correct, for example $12 \div 3 > \bigcirc \div 3$. What patterns can they see on a 100-square if they shade the numbers from the 3 times-table? What do they notice if they add up these digits?

KEY LANGUAGE

In lesson: times-table, multiply (×), **divide** (÷), group, multiplication fact, division, total, method

Other language to be used by the teacher: recall, number line, pattern

STRUCTURES AND REPRESENTATIONS

Number lines, arrays

RESOURCES

Mandatory: cubes, counters, number lines

Optional: times-tables written on stairs

 In the eTextbook of this lesson, you will find interactive links to a selection of teaching tools.

Before you teach 

- Can children count on and back in 3s?
- Do they understand the link between multiplication and division?

Discover

 Pair work

ASK

- Question ① a): *What times-table is shown here? How do you know? What do you notice about the answers to the multiplications? How do they link to the divisions? Can you convince me that 5 × 3 = 15?*
- Question ① b): *What different 3 times-table facts can you see here? What can the fact help you work out? Can you draw or write a statement to show 10 × 3? How can you work out how many candles in total?*

IN FOCUS Question ① a) asks children to work out missing multiplication and division facts. Children may use their knowledge from the previous two lessons where they were multiplying and dividing by 3. Question ① b) asks children to identify the relevant times-table fact shown by the image. This is important in helping children see that times-table facts are not just abstract representations.

PRACTICAL TIPS If you have the times-table facts on a poster in your classroom cover them up. Times-tables written on stairs are becoming increasingly common and could be something that your school considers.

ANSWERS

Question ① a): 18 ÷ 3 = 6   11 × 3 = 33
               9 ÷ 3 = 3   4 × 3 = 12

Question ① b): 5 × 3 = 15

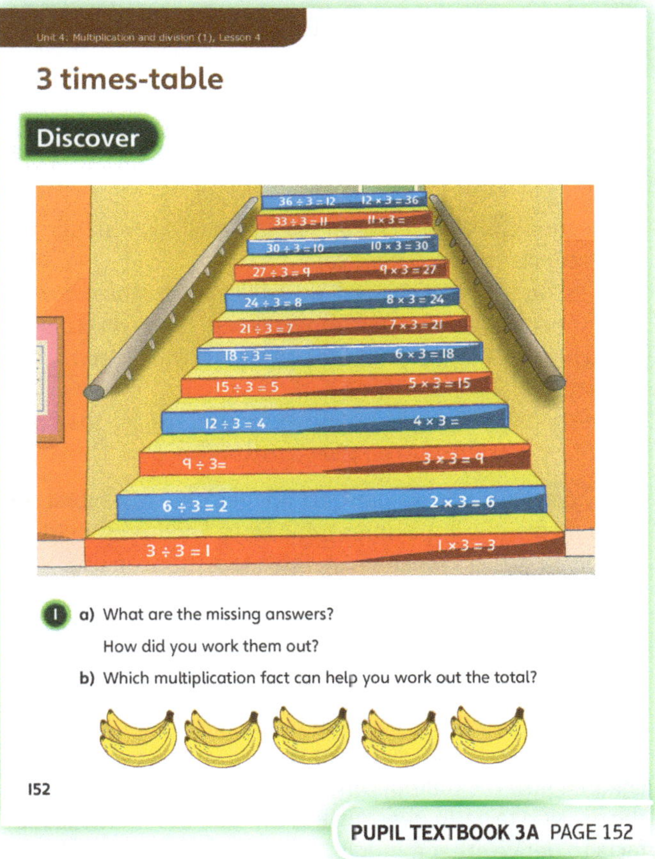

PUPIL TEXTBOOK 3A PAGE 152

Share

 Whole class teacher led

ASK

- Question ① a): *How can you show 4 × 3? What does 4 × 3 look like as an array? What method did you use to work out the answer? How did you work out the divisions?*
- Question ① b): *What times-table fact is shown? What does it help you work out? Could you write this another way?*

IN FOCUS In question ① a) arrays are shown as a visual representation of each question. To work out the answer, children may need to count up in 3s. For the divisions, this is shown as grouping. Most children understand division as grouping when it is read out, as teachers often say 'how many 3s go into 15?' Dexter shows how many groups of 3 are in each number and how this leads to the answer. For question ① b), children interpret the visual representation as a multiplication fact. Discuss with children that they can use the 3 times-table to work out totals if you know there are 3 in each group. Children should be encouraged to start to remember the times-table facts.

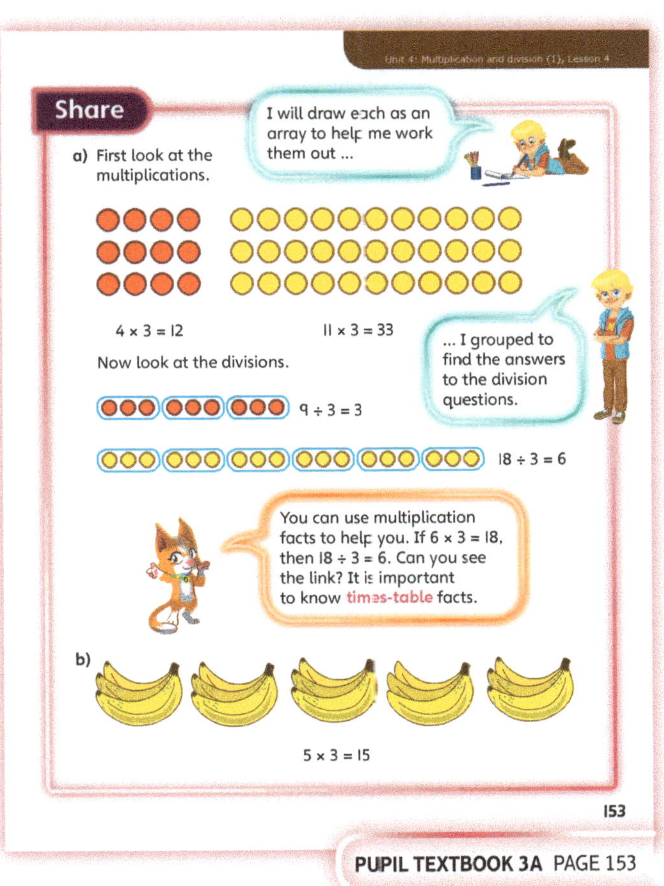

PUPIL TEXTBOOK 3A PAGE 153

Think together

 Whole class teacher led (I do, We do, You do)

ASK

- Question **1**: *What facts can you see? How can you use these facts to work out the total?*
- Question **2**: *What answers did you know straight away? How many did you complete in a minute? Which ones did you need help with? What can you do to help you learn them?*
- Question **3**: *How can you use the answer to 6 × 3 to work out 12 × 3? What else can you do to work out 12 × 3?*

IN FOCUS Question **1** presents images where the total can be worked out using 3 times-table facts. Children should be able to explain which times-table fact they can use. Encourage them to talk about the number of groups and the number of objects in each group. Question **2** aims to develop fluency with times-table facts by asking how fast children can find the answers. Explain that it is important for children to develop speed of recall with times-table facts and that they should also learn the division facts. Show children how the multiplication facts can be used to derive the division facts. In question **3**, children use multiplication facts to derive other multiplication facts such as using the fact that 12 × 3 is 10 × 3 plus 2 × 3, or that 12 × 3 is double 6 × 3. This might help children with strategies for working out times-tables facts that they do not know.

STRENGTHEN To strengthen understanding of times-table facts, show them as a visual or concrete representation such as 1 × 3 represented by a tower of 3 cubes, or 2 × 3 by 2 towers of 3 cubes. It is important for children to see what times-table facts are. Reinforce that they need to develop rapid recall of the multiplication and division facts, but also need to understand what they are.

DEEPEN For question **3**, ask: *If you know that 8 × 3 = 24, how many division facts can you show? What other multiplication facts can you work out from this?*

ASSESSMENT CHECKPOINT Have children begun to know the times-table multiplication and division facts using rapid recall? Can they work out which fact will help them work out the total, or the answer to a division question?

ANSWERS

Question **1** a): 4 × 3 = 12

Question **1** b): 11 × 3 = 33

Question **2**:

33	27	0	4
21	24	12	6
10	6	8	12

Question **3** a): Double 6 × 3
10 × 3 + 2 × 3

Question **3** b): 3 × 3 × 3 = 27 (use 9 × 3)
13 × 3 = 39 (add 3 to the answer to 12 × 3)
3 × 20 = 60 (double the answer to 10 × 3)

Think together

1 Use multiplication facts to work out how many items there are in each picture.

Which fact did you use to find each total?

a)   b)

☐ ○ ☐ = ☐   ☐ ○ ☐ = ☐

2 How many of these can you work out in a minute?

11 × 3	9 × 3	0 × 3	12 ÷ 3
3 × 7	8 multiplied by 3	☐ × 3 = 36	multiply 2 by 3
divide 30 by 3	3 × ☐ = 18	24 ÷ 3	number of questions in this grid

It is not always important to be quick, but knowing your times-table facts can help save time.

154

PUPIL TEXTBOOK 3A PAGE 154

3

a) Find two different methods for Luis to use to work out 12 × 3.

Which method do you prefer?

Which method is quicker?

b) How can you work these out using the 3 times-table?

3 × 3 × 3   13 × 3   3 × 20

155

→ Practice book 3A p112

PUPIL TEXTBOOK 3A PAGE 155

Practice

WAYS OF WORKING Independent thinking

IN FOCUS Question **2** checks if children know their times-table facts. Children should try to do these without referring back and should rely on mental recall. For those who are unable to do this, ask if they can work any out using other times-table facts. If they are still unable to work them out, then ask them to go back to counting up in 3s. In question **3**, children might need to work out the answers to compare them to 21, or they may develop a reasoning strategy. For example, if they know 8 × 3 is greater than 21 then 9 × 3 must also be. Question **5** asks children to put in inequality or equals signs to make statements correct. Encourage children to use reasoning instead of working out all the answers. Question **6** looks at the pattern that the 3 times-table makes on a 100 square.

STRENGTHEN To strengthen understanding of times-table facts, show them as a visual or concrete representation. Question **1** will help to strengthen this. As children move through the rest of the exercise, the questions become more abstract so they may still need to use cubes or counters to represent multiplications and divisions. It is important for children to see what times-table facts are. Reinforce that they need to develop rapid recall.

DEEPEN Ask children if they had to work out each answer in question **3**. If so, ask: *Is there another way you can do it? Which times-table fact do you need to know to start?* For question **4**, ask children to explain how and why they chose the symbol for each statement. After shading the 100-square in question **6**, can children come up with a rule for numbers that are in the 3 times-table? What do they notice if they add up all the digits?

ASSESSMENT CHECKPOINT Can children recognise 3 times-table facts from images? Children should start to develop a secure recall of the 3 times-table (both multiplication and associated division facts). Encourage children to practise at home, and build in daily practice in school if necessary.

ANSWERS Answers for the **Practice** part of the lesson appear in the separate **Practice and Reflect answer guide**.

Reflect

WAYS OF WORKING Pair work

IN FOCUS Children should try to work out the answers mentally and as quickly as possible to encourage rapid recall. Ask children to check each other's answers. Give support to any who are struggling. Ask children to make a note of any facts that they are consistently getting incorrect and focus on them.

ASSESSMENT CHECKPOINT Check if children have instant recall of 3 times-table facts up to 12 × 3. Look to see if there are any that they are consistently getting incorrect.

ANSWERS Answers for the **Reflect** part of the lesson appear in the separate **Practice and Reflect answer guide**.

After the lesson 

- Can children recall division and multiplication facts from the 3 times-table?
- Which facts are children struggling with?
- Do children know which multiplication or division fact they need to know in order to work out a total?

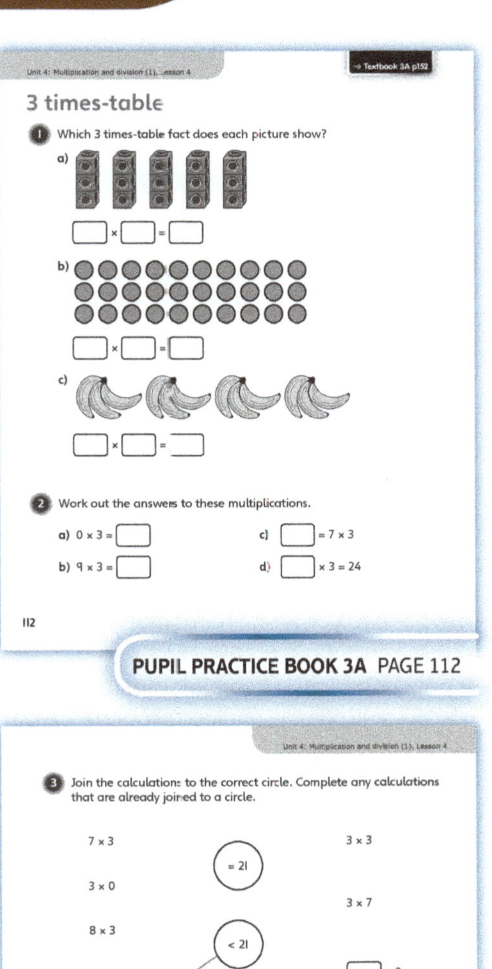

PUPIL PRACTICE BOOK 3A PAGE 112

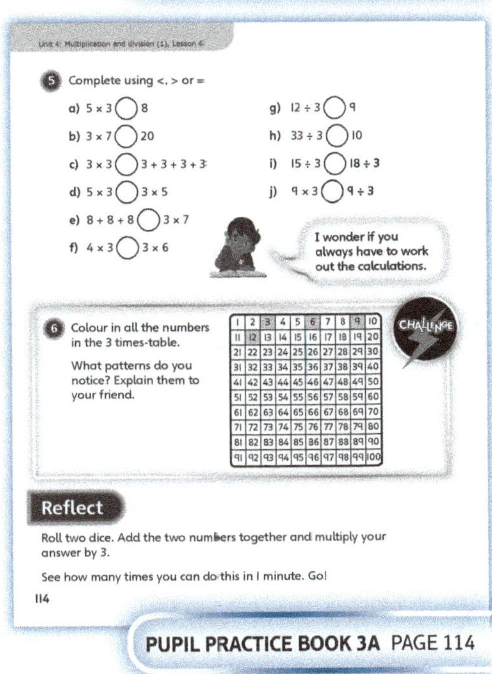

PUPIL PRACTICE BOOK 3A PAGE 113

PUPIL PRACTICE BOOK 3A PAGE 114

Multiplying by 4

Learning focus

In this lesson, children will start to understand what it means to multiply by 4. Children will see the link between repeated addition, counting up in 4s and multiplying by 4.

Small steps

→ Previous step: 3 times-table
→ **This step: Multiplying by 4**
→ Next step: Dividing by 4

NATIONAL CURRICULUM LINKS

Year 3 Number – Multiplication and Division

- Recall and use multiplication and division facts for the 3, 4 and 8 multiplication tables.
- Write and calculate mathematical statements for multiplication and division using the multiplication tables that they know, including for two-digit numbers times one-digit numbers, using mental and progressing to formal written methods.
- Solve problems, including missing number problems, involving multiplication and division, including positive integer scaling problems and correspondence problems in which *n* objects are connected to *m* objects.

ASSESSING MASTERY

Children can form a multiplication sentence involving multiplying by 4. Children can work out the answer to a multiplication sentence by knowing its link with repeated addition and using a number line to count up in 4s.

COMMON MISCONCEPTIONS

When children are presented with four groups of objects as opposed to 4 objects in groups, they may not see this as multiplying by 4. For example, the language used so far is that 12 × 4 is linked to 12 groups of 4. When children are presented with 4 groups of 8, they may only see this as 4 × 8 and so may not see the connection with 8 × 4. Show an array with 4 rows of 8. Work out the total and then rotate the array. Ask:
- *What is the total now? What has changed? What is still the same?*

Some children start the count from 0 when working out 11 × 4, for example, even if they know 10 × 4. Ask:
- *What × 4 fact do you already know? How could that help you here?*

STRENGTHENING UNDERSTANDING

Ensure that you show the direct link between the objects in a picture, a repeated addition, counting up in 4s on a number line and the multiplication statement. As children count up in 4s, put counters on the number line to show each group.

GOING DEEPER

Can children use multiplication facts that they know in order to work out other multiplication facts? For example, can they use 12 × 4 to work out 13 × 4, or 24 × 4? They could start to explore how to work out 2 × 4 + 5 × 4 as one multiplication, showing by adding arrays together that this is the same as 7 × 4.

KEY LANGUAGE

In lesson: multiply (×), multiplication statement, total, double, count up

Other language to be used by the teacher: multiplication sentence, equal groups, *x* groups of *y*, times-table

STRUCTURES AND REPRESENTATIONS

Number lines, arrays

RESOURCES

Mandatory: cubes, counters, number lines

Optional: plastic animals

 In the eTextbook of this lesson, you will find interactive links to a selection of teaching tools.

Before you teach 

- Can children count in 2s?
- Do children know their 2 times-table?
- Do children know the link between repeated addition and the answer to a multiplication?

Discover

WAYS OF WORKING Pair work

ASK

- Question **1** a): *How many groups (donkeys) do you have? How many (legs) are there in each group? How can we write this as a multiplication statement? How can we work out the total? How do you get from 6 × 2 to 6 × 4? What is the connection?*
- Question **1** b): *How much does each person pay? How much do they pay in total? What is the multiplication statement you have worked out?*

IN FOCUS Question **1** a) asks children to find the number of legs on 6 donkeys. Children should be able to write this as a multiplication calculation and should start to understand this as 6 groups of 4, which can be worked out using 6 × 4. Children may use different methods for working this out; try to encourage efficient methods as much as possible. Some children may work out 6 × 4 by counting in 2s and will see the connection between multiplying by 2 and by 4. Children should start to realise that an effective way for working out the total is to count up in 4s or to double the number and double again. Question **1** b) asks a similar question, but this time an answer is given and children have to reason whether it is correct. They could use counters to represent the coins.

PRACTICAL TIPS For this activity, you can use plastic animals to represent the donkeys. Ensure there are counters and cubes available for each child.

ANSWERS

Question **1** a): 6 × 4 = 24
There are 24 donkey legs.

Question **1** b): 5 × 4 = 20
This is the correct amount.

Share

WAYS OF WORKING Whole class teacher led

ASK

- Question **1** a): *Why do we count in 4s and not in 1s? Can we count in 2s to get the answer? Why can this be seen as a multiplication?*
- Question **1** b): *How does the multiplication relate to the array?*

IN FOCUS For question **1** a) encourage children to share their methods with the rest of the class. Astrid encourages children to find a more efficient way of finding the total rather than counting in 1s. Counting in 4s is used to show how children might have arrived at the answer. Explain that the situation can be seen as 6 groups of 4 and therefore can be written as a multiplication calculation. The method of multiplying by 2 and by 2 again is not covered until later, but you may find that you want to discuss it here. Question **1** b) shows a multiplication in the form of a 5 × 4 array and shows how the total is 20. Children should understand how an array links to a multiplication.

PUPIL TEXTBOOK 3A PAGE 156

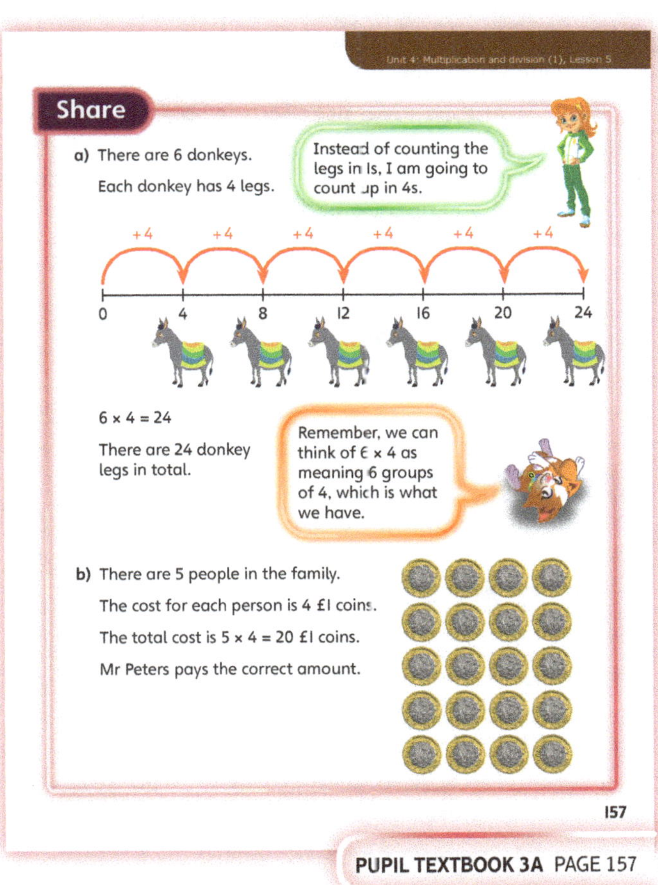

PUPIL TEXTBOOK 3A PAGE 157

Think together

WAYS OF WORKING Whole class teacher led (I do, We do, You do)

ASK

- Question **1**: *How many groups are there? How many are in each group? If we count from 0, where do we get to? How many 4s are we adding together? What multiplication is this? What is the answer? Could we have used an answer from earlier? How?*
- Question **2**: *How can we work out the number of donkeys? Do we count in 4s again? Where do we stop? Why is it 8 donkeys? Could we have got the answer quicker by using the answer to question* **1***?*
- Question **3**: *Is Ebo right? Can you use Ebo's method to work out the other number multiplied by 4? Why does Ebo's method work?*

IN FOCUS Question **1** builds on the **Discover** section by adding an extra donkey. This reinforces the link between counting up in 4s (repeated addition) and multiplying by 4. The question starts by counting in 4s from 0 again. Discuss with children how they can use their answer from **Discover** to get the answer by simply counting on another 4. In question **2**, children need to see this as counting up in 4s to 32, as opposed to a division. They should point to the number line and count up in 4s as they go. Encourage them to show why it is 8 donkeys. Question **3** draws out the method of multiplying by 2 and then multiplying by 2 again to multiply by 4. Children should follow Ebo's method and then apply it to other multiplications. Children are asked to use diagrams or equipment to explain why multiplying by 2 and 2 again is the same as multiplying by 4.

STRENGTHEN At each stage, use concrete objects alongside a number line to help children see the link with repeated addition, counting up in 4s and the eventual multiplication. In question **1**, you might want to encourage children to put 4 counters at each point on the number line to show that this represents a donkey. You may find it easier to support multiplying by 4 initially by multiplying by 2 and then multiplying by 2 again.

DEEPEN In question **3**, can children explain why multiplying by 2 and then by 2 again is equal to multiplying by 4? Ask: *How can you use 10 × 4 to work out 20 × 4, or 14 × 4?*

ASSESSMENT CHECKPOINT Can children understand what it means to multiply by 4 and use their knowledge of counting in 4s to work out the answers to the repeated addition and multiplication statements that they form?

ANSWERS

Question **1** a): 7 × 4 = 28
There are 28 donkey legs.

Question **1** b): 4 × 4 = 16
It costs 16 coins in total to go donkey trekking.

Question **2**: There are 8 donkeys because 8 × 4 = 32.

Question **3**: 9 × 4 = 36
Yes, Ebo's method always works.

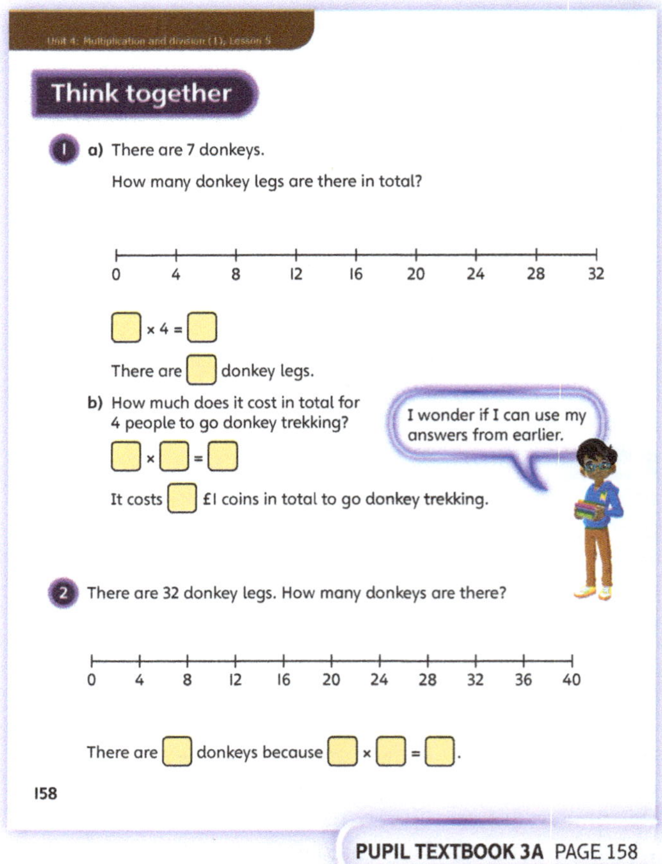

PUPIL TEXTBOOK 3A PAGE 158

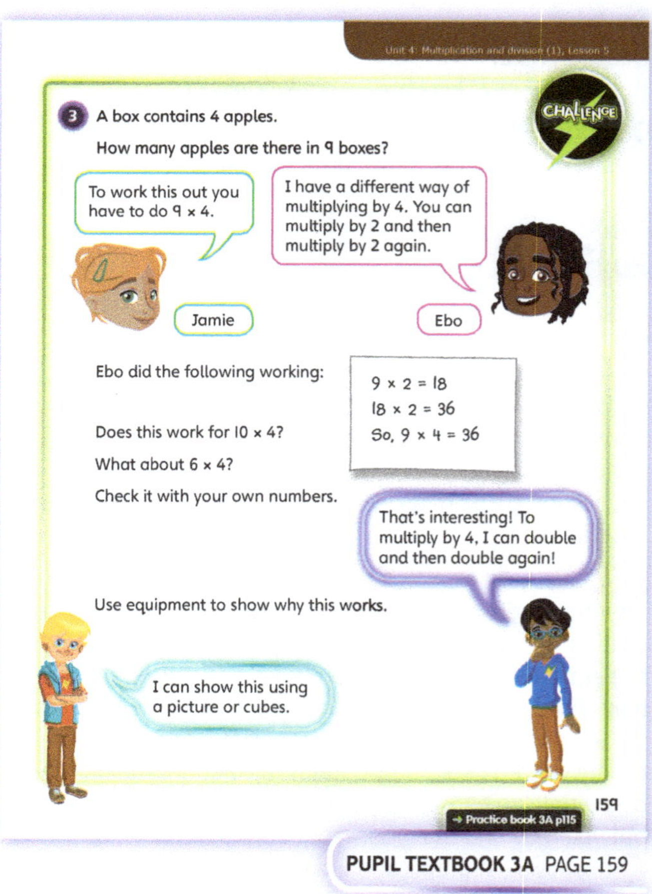

PUPIL TEXTBOOK 3A PAGE 159

Practice

WAYS OF WORKING Independent thinking

IN FOCUS Question ❶ provides practice for children to form the correct multiplication statements from given situations. Encourage children to use a number line and to count up in 4s to help them work out the answers. Some children may use the answer to part a) to help them with part b). Questions ❸ and ❹ provide contexts and the units involved are for measure and money. Children should write down the correct multiplication statement from the context given. Children may already know some of the 4 times-table facts off by heart and so will be able to just write down the answer. You might want to encourage checking of answers.

STRENGTHEN To support multiplying by 4, use concrete objects alongside a number line. Count aloud with children in 4s from 0, each time showing the jump on the number line. If children are confident with multiplying by 2, or can double, then you might want to encourage children to double and double again in order to multiply by 4.

DEEPEN In question ❹ c), how many different ways can children write their answer? For example, do they write it as an addition or can they write it as a single multiplication? Question ❻ is similar, but this time the problem is not separated into parts. Encourage children to attempt the question in their own way. If children have worked out two separate multiplications and added them, ask if there is a quicker way.

THINK DIFFERENTLY Question ❺ asks children to use the method of doubling and doubling again to multiply by 4. Children have already looked at this method in the **Think together** section. Children may not be able to deal with the numbers presented here mentally and so you may want to ask them to use a written method.

ASSESSMENT CHECKPOINT Can children represent a question as a multiplication sentence involving × 4 and can they use counting in 4s to work out the correct answer to the multiplication? Do children know that multiplying by 2 and multiplying by 2 again is the same as multiplying by 4?

ANSWERS Answers for the **Practice** part of the lesson appear in the separate **Practice and Reflect answer guide**.

Reflect

WAYS OF WORKING Independent thinking

IN FOCUS They may approach this by explaining a specific example or by explaining a generic method. Children may discuss how they could use a number line to count up in 4s or may use the method of doubling and doubling again. After explaining a method, ask them to apply it.

ASSESSMENT CHECKPOINT Check if children have an appropriate method for multiplying by 4.

ANSWERS Answers for the **Reflect** part of the lesson appear in the separate **Practice and Reflect answer guide**.

After the lesson ⏸

- Can children form multiplication statements involving × 4 from a word problem?
- Do children know how to work out an answer to a multiplication by counting in 4s?
- Do children know the method of doubling and doubling again?

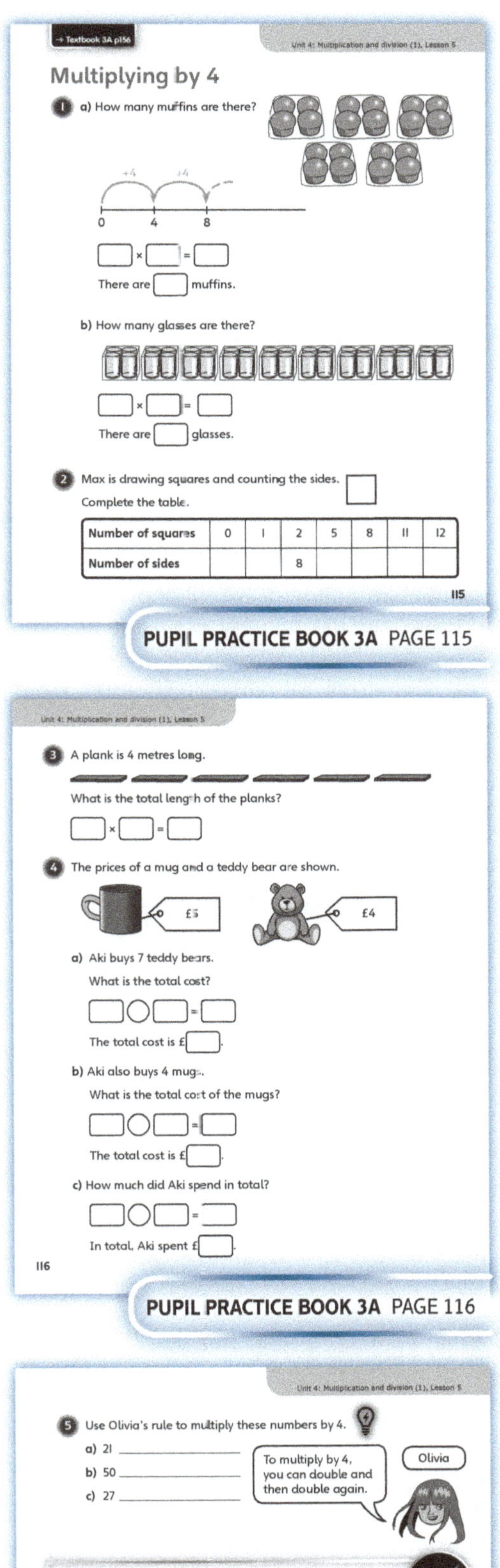

Number of squares	0	1	2	5	8	11	12
Number of sides			8				

PUPIL PRACTICE BOOK 3A PAGE 115

PUPIL PRACTICE BOOK 3A PAGE 116

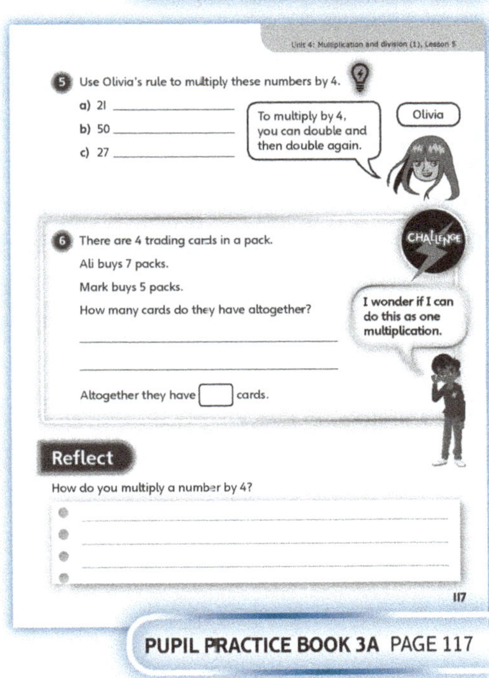

PUPIL PRACTICE BOOK 3A PAGE 117

Dividing by 4

Learning focus

In this lesson, children will start to understand what it means to divide by 4. Children will understand that division statements can be used to represent either equal grouping or equal sharing.

Small steps

→ Previous step: Multiplying by 4
→ **This step: Dividing by 4**
→ Next step: 4 times-table

NATIONAL CURRICULUM LINKS

Year 3 Number – Multiplication and Division

- Write and calculate mathematical statements for multiplication and division using the multiplication tables that they know, including for two-digit numbers times one-digit numbers, using mental and progressing to formal written methods.
- Recall and use multiplication and division facts for the 3, 4 and 8 multiplication tables.
- Solve problems, including missing number problems, involving multiplication and division, including positive integer scaling problems and correspondence problems in which *n* objects are connected to *m* objects.

ASSESSING MASTERY

Children can form a division sentence from either a grouping or a sharing situation as they know the difference between grouping and sharing. Children are able to use repeated subtraction and counting back in 4s to work out the result of a division; and they know that a method for dividing by 4 is to halve and halve again.

COMMON MISCONCEPTIONS

Children may think that ÷ always means sharing. Present children with examples to show that it can also mean grouping. Ask:
- *Can you tell me a story where you share and another where you group? What is the same in your stories? What is different?*

When dividing by 4, children may lose track of the count forwards or backwards and therefore arrive at the wrong answer. Explain that children need to commit the multiplication facts and associated division facts to memory. Ask:
- *Have you checked your answer? Do you know where your mistake came from? Is there a quicker way to find the answer?*

STRENGTHENING UNDERSTANDING

Arrays may help children find the answers to divisions. If the problem asks to divide by 4 and they are sharing, children can put objects into 4 columns; if they are grouping, they should put 4 in a column. The number of rows and columns formed will give the answer to the division.

GOING DEEPER

Ask children to show two different word problems for 12 ÷ 4 = 3. What is the same about their stories? What is different? Can children use 48 ÷ 4 to work out 52 ÷ 4? Ask children to solve missing number problems such as, ___ × 4 = 44.

KEY LANGUAGE

In lesson: divide (÷), division statement, group, share, shared equally, array, left over

Other language to be used by the teacher: equal, count back

STRUCTURES AND REPRESENTATIONS

Number lines, arrays

RESOURCES

Mandatory: cubes, counters, number lines
Optional: 48 playing cards

 In the eTextbook of this lesson, you will find interactive links to a selection of teaching tools.

Before you teach 

- Can children count back in 4s from 40 to 0?
- Can they share a set of objects between a given number of groups?
- Can they put a set of objects into groups of a given size?

Discover

 Pair work

ASK

- Question **1** a): *How many cards are there? How many cards are shared? How many people are we sharing them between? How many does each person receive? Is this equal grouping or sharing? How can you write it as a division?*
- Question **1** b): *How many cards are left over? How did you work this out? How many cards are going into each pile? Is this equal sharing or grouping? How many piles do you get? Can you write this as a division?*

IN FOCUS Question **1** a) is about showing division as equal sharing. Children may share cards, cubes or counters out equally to imitate sharing the 20 cards between 4 people. Ask them how they could record their working. For example, they may draw 20 cards and 4 people and draw arrows from each card to a person. They may share 1 at a time or they may take 4 cards at a time and give 1 to each person. Question **1** b) provides a contrasting example and this time looks at grouping. There is an additional step in this question as children first have to work out how many left-over cards there are. They may approach this by counting the ones remaining or by doing a subtraction, which is the method that should be encouraged. This activity should reinforce the different instances where the division sign is used (grouping and sharing).

PRACTICAL TIPS Put children into groups of 4 and play a game using 48 cards to help them understand the context and the problem.

ANSWERS

Question **1** a): $20 \div 4 = 5$
Each player gets 5 cards.

Question **1** b): $48 - 20 = 28$
$28 \div 4 = 7$
There are 7 piles.

Share

 Whole class teacher led

ASK

- Question **1** a): *How will Dexter share out the cards? Is there another way? Do they have equal groups? How do you know?*
- Question **1** b): *How can you work out how many cards are left over? What is the difference between this question and the one before? Why do you still use the division sign? What does each number in the division mean?*

IN FOCUS For question **1** a), ensure that the method of sharing is demonstrated and explain each number in the division calculation. It is important to link each number in the calculation to what it means in the image. The most efficient method for sharing would be to take 4 cards at once, with 1 given to each child. Explain that children could also have shared the cards 1 by 1, like Dexter. Question **1** b) shows a method for grouping. Children first need to find the left-over cards. Discuss methods that they may have used for this. Children need to understand the methods for grouping and sharing.

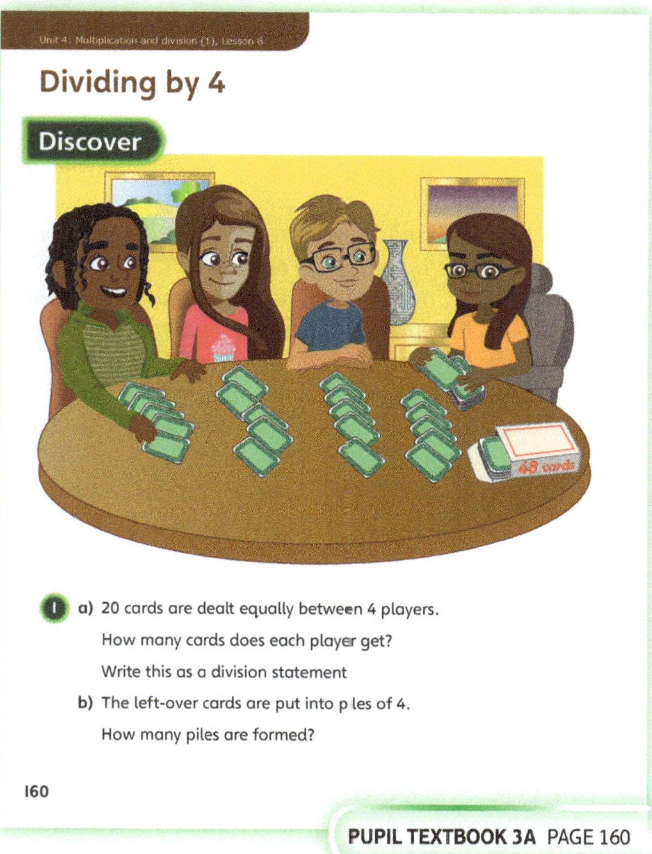

PUPIL TEXTBOOK 3A PAGE 160

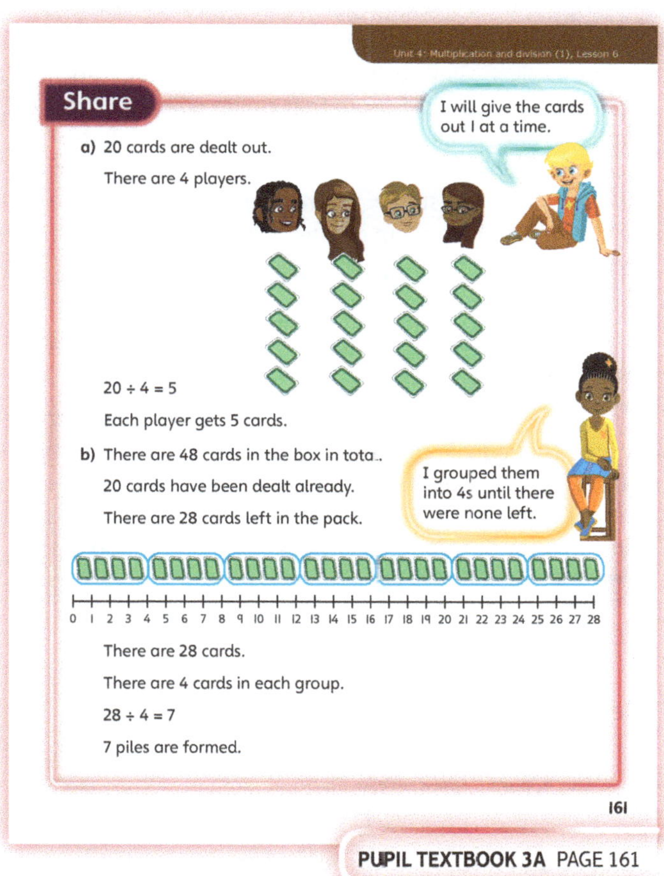

PUPIL TEXTBOOK 3A PAGE 161

Think together

WAYS OF WORKING Whole class teacher led (I do, We do, You do)

ASK

- Question **1** : *How many coins are there? How many money boxes are there? Is this grouping or sharing? Can you explain each step of your method? What does the division look like? What does each number in the division statement mean?*
- Question **2** : *How many flowers are there? How many flowers are put into a bunch? Is this grouping or sharing? What division do you do? Can you explain what each number in the division means?*
- Question **3** : *What is Jamie's method? Can you explain why Jamie's method works? Can you share 44 marbles equally? Why can you not share 22 marbles equally between 4 boxes? How many have you got left over? How many more would you need to share equally?*

IN FOCUS Question **1** is an example of equal sharing. Children may share out 1 at a time or take a group of 4. Children should be confident with sharing and may go straight to the division. Question **2** is an example of equal grouping. Children may circle the flowers with their finger to work out how many groups of 4. As a class, discuss that the flowers may be grouped in different ways, but the same number of groups remains. In both questions, link the numbers in the division to the context. Question **3** explores the method of halving and halving again as an alternative for dividing by 4: children may find this easier.

STRENGTHEN For questions **1** and **2**, carry out the grouping and sharing with counters to represent the objects. Explicitly link each action with the steps in repeated subtraction or sharing. Ensure that children know what each number in each division statement represents. Children may find making an array with 4 counters in each row an effective way. In question **3**, some children may need to use counters to help them do the division.

DEEPEN Ask children how they could use division to work out the missing number problem ___ × 4 = 36? Ask: *Can you explain why you use division? Can you make up your own problems similar to this?*

ASSESSMENT CHECKPOINT Children understand that division can be either equal grouping or equal sharing and are able to form a division statement to answer a question. They can also work out the answer to division statements, practically, using an array or through knowledge of multiplication facts.

ANSWERS

Question **1** : 12 ÷ 4 = 3
There are 3 coins in each money box.

Question **2** : 32 ÷ 4 = 8
There are 8 bunches.

Question **3** a): 44 ÷ 4 = 11
There are 11 marbles in each box.

Question **3** b): You cannot share 22 marbles equally between 4 boxes. There are 2 marbles too many.
You would need 2 more marbles to share equally.

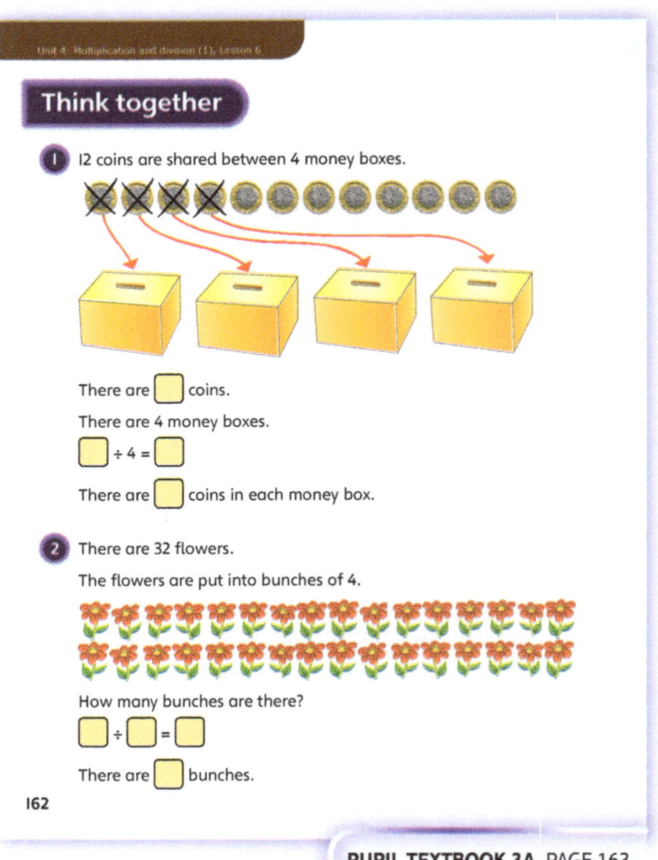

PUPIL TEXTBOOK 3A PAGE 162

PUPIL TEXTBOOK 3A PAGE 163

Practice

WAYS OF WORKING   Independent thinking

IN FOCUS   Questions ❶ and ❷ provide a mixture of grouping and sharing problems. Encourage children to link each number in the division statement with the context. Questions ❸ and ❹ focus on using an array. Children should start to see that if a group of objects can be shown as either 4 rows or 4 columns then they can work out the answer to the division.

STRENGTHEN   To support dividing by 4, ensure children understand the difference between equal grouping and equal sharing. Provide simple contexts and ask children to show with counters. For example, in question ❶, they could take 24 counters to represent the jelly beans. They should make 4 piles to show the division or put them into an array with 4 columns (or rows). The number of rows or columns gives the answers to the division: make sure they see the link. In question ❺, children can replace the £20 note with twenty £1 coins, which may help them with grouping. If children struggle with division, they may find halving and halving again easier.

DEEPEN   Question ❻ provides an opportunity to lead into questions such as 12 ÷ 4 ◯ 12 ÷ 3. Ask children to reason which sign goes between the expressions. They may say that 12 ÷ 4 is smaller as you are sharing the 12 between more 'people'.

THINK DIFFERENTLY   Question ❻ asks children to consider which is bigger, 24 divided by 4 or 24 divided by 3. Some children may work out the answers and make a decision based on the answers. However, children should be encouraged to reason that the answer to 24 ÷ 4 will be smaller as they are sharing amongst more groups.

ASSESSMENT CHECKPOINT   Can children represent a question as a division sentence involving ÷ 4 and can they use their knowledge of counting back in 4s to work out the answer? Do children understand the difference between sharing and grouping, and the division sign can represent both? Do children understand that one method to divide by 4 is to halve and halve again?

ANSWERS   Answers for the **Practice** part of the lesson appear in the separate **Practice and Reflect answer guide**.

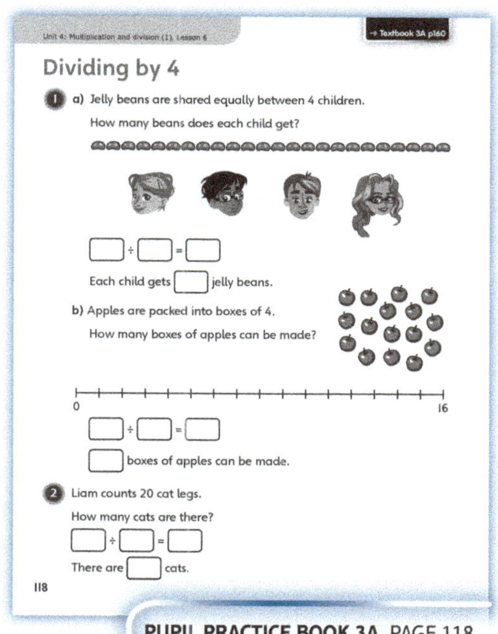

PUPIL PRACTICE BOOK 3A PAGE 118

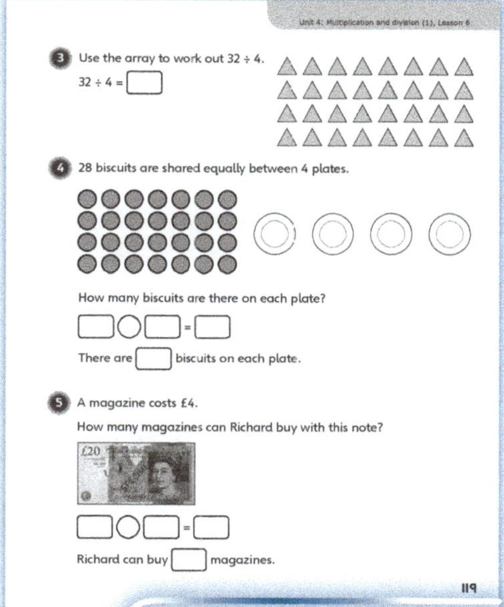

PUPIL PRACTICE BOOK 3A PAGE 119

Reflect

WAYS OF WORKING   Pair work

IN FOCUS   Children may draw a shape and show that if they divide the shape into 2 equal parts and then divide each half into half again, they end up with 4 equal parts.

ASSESSMENT CHECKPOINT   Check whether children understand that a method for dividing by 4 is to halve and halve again.

ANSWERS   Answers for the **Reflect** part of the lesson appear in the separate **Practice and Reflect answer guide**.

After the lesson ⏸

- Can children form division statements involving 4 from a word problem?
- Do children know how to use repeated subtraction and counting down in 4s to divide by 4?
- Do children know that a method for dividing by 4 is to halve and halve again?

PUPIL PRACTICE BOOK 3A PAGE 120

195

4 times-table

Learning focus

In this lesson, children will focus on learning the 4 times-table and should be able to recall associated division facts from multiplication facts. They will know how the 4 times-table can be derived from the 2 times-table.

Small steps

→ Previous step: Dividing by 4
→ **This step: 4 times-table**
→ Next step: Multiplying by 8

NATIONAL CURRICULUM LINKS

Year 3 Number – Multiplication and Division

- Recall and use multiplication and division facts for the 3, 4 and 8 multiplication tables.
- Write and calculate mathematical statements for multiplication and division using the multiplication tables that they know, including for two-digit numbers times one-digit numbers, using mental and progressing to formal written methods.
- Solve problems, including missing number problems, involving multiplication and division, including positive integer scaling problems and correspondence problems in which *n* objects are connected to *m* objects.

ASSESSING MASTERY

Children can recognise a multiplication fact from the 4 times-table in a given image. Children should be developing a rapid recall of multiplication facts and associated division facts from the 4 times-table.

COMMON MISCONCEPTIONS

Children may think that $0 \times 4 = 4$. This is a common mistake, mainly because children have answered too quickly and not thought about their response. Remind children that 0 × anything always gives the answer 0. Ask:
- *What is 0×3? Can that help you work out 0×4? How many objects or groups are there when you multiply by 0?*

To work out if 9×4 is greater than or equal to 7×4, children may think you have to work out each multiplication fact, but this is not the case. Children can draw the groups in a straight line to show this. Ask:
- *Do you need to work out both calculations? Why not? How else could you work it out?*

STRENGTHENING UNDERSTANDING

To strengthen understanding of times-table facts, show them as a visual or concrete representation, for example, towers or blocks of 4 cubes. Link this to children's work on counting in 4s on a number line. It is fine at this stage if some children still need to use cubes or counters to represent multiplications and division. What is important is that children can see what times-table facts are. Reinforce that they need to develop rapid recall of multiplication and division facts, but also need to understand what they are.

GOING DEEPER

Ask children to reason why $6 \times 4 = 8 \times 3$ without working out the two multiplications. Can they also work out missing numbers to complete statements such as $12 \div 4 < \_\_\_ \div 3$?

KEY LANGUAGE

In lesson: times-table, multiplication fact, array, pattern

Other language to be used by the teacher: divide, division, grouping, multiply, recall

STRUCTURES AND REPRESENTATIONS

number lines, arrays

RESOURCES

Mandatory: cubes, counters, number lines, 100 square

Optional: 4 times-table flashcards

 In the eTextbook of this lesson, you will find interactive links to a selection of teaching tools.

Before you teach 

- Can children multiply by 4?
- Can children divide by 4?
- Do children know the 2 times-table?

Discover

 Pair work

- Question **1** a): *What times-table is shown here? How do you know? What do you notice about the answers to the multiplications? What numbers are being covered? How did you work out the answers?*
- Question **1** b): *What multiplication can you use to work out 4 × 7? Why? How can you show this using an array? What about a division? Do you have to share or group? Is there a quicker way of doing this?*

 Question **1** a) asks children to work out the two multiplication facts that children in the image are standing on. Children may use their knowledge of multiplying and dividing by 4 from previous lessons to answer the questions. Some children may already know their times-table facts; you could ask them to convince you of their answers. Others may also link this to their knowledge of the 2 times-table, but some may just count on from previous answers. Question **1** b) asks children to work out which fact they can use to work out other multiplication facts. Encourage children to use an array to show that 4 × 7 is the same as 7 × 4. For 48 ÷ 4, they may need to use equal grouping or sharing. Try to encourage these children to use one of the multiplication facts to help them.

 Ensure any times-table facts on a poster in your classroom are covered up. You could consider using flashcards with 4 times-table multiplication facts on them.

Question **1** a): 6 × 4 = 24, 8 × 4 = 32
Question **1** b): Use 7 × 4 to work out 4 × 7 = 28.
 Use 12 × 4 to work out 48 ÷ 4 = 12.

Share

 Whole class teacher led

- Question **1** a): *What does each array show? Could you use something other than an array?*
- Question **1** b): *Which fact did you use? Why do you think it is important to know your 4 times-table off by heart? Can you work out the 4 times-table from the 2 times-table? How?*

 In question **1** a) an array is used to give a visual understanding of the two multiplication facts. There are multiple ways that children may get the correct answer. For example, they may just count on 4 or double and double again. All are valuable methods, but children need to be moving towards instant recall of these facts. In question **1** b), an array is used to show that 4 × 7 and 7 × 4 are the same. Discuss that using the fact 12 × 4 = 48 makes it easier to work out 48 ÷ 4, rather than going through the process of grouping or sharing.

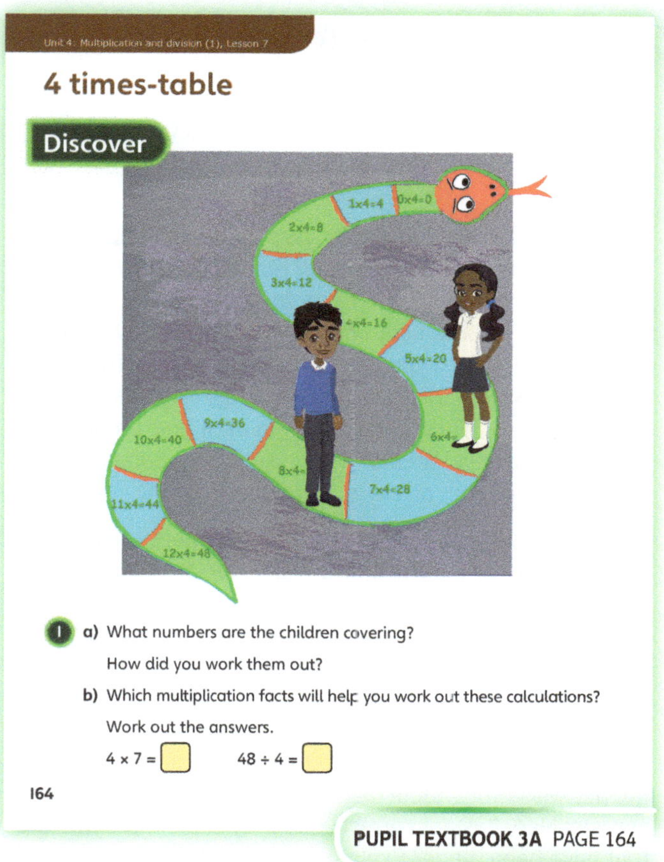

PUPIL TEXTBOOK 3A PAGE 164

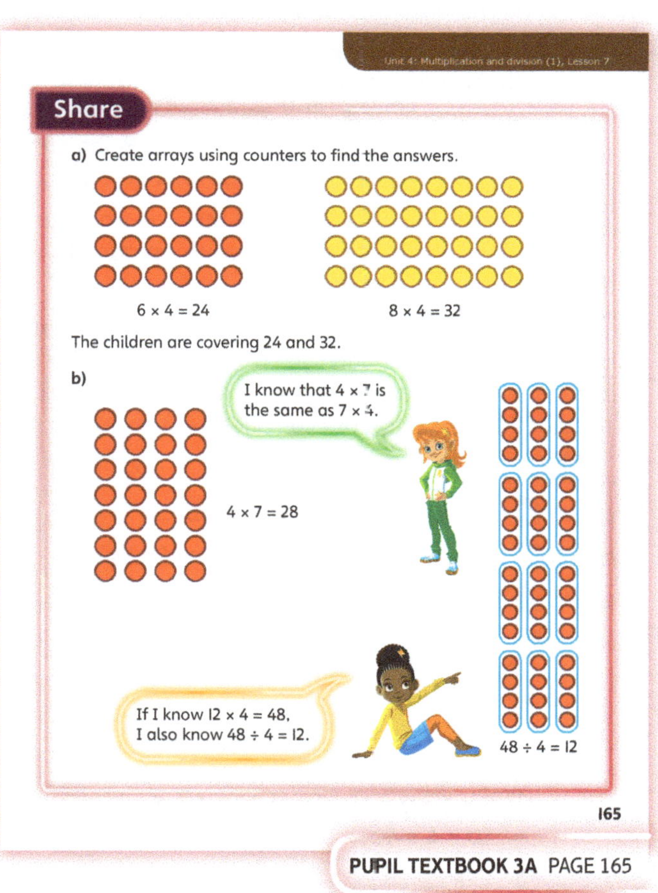

PUPIL TEXTBOOK 3A PAGE 165

Think together

WAYS OF WORKING Whole class teacher led (I do, We do, You do)

ASK

- Question ❶: *What facts can you see? How can you use these facts to work out the total? How can you work out 4 × 6 if you do not know the 6 times-table yet?*
- Question ❷: *How many answers has Mary got right? Which one can you tell is wrong straight away? How do you know? Which ones did Mary get wrong? What mistakes has she made?*
- Question ❸: *Which numbers are in the 4 times-table? Can you see any patterns? What do you notice about the numbers? Are they all even?*

IN FOCUS Question ❶ presents images where the total can be worked out using the 4 times-table facts. Children should be able to explain which times-table fact they can use. Encourage them to talk about the number of groups and the number of objects in each group. Children may notice that 4 × 6 can be worked out using 6 × 4. Again, children may use methods they have explored in previous lessons if they do not have recall of the facts. Question ❷ aims to develop fluency with times-table facts. Can they work out how many answers Mary has got correct? Can they explain the mistakes that have been made? Explain that they should also learn the division facts and show them how multiplication facts can be used to derive division facts. In question ❸ children start to explore times-table patterns on a 100 square. They should see that the numbers are all even and the pattern repeats every two rows.

STRENGTHEN To strengthen understanding of times-table facts, show them as a visual or concrete representation, for example with towers of cubes. It is important that children can visualise and understand times-table facts. Reinforce that eventually they will need to develop rapid recall.

DEEPEN In question ❸, ask children if they can work out the pattern of the 1s digits. Can they use this pattern to work out if 192 can be divided by 4?

ASSESSMENT CHECKPOINT Children start to know and remember their times-table multiplication and division facts using rapid recall. They can work out which fact will help them to work out the total or the answer to a division question.

ANSWERS

Question ❶ a): 5 × 4 = 20; There are 20 cubes.

Question ❶ b): 4 × 6 = 24; There are 24 bread rolls.

Question ❶ c): 11 × 4 = 44; There are 44 boxed pens.

Question ❷: Mary has got five answers right; they should be:

7 × 4 = 28	12 ÷ 4 = 3
4 × 9 = 36	**4 ÷ 4 = 1**
4 × 1 = 4	**8 ÷ 4 = 2**
0 × 4 = 0	**24 ÷ 4 = 6**
10 × 4 = 40	44 ÷ 4 = 11

Question ❸: 4, 8, 12, 16, 20, 24, 28, 32, 36, 40, 44, 48, 52, 56, 60, 64, 68, 72, 76, 80, 84, 88, 92, 96, 100
They are all even numbers, and the pattern repeats every two rows.

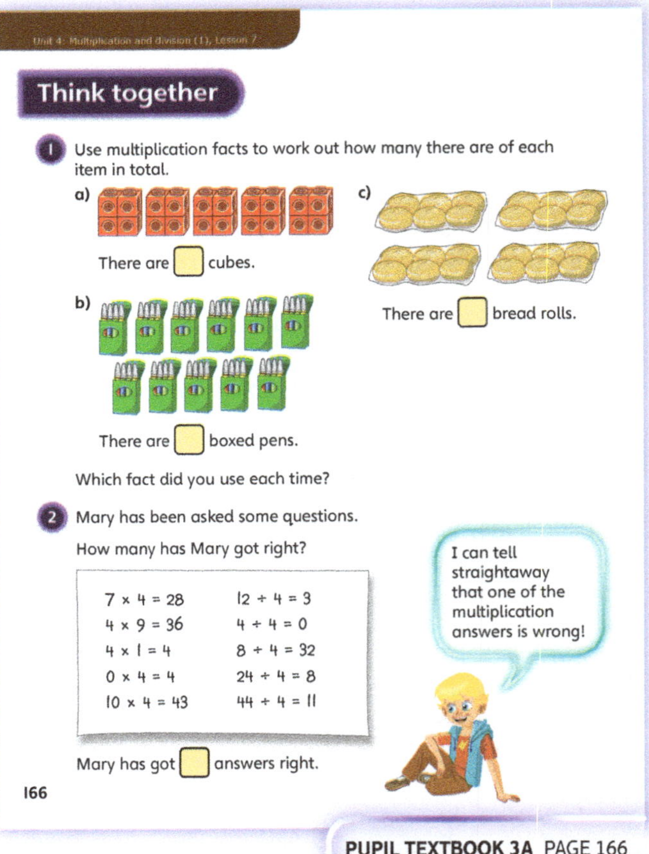

PUPIL TEXTBOOK 3A PAGE 166

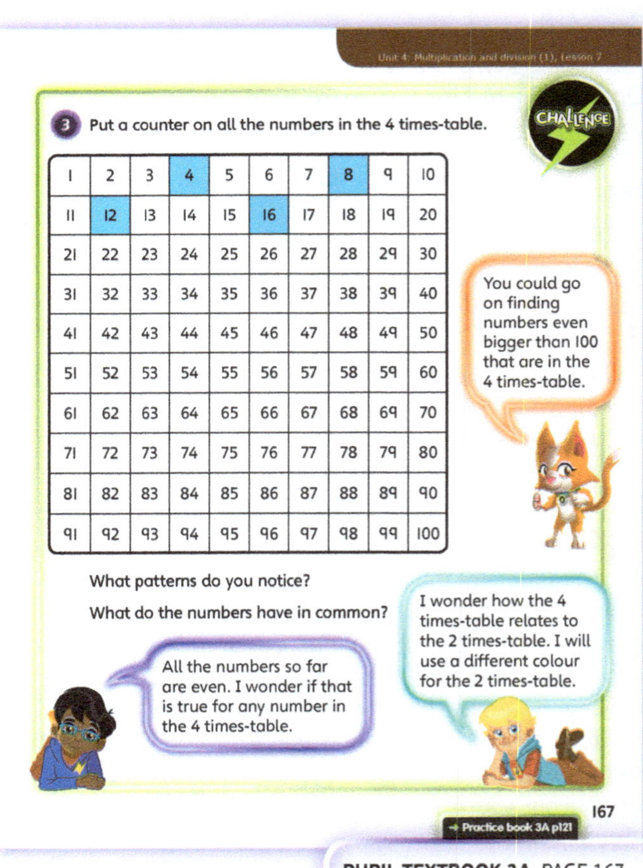

PUPIL TEXTBOOK 3A PAGE 167

Practice

WAYS OF WORKING Independent thinking

IN FOCUS Question ① reinforces visual images of several multiplication facts from the 4 times-table. Questions ② and ④ check whether children know the multiplication and division facts. Children should try to do these without referring back, and should be relying on mental recall. For any that they are still unable to work out, ask them to think about 2 times-table facts or their work from the previous two lessons. In question ③, children identify numbers that are in the 4 times-table. They use reasoning to try to explain the two children's statements. Question ⑤ asks children to put in inequality or equals signs to make statements correct. Encourage children to use reasoning instead of working out all the answers and then comparing.

STRENGTHEN To strengthen understanding of times-table facts, show them as a visual or concrete representation; question ① will help with the strengthening of this. As children move through the rest of the exercise, the questions become more abstract. Children may still need to use cubes or counters to see what times-table facts are, although they will also need to develop instant recall.

DEEPEN In question ⑤, ask children to explain how they can work out which sign completes the sentence without having to work out the answers. Which ones do they need to work out in order to compare? Can children find the pattern in question ⑥? Can they work out how the missing numbers are formed?

ASSESSMENT CHECKPOINT Can children recognise 4 times-table facts from images? They should be starting to develop a secure recall of their 4 times-table (both multiplication and associated division facts). Help children to develop quick recall by setting daily practice at home and at school.

ANSWERS Answers for the **Practice** part of the lesson appear in the separate **Practice and Reflect answer guide**.

Reflect

WAYS OF WORKING Pair work

IN FOCUS This brings together work on the 3 times-table and the 4 times-table. Children work out which numbers appear in both the 3 and the 4 times-table. Encourage children to take a strategic approach rather than just guessing. Ask children to list the numbers in the times-tables and to circle those numbers that appear in both. They should start to notice that the numbers go up in 12s.

ASSESSMENT CHECKPOINT Check whether children have instant recall of the times-table facts up to 12 × 3 and 12 × 4. Check if there are any facts that they are consistently getting wrong.

ANSWERS Answers for the **Reflect** part of the lesson appear in the separate **Practice and Reflect answer guide**.

After the lesson ⏸

- Can children recall division and multiplication facts from the 4 times-table?
- Which facts are children struggling with? How can these be reinforced further at school?

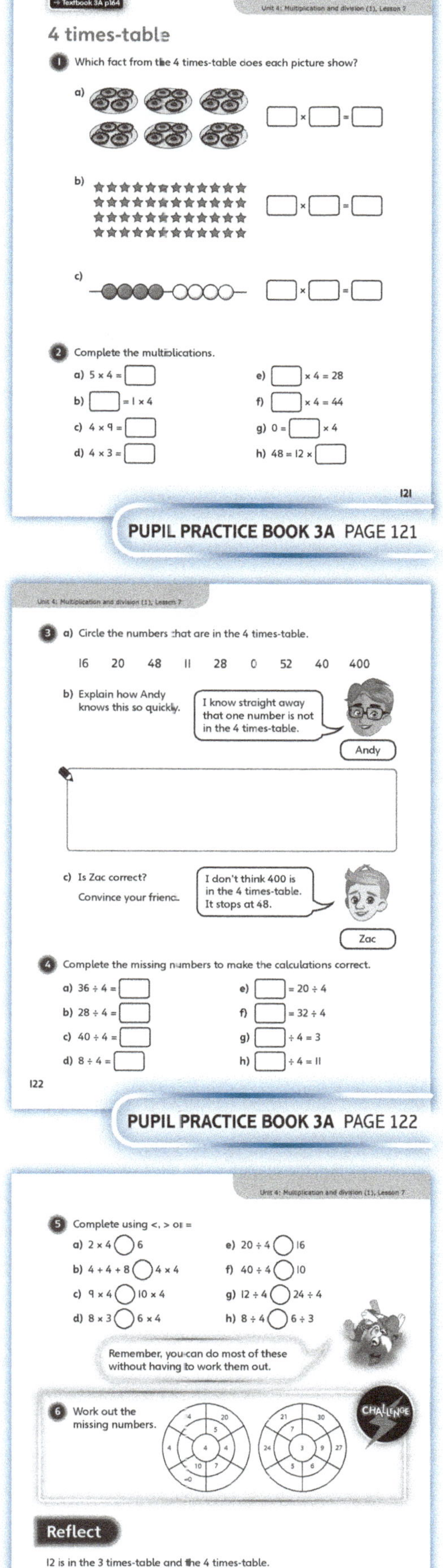

Multiplying by 8

Learning focus

In this lesson, children will start to understand what it means to multiply by 8. Children will use counting up in 8s to work out the answers to multiplications.

Small steps

→ Previous step: 4 times-table
→ **This step: Multiplying by 8**
→ Next step: Dividing by 8

NATIONAL CURRICULUM LINKS

Year 3 Number – Multiplication and Division

- Recall and use multiplication and division facts for the 3, 4 and 8 multiplication tables.
- Write and calculate mathematical statements for multiplication and division using the multiplication tables that they know, including for two-digit numbers times one-digit numbers, using mental and progressing to formal written methods.
- Solve problems, including missing number problems, involving multiplication and division, including positive integer scaling problems and correspondence problems in which *n* objects are connected to *m* objects.

ASSESSING MASTERY

Children can form a multiplication sentence involving multiplying by 8 and can work out the answer to a multiplication sentence by knowing its link with repeated addition and using a number line to count up in 8s. They will start to remember some of the multiplication facts for multiplying by 8 and they will know the link between multiplying by 2, by 4 and by 8.

COMMON MISCONCEPTIONS

Children often start the counting in 8s from 0 when working out, for example 11×8, even if they already know 10×8. Encourage children to start counting on from known facts. Ask:

- *How will you work this one out? Is there a quicker way?*

Children may lose count, for example when working out 6×8 in their head; they may count up too many or too few 8s. This should reinforce the need for them to know their multiplication facts. Ask:

- *How can you be more efficient? Could a number line help you to keep count? Is there a quicker way to find the answer?*

STRENGTHENING UNDERSTANDING

Children often find it easier to multiply by 8 by doubling the number, then doubling again and doubling again. You can show children this visually by using simple arrays: the array gets twice as big each time. This is a useful method, although children will eventually need to know their 8 times-table facts.

GOING DEEPER

Can children use multiplication facts they know in order to work out other multiplication facts? For example, can they use 12×8 to work out 13×8, or 24×8? Ask children to explore how to work out $4 \times 8 + 6 \times 8$ as one multiplication. They could show, by joining arrays together, that this is the same as 10×8.

KEY LANGUAGE

In lesson: multiply (×), total, multiplication statement, group, double, method

Other language to be used by the teacher: multiplication sentence, equal groups, count in 8s, times-table, *x* groups of *y*, altogether

STRUCTURES AND REPRESENTATIONS

Number lines, arrays

RESOURCES

Mandatory: cubes, counters, number lines

Optional: circles divided into 2, 4 and 8 equal pieces

 In the eTextbook of this lesson, you will find interactive links to a selection of teaching tools.

Before you teach 

- Can children count in 2s and 4s?
- Do children know the 2 and 4 times-tables?
- Do children know the link between repeated addition and the answer to a multiplication?

Discover

WAYS OF WORKING Pair work

ASK

- Question ❶ a): *How many groups or pies do you have? How many (slices) are there in each group? How can we write this as a multiplication statement? How can we work out the total? Do you know how to count in 8s?*
- Question ❶ b): *What is the connection between multiplying by 2, 4 and 8? Can you come up with a rule to help other children multiply by 8? Can you show why the rule works?*

IN FOCUS Question ❶ a) asks children to find the total number of slices. Each pie is cut into 8 slices. Children should be able to write this as a multiplication calculation. They should start to understand this as 4 groups of 8, which you can work out using 4×8. Children may use different methods for working this out. For example, some children may use a number line to count up in 8s, or they may link multiplying by 8 with multiplying by 2 and/or 4. However, try to encourage the use of efficient methods as much as possible. Question ❶ b) asks children to see the connection between multiplying by 2, 4 and 8. Children should notice that the answer doubles each time. Discuss how this can be used to help form a rule to multiply by 8.

PRACTICAL TIPS For this activity, you could provide circles that are divided into 2, 4 and 8 equal pieces to represent the pies.

ANSWERS

Question ❶ a): $4 \times 8 = 32$

Question ❶ b): $5 \times 2 = 10$, $5 \times 4 = 20$, $5 \times 8 = 40$
The answer doubles each time.

Share

WAYS OF WORKING Whole class teacher led

ASK

- Question ❶ a): *Why do we count in 8s and not in 1s? Can we count in 2s or 4s to get the answer? Do you know the answer to 4×8 from your knowledge of the 4 times-table?*
- Question ❶ b): *What is the connection between the answers? How can you use the answer to 5×4 to find the answer to 5×8?*

IN FOCUS In question ❶ a) counting in 8s on a number line is used to show how children might have arrived at the answer. Use what Sparks says to explain that the situation can be seen as 4 groups of 8 and can therefore be written as a multiplication calculation: $4 \times 8 = 32$. You can also link this to the 4 times-table: $8 \times 4 = 32$. In question ❶ b), draw out the connection between the three multiplications. Children should see from the arrays that the answer doubles each time. Discuss the implications of using this method to help multiply by 8. For example, you can multiply by 4 and double the answer, or you could double, double and double again.

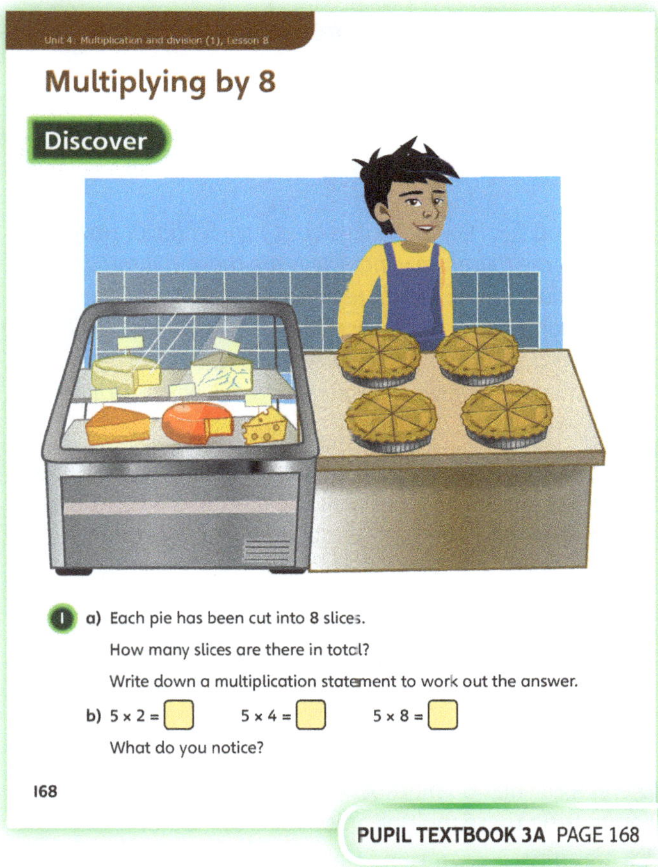

PUPIL TEXTBOOK 3A PAGE 168

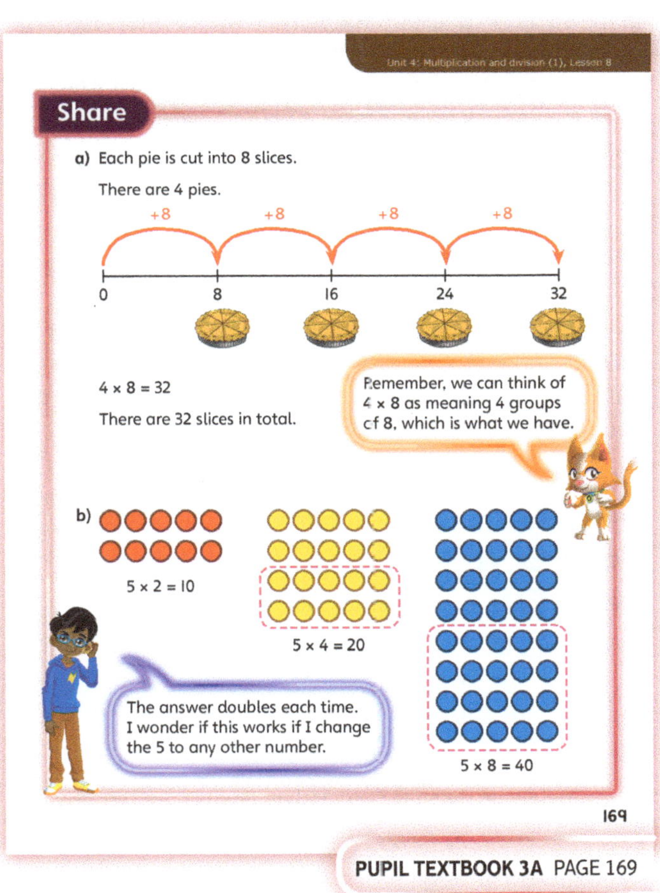

PUPIL TEXTBOOK 3A PAGE 169

Think together

WAYS OF WORKING Whole class teacher led (I do, We do, You do)

ASK

- Question ❶: *How many legs does 1 spider have? How many spiders are there? How many legs are there in total? What multiplication can you do to work this out?*
- Question ❷: *How many people are in the queue? How much is a ticket? How can you work out the total cost?*
- Question ❸: *Can you explain to a friend why Isla's method works?*

IN FOCUS Question ❶ asks children to work out how many legs are on six spiders. Children use a number line to find the total, with the first few jumps given. In question ❷, children have to find 11 × 8. This time there is a unit of money in the answer. Question ❸ aims to get children to think about different methods for multiplying by 4 and 8. Children use the different methods to multiply numbers by 4 and 8. Discuss if there is a particular method that children prefer for multiplying by 8. For example, children who struggle to remember the count in 8s may prefer a calculation method. Emphasise the importance of committing these multiplication facts to memory.

STRENGTHEN At each stage, use concrete objects alongside a number line to help children see the link with repeated addition, counting up in 8s and multiplication. In question ❶, you can encourage children to put 8 counters at each point on the number line to show that this represents the legs on the spider. In question ❷, you can use counters to represent the coins and put them into an array. Children who struggle to remember the count sequence for 8s may find it easier to double, double and double again.

DEEPEN Ask: *Why is multiplying by 2 and then by 2 and then by 2 again equal to multiplying by 8? How can you use 12 × 8 = 96 to work out 13 × 8, 14 × 8 or 15 × 8?*

ASSESSMENT CHECKPOINT Children understand what it means to multiply by 8 and use their knowledge of counting in 8s to work out the answers to the repeated addition and multiplication statements that they form. Children know that to multiply by 8 they can multiply by 4 and double, or multiply by 2, then by 2 and then by 2 again.

ANSWERS

Question ❶: 6 × 8 = 48 There are 48 legs altogether.

Question ❷: 11 × 8 = £88 The total cost is £88.

Question ❸ a): 9 × 2 = 18, 18 × 2 = 36

Question ❸ b): 9 × 2 = 18
  18 × 2 = 36
  36 × 2 = 72

Question ❸ c): 15 × 4 = 60, 15 × 8 = 120

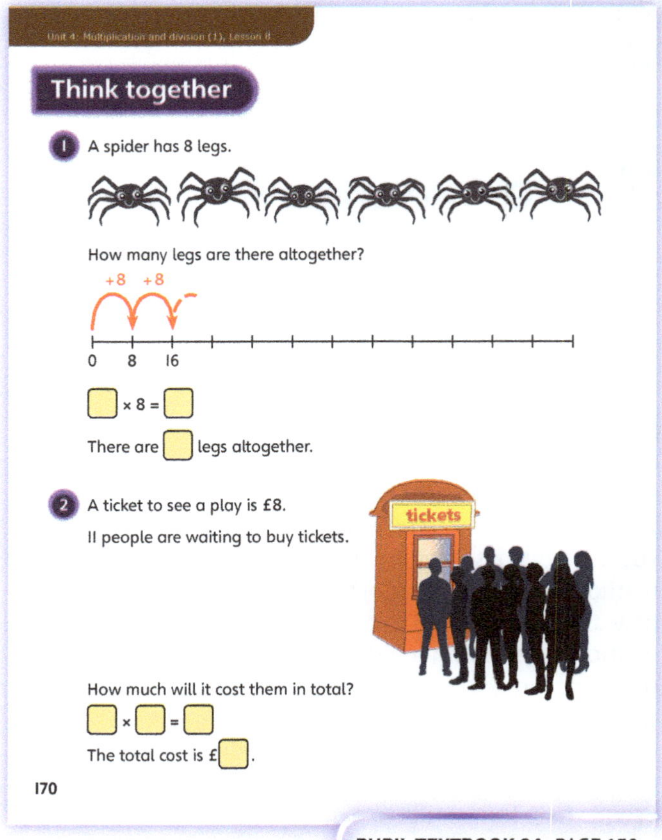

PUPIL TEXTBOOK 3A PAGE 170

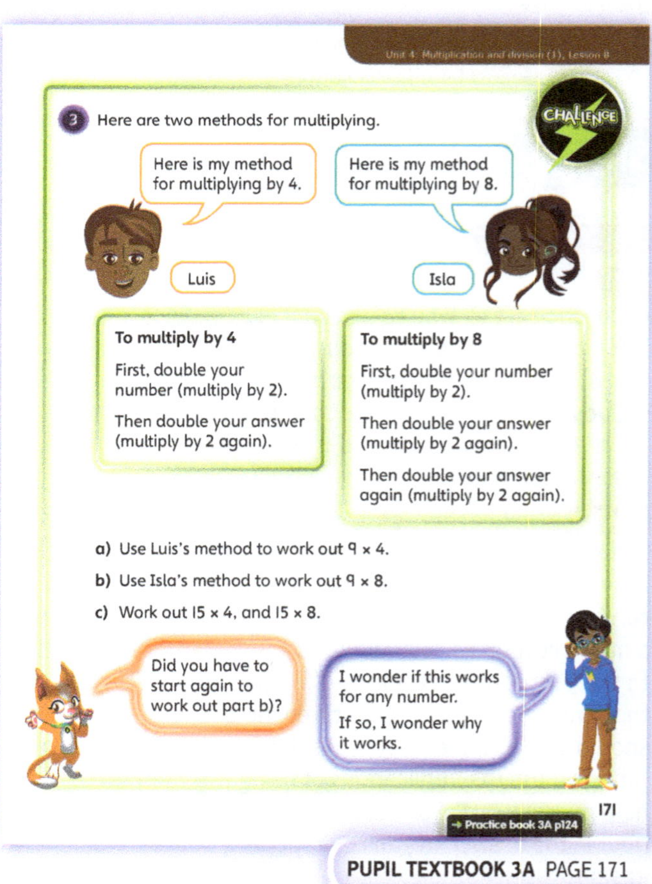

PUPIL TEXTBOOK 3A PAGE 171

Practice

WAYS OF WORKING Independent thinking

IN FOCUS Questions ① to ④ provide practice for children to form the correct multiplication statements from given situations. Encourage children to use a number line and counting up in 8s to help them work out the answers to the calculations; or they may use the rule of doubling, doubling and doubling again. Children may also use their knowledge of other times-tables to work out some of the answers: for example, they may use 8 × 4 to work out 4 × 8. They should write down the correct multiplication statement from the context given. Children may already know some of the 8 times-table facts off by heart and will simply write down the answer, so encourage checking.

STRENGTHEN To support multiplying by 8, use concrete objects alongside a number line. Count aloud with children in 8s from 0, each time showing the jump on the number line. If children are confident with multiplying by 2, or can double, encourage them to double, double and double again.

DEEPEN Can children make up their own questions similar to the ones in question ⑤? For example, if I know that 2 × a number is 18, what is 4 × the number and 8 × the same number? Can children explain what they did and why it works? Question ⑥ requires children to use a rule to multiply by 8. Encourage them to come up with other rules for multiplying by 8. For example, they may multiply by 10 and then subtract 8 from the number twice.

THINK DIFFERENTLY Question ⑤ gives children a multiplication fact from the 4 times-table and requires them to work out what the number is when multiplied by 8. They may think that they need to do a division to work out the missing number, but instead they should be encouraged to think of a way of doing it using their knowledge of the connection between multiplying by 4 and multiplying by 8.

ASSESSMENT CHECKPOINT Can children represent a question as a multiplication sentence involving × 8 and use counting in 8s to work out the correct answer? Do children know that multiplying by 2, then multiplying by 2 and multiplying by 2 again, is the same as multiplying by 8?

ANSWERS Answers for the **Practice** part of the lesson appear in the separate **Practice and Reflect answer guide**.

Reflect

WAYS OF WORKING Independent thinking

IN FOCUS Children should be aware of the connection between multiplying by 2, 4 and 8. Children need to use this knowledge to work out 6 × 8 from 6 × 4, which they can do by doubling. Children may use other valid methods, such as starting from 0 and counting up in 8s, or counting up in 6s from 24.

ASSESSMENT CHECKPOINT Check whether children can use a 4 times-table fact to multiply the same numbers by 8.

ANSWERS Answers for the **Reflect** part of the lesson appear in the separate **Practice and Reflect answer guide**.

After the lesson ⏸

- Can children form multiplication statements involving × 8 from a word problem?
- Can they work out multiplications by counting in 8s?
- Do children know the connection between multiplying by 2, 4 and 8?

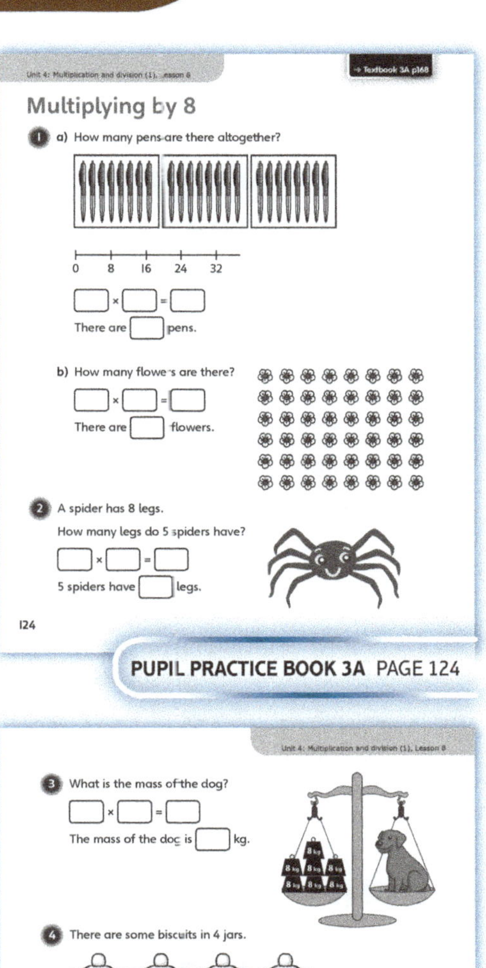

PUPIL PRACTICE BOOK 3A PAGE 124

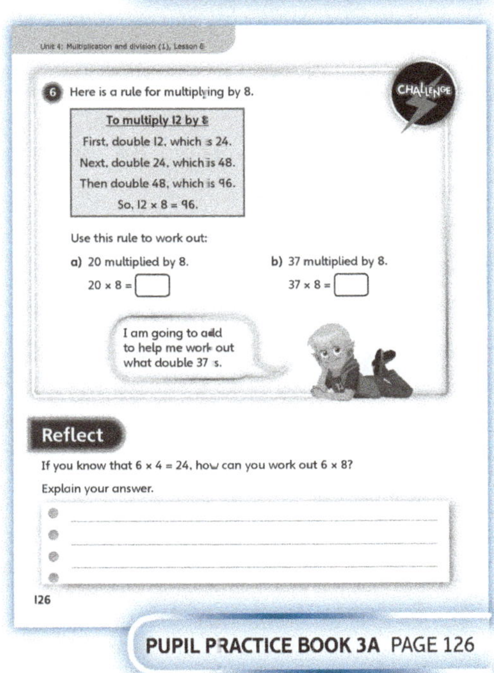

PUPIL PRACTICE BOOK 3A PAGE 125

PUPIL PRACTICE BOOK 3A PAGE 126

Dividing by 8

Learning focus

In this lesson, children will understand how they can divide a number by 8.

Small steps

→ Previous step: Multiplying by 8
→ **This step: Dividing by 8**
→ Next step: 8 times-table

NATIONAL CURRICULUM LINKS

Year 3 Number – Multiplication and Division

- Recall and use multiplication and division facts for the 3, 4 and 8 multiplication tables.
- Write and calculate mathematical statements for multiplication and division using the multiplication tables that they know, including for two-digit numbers times one-digit numbers, using mental and progressing to formal written methods.
- Solve problems, including missing number problems, involving multiplication and division, including positive integer scaling problems and correspondence problems in which *n* objects are connected to *m* objects.

ASSESSING MASTERY

Children can form a division sentence from either a grouping or a sharing situation and know the difference between grouping and sharing. Children can use repeated subtraction and counting forwards or backwards in 8s to work out the result of a division, as well as know that a method for dividing by 8 is to halve the number, halve the answer and then halve again.

COMMON MISCONCEPTIONS

When dividing by 8, children may lose track of the count forwards or backwards. Explain that the sooner they can commit the multiplication facts (and associated division facts) to memory, the better. Ask:
- *Is there a quicker way to work this out? What multiplication fact could have helped you with this?*

When using the method where they have to halve, halve again and halve again, often children may only halve twice (and therefore only divide by 4). Ask:
- *What is the link between the 2, 4 and 8 times-tables? Can you use the 4 times-table to check your answer?*

STRENGTHENING UNDERSTANDING

Work with concrete objects to show the difference between grouping and sharing. Explicitly link each action in grouping with the steps in repeated subtraction. Use a number line to show how grouping can be seen as repeated subtraction by counting back in 8s (or subtracting 8).

To understand sharing, first ask children to share the objects one by one and then ask children to take 8 at a time and allocate to each group. Ensure children know what each number in the division statement represents.

GOING DEEPER

Ask children to show two different word problems for $32 \div 8 = 4$. What is the same about their stories? What is different? Can children use $96 \div 8$ to work out $104 \div 8$?

KEY LANGUAGE

In lesson: divide (÷), division, division statement, share equally

Other language to be used by the teacher: equal, grouping, array, halve, count in 8s, method, multiply, multiplication fact

STRUCTURES AND REPRESENTATIONS

Number lines, arrays

RESOURCES

Mandatory: cubes, counters, number lines

Optional: ice lolly moulds, fruit juice, paper circles

 In the eTextbook of this lesson, you will find interactive links to a selection of teaching tools.

Before you teach 

- Can children count back in 8s from 80 to 0?
- Can children share a set of objects between a given number of groups?
- Can children put a set of objects into groups of a given size?

Discover

WAYS OF WORKING Pair work

ASK

- Question **1** a): *How many lollies are in each mould? How many complete moulds can you make? What division calculation should you do? What does each number and sign represent?*
- Question **1** b): *What is different about this question? Does this question work? Do you have enough lollipop sticks to fill up the last mould? How can you write the division? What do you think your answer could look like?*

IN FOCUS Question **1** a) focuses on equal grouping. Children must work out how many 8s are in 24. Children may count on in 8s until they get to 24, or use repeated subtraction of 8. Those children who group counters in 8s should be encouraged to think about the mathematical process that is going on by saying, for example, 'Each time I group 8, I reduce the original amount by 8'. Encourage children to show this on a number line. Question **1** b) is still about grouping, however this time children have some lollipop sticks left over. Some children may think that the question is wrong, because they do not have enough lollipop sticks to complete the last container. At this stage, we are not writing formal remainders; we are just interested in how many complete moulds can be filled. For both questions, encourage children to write down the correct division calculation and ensure that children know what each number and sign represents.

PRACTICAL TIPS You could make ice lollies out of fruit juice in groups of 8, as in the image.

ANSWERS

Question **1** a): 24 ÷ 8 = 3; 3 moulds can be filled.

Question **1** b): She can fill 4 moulds.

Share

WAYS OF WORKING Whole class teacher led

ASK

- Question **1** a): *Can you explain each of Flo's steps? Why does the number line help? What other methods could you have used? Could an array help you?*
- Question **1** b): *What is the same about this question and the way you can answer it? What is different about what you do? How could you record your answer?*

IN FOCUS For question **1** a), ensure children demonstrate the method of grouping. Make each step of the subtraction clear alongside a demonstration of grouping. You can use counters to represent the lollipop sticks. It is important to link any numbers in the calculation to what it means in the context of the image. Children should notice that there are three groups of 8.

Question **1** b) is similar, but this time children may notice that there are some left over. At this point, we are only interested in the complete number of containers that can be filled. Flo points out that there are not enough lollipop sticks to fill another container. Discuss this with children.

PUPIL TEXTBOOK 3A PAGE 172

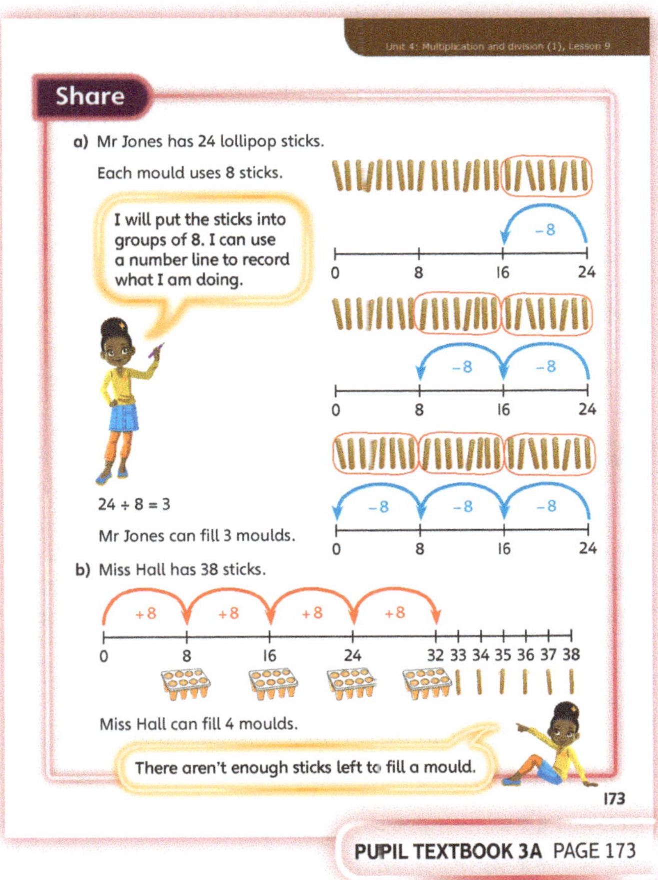

PUPIL TEXTBOOK 3A PAGE 173

Think together

WAYS OF WORKING Whole class teacher led (I do, We do, You do)

ASK

- Question **1** : *How many cupcakes are there? How many chocolate chips are there? Is this grouping or sharing? Can you explain each step of your method? What does the division look like? What does each number in the division statement mean? Could you find the answer more quickly?*
- Question **2** : *What array is shown? How can you use the array to work out the answer to the division?*
- Question **3** : *Can you explain Lexi's method? How do the candles help you work out the answer to the problem? What divisions can Lexi work out from her demonstration?*

IN FOCUS Question **3** explores the method of halving, halving again and halving again as an alternative to dividing by 8. Children should come up with this method themselves as they see a birthday cake gradually cut into 8 pieces. Draw children's attention to the number of candles on the cake at the start, and the number on each part of the cake as they go through each of the divisions, to help them see the connection.

STRENGTHEN In question **1**, children can share practically using counters to represent the chocolate chips. Some children may group these into 8s in order to share them more quickly. Ask children what each number in the division represents and how they know they are doing a division. What clues tell them it is a division?

Children may find making an array with 8 counters in each row an effective way of finding the answers to the divisions. They may find it easier to understand dividing by 8 as dividing by 2, dividing by 2 again and then dividing by 2 again. Question **3** provides a demonstration of why this works. Also, talk through the steps and consider modelling it with a paper circle and counters.

DEEPEN To deepen understanding, discuss with children how they could use division to work out the missing number problem ___ × 8 = 72? Ask: *Can you explain how you can do this? Can you make up your own problems similar to this?* Follow on from question **3** by asking children how they can use this method to divide 288 by 8.

ASSESSMENT CHECKPOINT Children understand that division can either be equal grouping or sharing and they are able to form a division statement to answer a question. They can use a variety of different methods to work out an answer to a division by 8, including counting forwards or backwards, using an array or through knowledge of multiplication facts.

ANSWERS

Question **1** : 40 ÷ 8 = 5.
She can use 5 chocolate chips on each cupcake.

Question **2** a): 72 ÷ 8 = 9

Question **2** b): 48 ÷ 8 = 6

Question **3** a): 16 ÷ 2 = 8, 16 ÷ 4 = 4, 16 ÷ 8 = 2

Question **3** b): Halve it, halve it again and halve it again.
88 ÷ 2 = 44,
44 ÷ 2 = 22,
22 ÷ 2 = 11

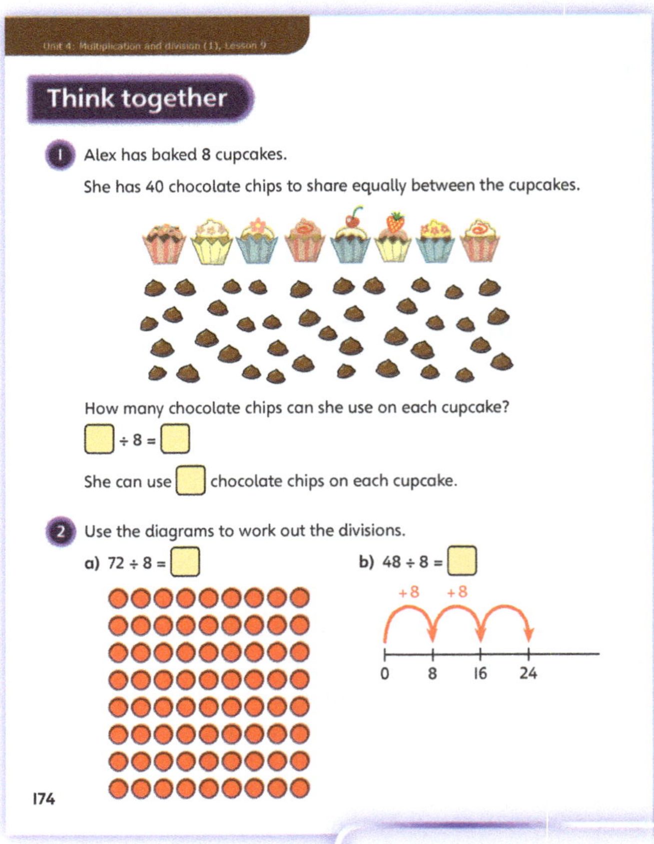

PUPIL TEXTBOOK 3A PAGE 174

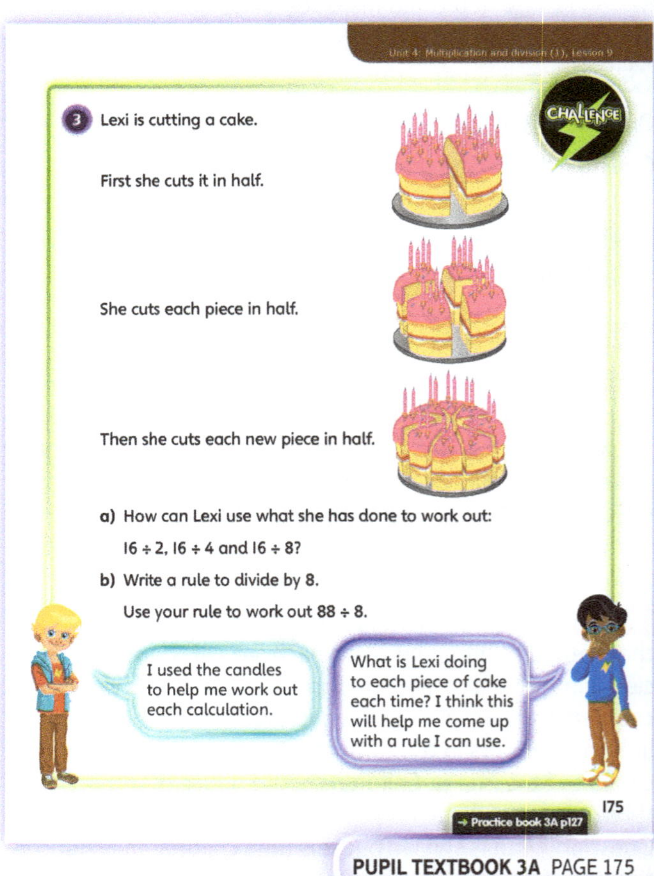

PUPIL TEXTBOOK 3A PAGE 175

Practice

WAYS OF WORKING Independent thinking

IN FOCUS Questions ❶, ❷ and ❸ provide a mixture of grouping and sharing problems. Encourage children to link each number in the division statement with the context. Ask them to consider whether the problem is grouping or sharing. Question ❹ a) asks children to use arrays with groups circled to find the answers to divisions. In question ❹ b), children may circle the groups in different ways. Question ❺ gives the answer to a division and asks children to use their related multiplication facts to work out the total. Children need to understand that 8 is the result of the division, rather than dividing by 8.

STRENGTHEN To support dividing by 8, ensure that children understand the difference between equal grouping and equal sharing. Provide simple contexts and ask children to show with counters what they are doing. Use counters or other materials to show the actions of grouping and sharing. For example, in question ❶, 24 counters can represent the pegs.

For grouping problems, ask children to take 8 counters at a time. For sharing problems, you may still want children to take 8 counters, but explain that by taking 8 you are giving one to each group.

DEEPEN Question ❻ provides an opportunity to look at the connection between multiplying and dividing by 2, 4 and 8. Give children a multiplication or division fact about a number, such as: *My number divided by 4 is 12. What is my number divided by 8? What is my number divided by 2?*

Ask questions such as *24 ÷ 8 ___ 24 ÷ 4: which sign goes between the expressions?* Ask children to reason what the answer might be and why.

ASSESSMENT CHECKPOINT Can children represent a question as a division sentence involving ÷ 8, and use their knowledge of counting on or back in 8s to work out the answer? Children should understand the difference between sharing and grouping, but also know that the division sign can be used to represent both. Do children understand that one method to divide by 8 is to halve the number, halve it again and then halve it again?

ANSWERS Answers for the **Practice** part of the lesson appear in the separate **Practice and Reflect answer guide**.

Reflect

WAYS OF WORKING Pair work

IN FOCUS Children explore the problem of dividing 16 by 8. Ask children to decide what method they will use to solve this and to share their methods. Children may count on, count back, or use practical equipment or pictures to show their thinking. Try to highlight that some children have grouped and others have shared. Draw out the connection between the numbers and what children may have drawn or done.

ASSESSMENT CHECKPOINT Check whether children understand the methods for dividing by 8.

ANSWERS Answers for the **Reflect** part of the lesson appear in the separate **Practice and Reflect answer guide**.

After the lesson ⏸

- Can children form division statements involving 8 from a word problem?
- Do children know the difference between division as grouping and division as sharing?
- Do children know at least one method to divide by 8?

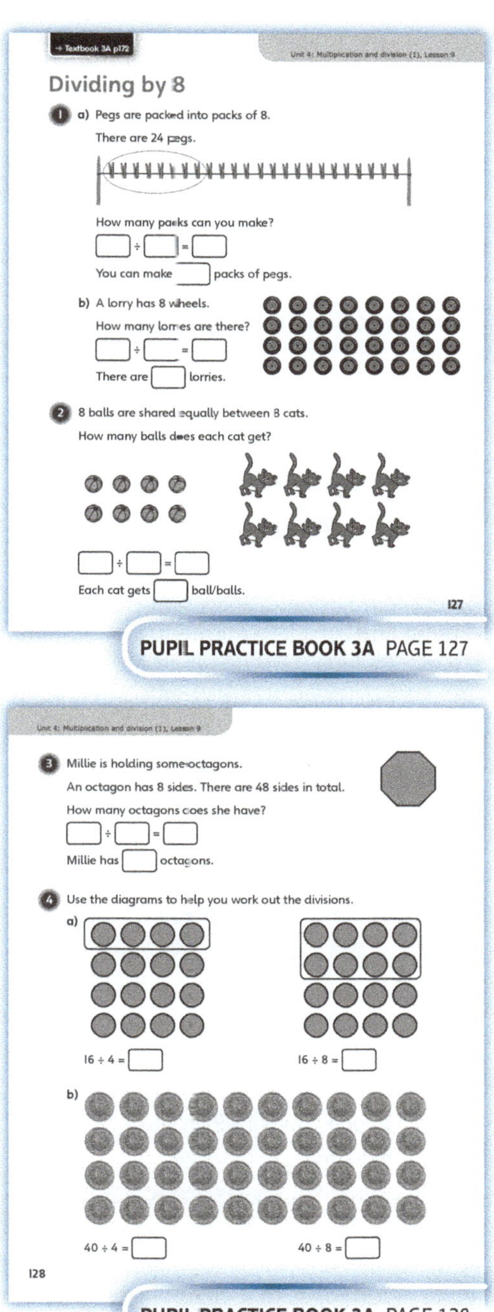

PUPIL PRACTICE BOOK 3A PAGE 127

PUPIL PRACTICE BOOK 3A PAGE 128

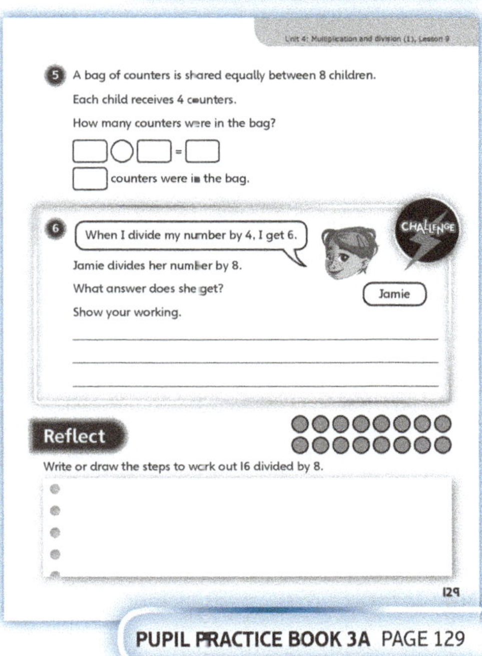

PUPIL PRACTICE BOOK 3A PAGE 129

8 times-table

Learning focus

In this lesson, children will focus on learning the 8 times-table and associated division facts from multiplication facts. They will show that they know how the 8 times-table can be derived from the 2 and 4 times-tables.

Small steps

→ Previous step: Dividing by 8
→ **This step: 8 times-table**
→ Next step: Problem solving – multiplication and division (1)

NATIONAL CURRICULUM LINKS

Year 3 Number – Multiplication and Division

- Recall and use multiplication and division facts for the 3, 4 and 8 multiplication tables.
- Write and calculate mathematical statements for multiplication and division using the multiplication tables that they know, including for two-digit numbers times one-digit numbers, using mental and progressing to formal written methods.
- Solve problems, including missing number problems, involving multiplication and division, including positive integer scaling problems and correspondence problems in which *n* objects are connected to *m* objects.

ASSESSING MASTERY

Children can recognise a multiplication fact from the 8 times-table given as an image. Children should be developing a rapid recall of multiplication facts and associated division facts from the 8 times-table.

COMMON MISCONCEPTIONS

Children may think that $0 \times 8 = 8$. This is a common mistake because children may answer too quickly and not think about their response. Ask:

- *Have you tried dividing by 0 before? What answer do you always get when you divide by 0? Why?*

To work out if 9×8 is greater than or equal to 7×8, children may think you have to work out each multiplication fact. Ask:

- *How do you know that 9 groups of 8 must be greater than 7 groups of 8? Why do you not need to work out both calculations to find the answer?*

STRENGTHENING UNDERSTANDING

At this stage, it is fine if some children still need to use cubes or counters to represent multiplications and divisions. The important thing is that children see what times-table facts are. Reinforce that they need to develop rapid recall of the multiplication and division facts, but also understand what they are.

Remind children that if they are unsure of how to work out an 8 times-table fact, they can use their knowledge from the previous lessons: for example, to work out 3×8 they could double 3, double again and double again.

GOING DEEPER

Ask children to reason why $6 \times 8 < 7 \times 8$ without working out the two multiplications, and to find missing numbers to complete statements such as $40 \div 8 > \bigcirc \div 8$. Children can explore the connection between the 2, 4 and 8 times-tables; for example, if they know that $2 \times \_\_\_ = 15$, can they find 4 × that number and 8 × the same number.

KEY LANGUAGE

In lesson: times-table, multiplication fact, array, double, pattern, total

Other language to be used by the teacher: multiply, divide, grouping, division fact, recall

STRUCTURES AND REPRESENTATIONS

Number lines, arrays

RESOURCES

Mandatory: cubes, counters, number lines

 In the eTextbook of this lesson, you will find interactive links to a selection of teaching tools.

Before you teach 

- Can children multiply by 8?
- Can children divide by 8?
- Do children know their 2 and 4 times-tables?

Discover

 Pair work

- Question **1** a): *What times-table facts are missing? How did you work them out? How many of the times-tables facts on the stairs do you know off by heart?*
- Question **1** b): *What is 5 × 2, 5 × 4 and 5 × 8? What do you notice about this? What is the connection?*

 Question **1** a) asks children to work out three missing multiplication facts. Children may use their knowledge from previous lessons of multiplying and dividing by 8. Some children may already know their times-table facts; you might want to ask these children to convince you of their answers. Some children may also link this to their knowledge of the 2 or 4 times-tables. Others may just count on from the previous values. Question **1** b) asks them to start finding the connection between the 2, 4 and 8 times-tables. Children have previously done work on doubles and so they should start to see that to get from one to the other you can double.

 If you have the times-table facts on a poster in your classroom, ensure that the relevant answers are covered up so that children do not just look at them.

Question **1** a): 2 × 8 = 16, 6 × 8 = 48, 9 × 8 = 72

Question **1** b): The 4 times-table is double the 2 times-table.
The 8 times-table is double the 4 times-table.

Share

 Whole class teacher led

- Question **1** a): *What does each array show? Could you use something other than an array?*
- Question **1** b): *What do you notice about the arrays? Can you explain, using the arrays, why you double each time?*

 In question **1** a) an array is used to give a visual understanding of the multiplication facts. There are multiple ways that children could find the correct answer. For example, they may have just counted on 8 from the previous multiplication fact; some may have doubled, doubled again and doubled again. All are valid methods, although children do need to be moving towards instant recall of these facts at this stage. Question **1** b) shows, using arrays, that to get from the 2 to the 4 times-table you double and then to get to the 8 times-table you double again.

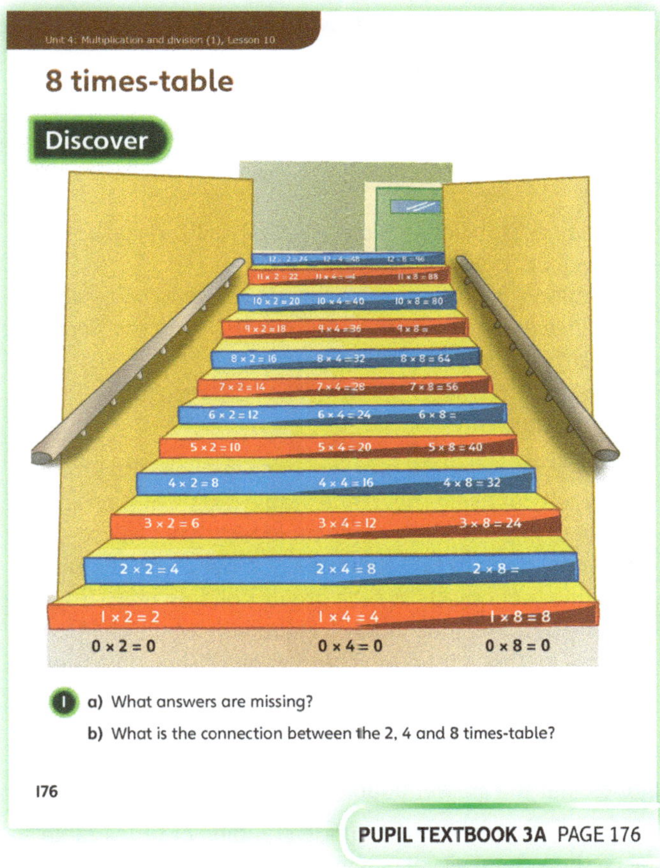

PUPIL TEXTBOOK 3A PAGE 176

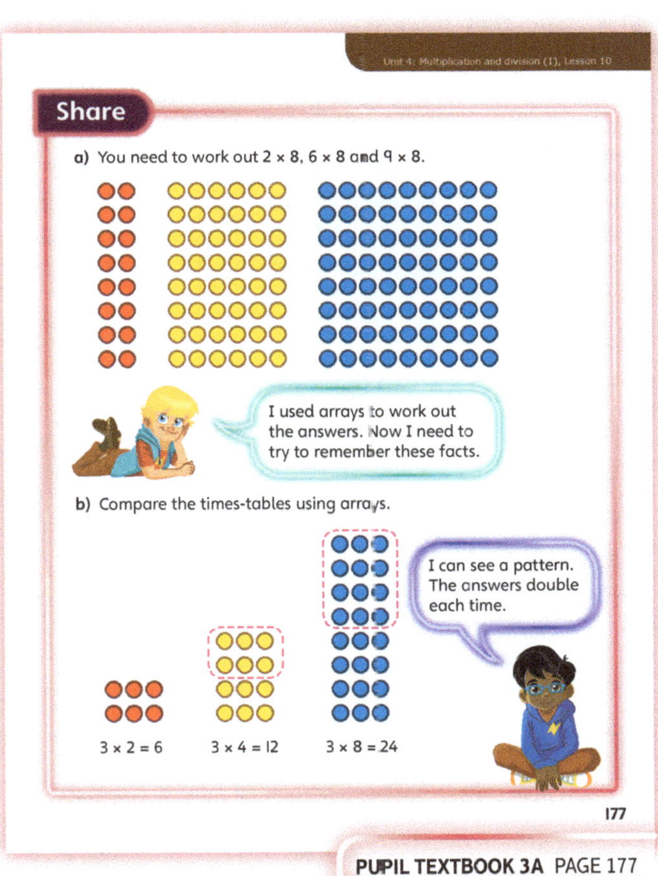

PUPIL TEXTBOOK 3A PAGE 177

Think together

WAYS OF WORKING Whole class teacher led (I do, We do, You do)

ASK

- Question **1**: *What facts can you see? How can you use these facts to work out the total?*
- Question **2**: *How many answers do you know off by heart? Which ones did you have to work out? How can you work out the divisions from the multiplications?*
- Question **3**: *What is the connection between the multiplications?*

IN FOCUS Question **1** presents images where the total can be worked out using the 8 times-table facts. Children should be able to explain which times-table fact they can use. Encourage them to talk about the number of groups and the number of objects in each group. Children may struggle with the problem that shows 1 group of 8. Again, children may use methods from previous lessons if they do not have rapid recall of the facts. In question **3**, children further explore the connection between the 2, 4 and 8 times-tables by matching facts that give the same answer. They should notice that the answer doubles each time.

STRENGTHEN To strengthen understanding of times-table facts, show them visual or concrete representation, for example towers of 8 cubes. It is important that children can visualise and understand what times-table facts are. Reinforce that, eventually, children will need to develop rapid recall of the multiplication and division facts, but also understand what they are.

DEEPEN Expand question **3** by asking children if they know that a number multiplied by 8 is 100, what is the same number multiplied by 4 and multiplied by 2.

ASSESSMENT CHECKPOINT Children should start to develop rapid recall of the 8 times-table multiplication and division facts. They should show an understanding of the connection between the 2, 4 and 8 times-tables and start to explain why this is the case.

ANSWERS

Question **1** a): $5 \times 8 = 40$ There are 40 bottles of water.

Question **1** b): $1 \times 8 = 8$ There are 8 cubes.

Question **1** c): $11 \times 8 = 88$ There are 88 eggs.

Question **2**: 24, 80, 1, 96, 5, 40, 16, 0

Question **3**:

5 × 8		2 × 4		12 × 2
3 × 8		4 × 20		32 × 2
1 × 8		10 × 4		2 × 40
8 × 11		4 × 22		4 × 2
8 × 10		6 × 4		2 × 44
8 × 8		16 × 4		20 × 2

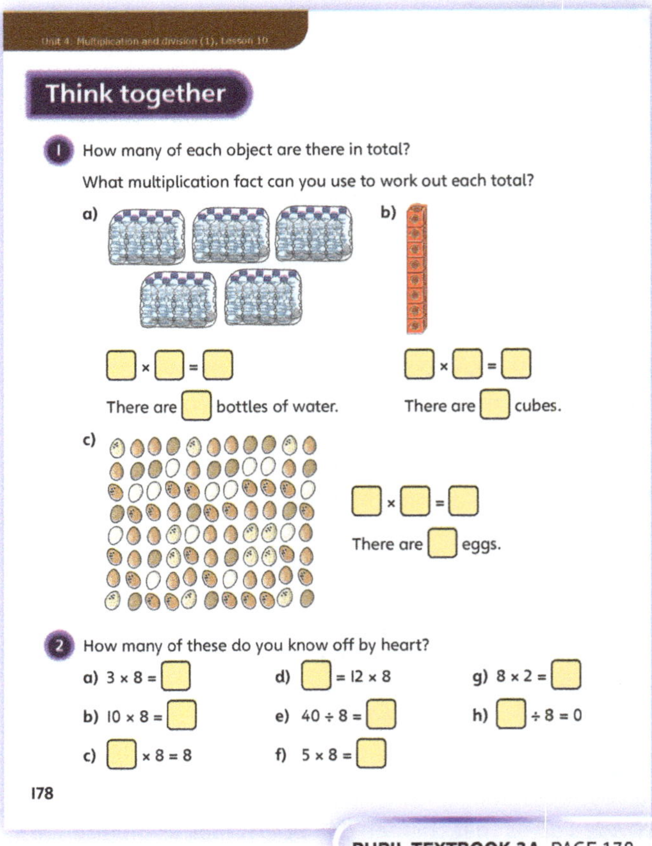

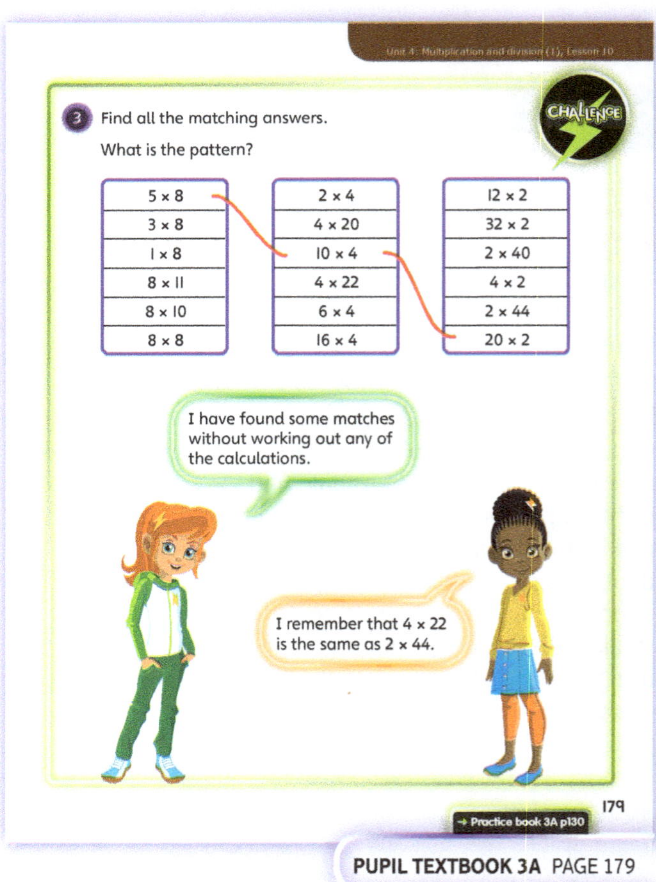

PUPIL TEXTBOOK 3A PAGE 178

PUPIL TEXTBOOK 3A PAGE 179

Practice

WAYS OF WORKING Independent thinking

IN FOCUS Question ① reinforces visual images of several multiplication facts from the 8 times-table. Questions ② and ④ check whether children know the multiplication and division facts: children should try to do them without looking back in the book. If there are any that they do not know off by heart, they could use methods from the previous two lessons. In question ③, children complete number tracks to show the multiples of 8. Question ⑤ asks children to add equals or inequality signs to make statements correct. Encourage children to use reasoning instead of working out all the answers and then comparing. Question ⑥ brings together their work on different times-tables.

STRENGTHEN To strengthen understanding of times-tables facts, show them as a visual or concrete representation; question ① will help with this. As children move through the rest of the exercise, the questions become more abstract, although some children may still need to use cubes or counters to represent multiplications and divisions. What is important is that they can see what times-table facts are. Reinforce that eventually they will need to develop rapid recall of the multiplication and division facts.

DEEPEN In question ⑤, ask children to explain how they can work out which sign completes the sentence without having to work out the answers. Which ones do they need to work out to compare? Can children make up similar problems to the ones in question ⑥, involving other times-tables?

ASSESSMENT CHECKPOINT Can children recognise 8 times-table facts from images? Children should start to be developing a secure recall of the 8 times-table multiplication and division facts.

ANSWERS Answers for the **Practice** part of the lesson appear in the separate **Practice and Reflect answer guide**.

Reflect

WAYS OF WORKING Pair work

IN FOCUS This brings together work on all the times-tables that children know so far, including those covered in Year 2. Children have to find times-table facts to write into the relevant boxes. They should reflect on the strategy that they use. For example, do they list out the times-table facts first or do they just know some answers? Do they notice any patterns in any of the columns?

ASSESSMENT CHECKPOINT Check if children have instant recall of times-table facts across a range of times-tables they have met in Year 2 and Year 3.

ANSWERS Answers for the **Reflect** part of the lesson appear in the separate **Practice and Reflect answer guide**.

After the lesson 

- Can children recall their division and multiplication facts from the 8 times-table?
- Which facts are children struggling with? Why?
- Do children know the connection between the 2, 4 and 8 times-tables?

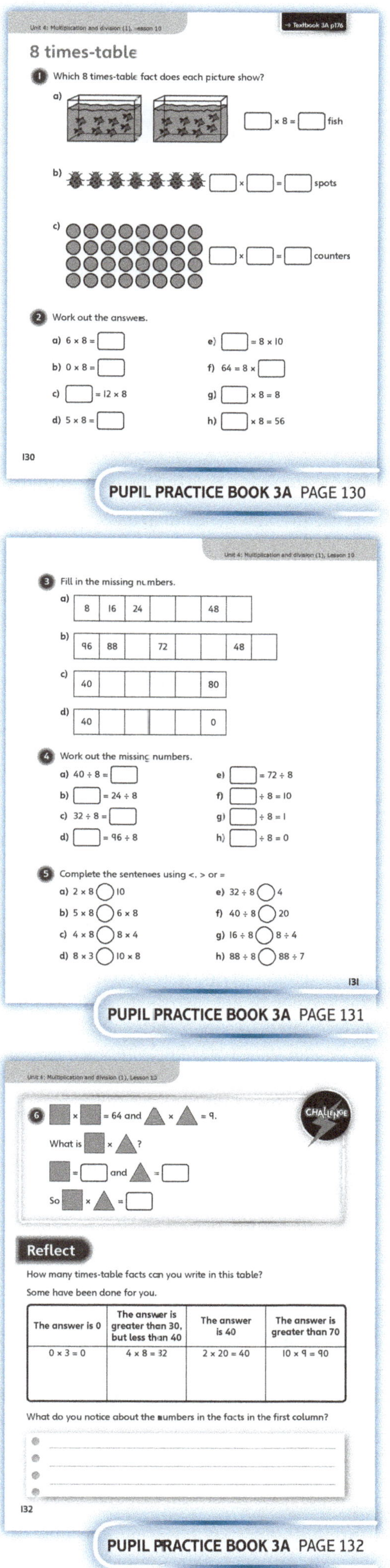

PUPIL PRACTICE BOOK 3A PAGE 130

PUPIL PRACTICE BOOK 3A PAGE 131

PUPIL PRACTICE BOOK 3A PAGE 132

211

Problem solving – multiplication and division ①

Learning focus

In this lesson, children will solve simple one-step multiplication and division problems. They will recognise when they need to multiply and divide and will draw a simple bar model to represent the problem.

Small steps

→ Previous step: 8 times-table
→ **This step: Problem solving – multiplication and division (1)**
→ Next step: Problem solving – multiplication and division (2)

NATIONAL CURRICULUM LINKS

Year 3 Number – Multiplication and Division

- Solve problems, including missing number problems, involving multiplication and division, including positive integer scaling problems and correspondence problems in which *n* objects are connected to *m* objects.
- Write and calculate mathematical statements for multiplication and division using the multiplication tables that they know, including for two-digit numbers times one-digit numbers, using mental and progressing to formal written methods.
- Recall and use multiplication and division facts for the 3, 4 and 8 multiplication tables.

ASSESSING MASTERY

Children can solve a range of simple multiplication and division one-step word problems. They are able to use key information in a question to draw a bar model and then work out whether they need to multiply or divide.

COMMON MISCONCEPTIONS

Children may not be sure whether they need to multiply or divide. Drawing a bar model to represent the situation may help them to see. Encourage children to look at the words in the question for clues. Ask:
- *What will your bar model look like? Do you know how many parts there should be, or the value of each part? What does each number in the question mean?*

STRENGTHENING UNDERSTANDING

Children may not understand the question and what they need to do. To support understanding, read the questions with children line by line. Ask them to identify key information in the questions. If necessary, get them to represent the objects with counters or cubes. This may help them understand the abstract questions better. Try to encourage children to draw a bar model, which may help them see whether they need to multiply or divide. Some children may not be able to recall times-table facts quickly so could use counting up, grouping or sharing strategies.

GOING DEEPER

Ask children to solve problems that involve a single multiplication but are in two parts, such as: *Mary has 4 boxes of 8 tarts and 3 more boxes of 8 tarts. How many are there in total?* Get children to find different ways of working out the total. Encourage them to try to find a solution that uses a single multiplication. How did they do it? Can they explain why it works?

KEY LANGUAGE

In lesson: multiplication, division, multiply, divide, group, shared equally, bar model, total

Other language to be used by the teacher: times-table, recall, grouping, repeated addition, equal, sharing, word problem

STRUCTURES AND REPRESENTATIONS

Bar models

RESOURCES

Optional: cubes, counters, number lines

 In the eTextbook of this lesson, you will find interactive links to a selection of teaching tools.

Before you teach 

- Do children know 2, 3, 4, 5, 8 and 10 times-table facts?
- Can children draw a bar model for addition?
- Do children know how to calculate a multiplication or division if they do not know the relevant times-table fact?

Discover

WAYS OF WORKING Pair work

ASK

- Question ❶ a): *How many plants are there? How many plants are in each row? Can you represent this as a bar model? What type of problem are you solving? Is it division or multiplication?*
- Question ❶ b): *How many bunches of flowers does Amal have? How many are in each bunch? Can you show this on a bar model? How do you work out the total? What calculation do you have to do? Do you know the answer mentally?*

IN FOCUS Question ❶ a) requires children to work out how many rows there are, if there are 24 plants. Ensure that children have at least 24 counters which they can group into 4s or put into an array with 4 columns. Children should be encouraged to show their working and write down a division calculation to work out the answer. Question ❶ b) is a multiplication, which children should notice. They need to look at the picture to realise that Amal has three bunches of flowers. To work out the answers, children should use mental recall where they can. Encourage children to use a bar model to represent the situations throughout. This will help them determine whether the problem they are trying to solve is a multiplication or a division.

PRACTICAL TIPS Use counters or cubes to represent the plants and flowers.

ANSWERS

Question ❶ a): $24 \div 4 = 6$
There will be 6 rows of 4 plants.

Question ❶ b): $3 \times 10 = 30$
Amal has 30 flowers in total.

Share

WAYS OF WORKING Whole class teacher led

ASK

- Question ❶ a): *How many groups of 4 plants are there? How has this been shown using a number line? How does the bar model show this?*
- Question ❶ b): *What does the bar model look like? Why are there 3 bars, each with 10 in?*

IN FOCUS For question ❶ a) show children how the grouping bar model is built up, starting with the brace of 24 and then adding bars using repeated addition until they get to 24. For the multiplication in question ❶ b), children can put the 3 bars of 10 and then use them to work out how many, in total. Showing multiplication as a bar model reinforces the fact that multiplication is repeated addition.

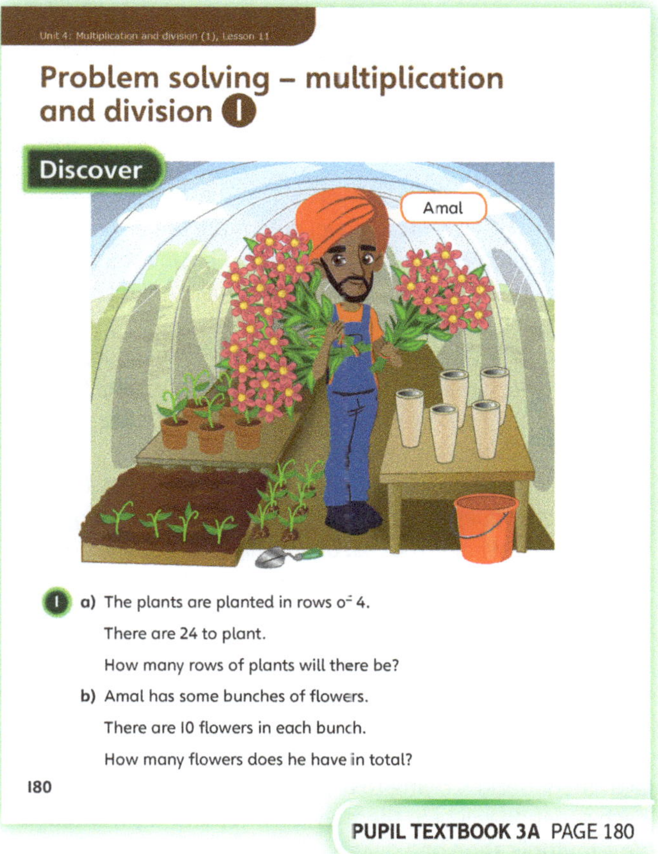

PUPIL TEXTBOOK 3A PAGE 180

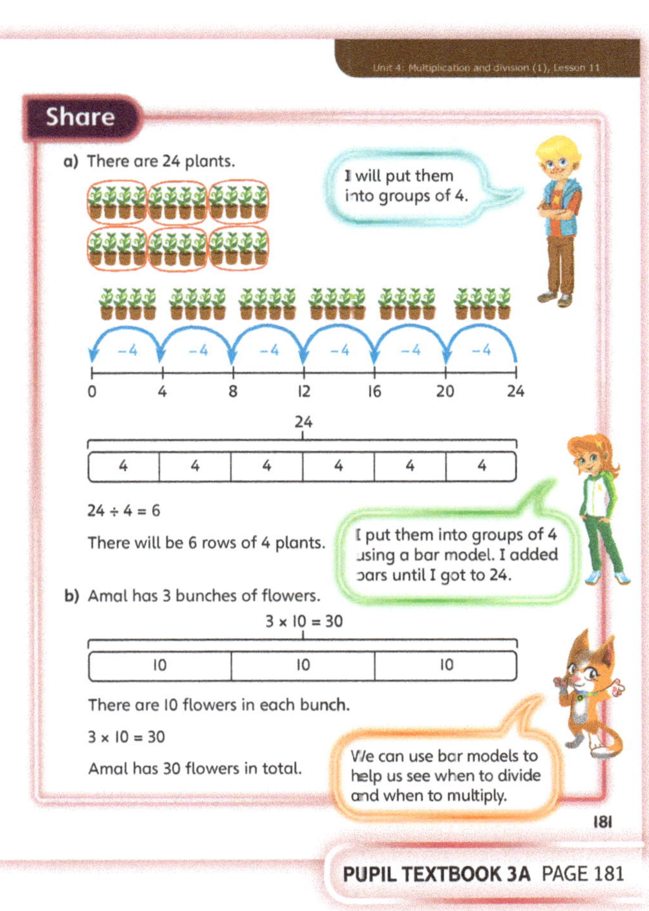

PUPIL TEXTBOOK 3A PAGE 181

Think together

WAYS OF WORKING Whole class teacher led (I do, We do, You do)

ASK

- Question **1**: *How many equal groups are there? How many in each group? How many flowers are there in total? How did you work it out? How does the arrangement help? Can you draw a bar model to show this?*
- Question **2**: *What is the key information in the question that you need to know? How many flowers are there? How many vases? Can you explain the bar model? Do you need to divide or multiply? How do you know?*
- Question **3**: *How much do the small plant pots cost in total? What does this tell you about the cost of the large plant pots? What is the cost of one large pot? Did you do two calculations? Can you represent each calculation on a bar model? Is there a way of working this out more easily?*

IN FOCUS This section takes children through a series of word problems. The key aim is for children to decide whether they need to multiply or divide. They may look at the words, such as 'how many are there in total' or 'shared' to help them decide. Trying to draw a bar model to represent each situation will help them to see whether they have to do a multiplication or a division. Most children are likely to approach question **3** as a two-step problem, first multiplying to find the total and then dividing to find the cost of a large pot. Some children may notice that a large plant pot must be double the price of a small plant pot because 4 large pots cost the same as twice as many small pots.

STRENGTHEN Encourage children to work through each problem line by line. Ask them to highlight the key information in the question. Have they been given the total to start with or do they have to work it out? Support children in drawing a bar model to help them. For question **1**, they will need to draw 3 bars, each with 8 in it. Ask children what the bars represent and how they can work out the total. If children are struggling with finding the answers, support them by using methods of repeated addition, subtraction, grouping and sharing.

DEEPEN For question **3**, ask children if they have to do both a multiplication and division. Ask: *How could 2 bars the same length help you work out the cost of the large plant pot?* This might help them see that they can just double the cost of a small plant pot. Challenge children to make up their own word problems. Can they make up a multiplication and then a division word problem with the answer 12?

ASSESSMENT CHECKPOINT Children can solve simple multiplication and division word problems. They can represent a simple problem as a bar model to help them see whether they need to do a multiplication or a division.

ANSWERS

Question **1**: 8 × 3 = 24
There are 24 plants.

Question **2**: 30 ÷ 5 = 6
There are 6 flowers in each vase.

Question **3**: 8 × £3 = £24; £24 ÷ 4 = £6
A large plant pot costs £6.

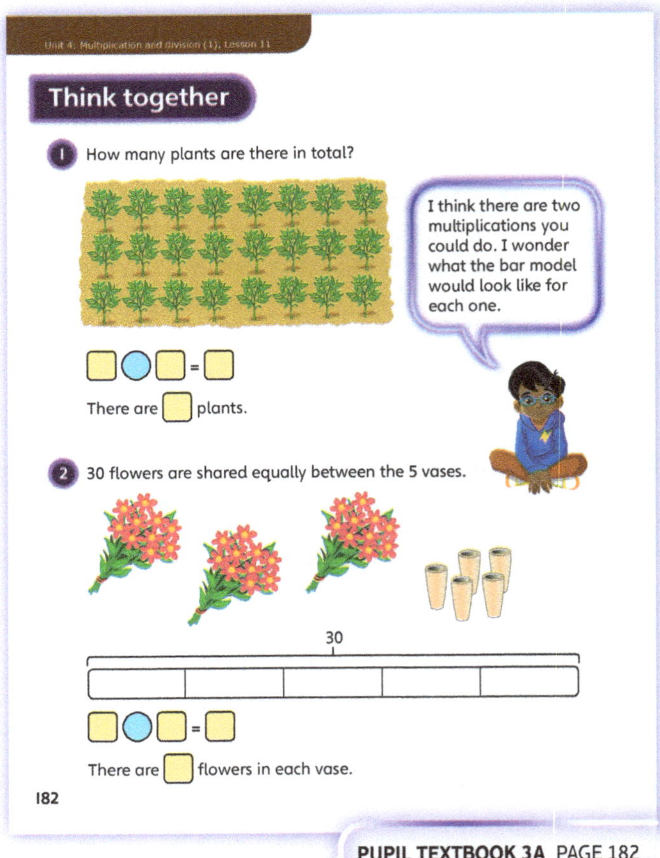

PUPIL TEXTBOOK 3A PAGE 182

PUPIL TEXTBOOK 3A PAGE 183

Practice

WAYS OF WORKING Independent thinking

IN FOCUS This practice exercise contains a series of multiplication and division word problems. Gradually, the structure is removed so children may find it increasingly difficult to work out whether they need to multiply or divide. For each question, a bar model or part of a model is drawn, except towards the end, when children are asked to draw their own to represent a situation. Children will start to realise that a bar model will help them determine the calculation that they need to do. Encourage children to highlight any key information in the question, including numbers and key words; for example, find a 'total' implies multiplication and 'each' often implies division. Questions **1** to **3** are largely fully structured. In question **4**, the structure is removed, which requires children to work out the calculation they need to do. All the questions feature answers from times-tables that they should know, so encourage recall before using other methods to work out an answer.

STRENGTHEN Read the questions line by line and ask them to identify key information. If necessary, ask children to represent the objects with counters or cubes, which may help them to understand the more abstract questions. Encourage children to draw a bar model to help them see whether they need to multiply or divide.

DEEPEN Question **5** is a more complex two-step problem, in the context of weighing. To further deepen understanding, give children multiplication and division bar models for different situations and ask them to write down a word problem that could be solved using this bar model. Can they write another? What is the same about their word problems? What is different?

ASSESSMENT CHECKPOINT Can children convert a word problem into a bar model? Can they use the bar model to work out whether they need to do a multiplication or a division to work out the answer?

ANSWERS Answers for the **Practice** part of the lesson appear in the separate **Practice and Reflect answer guide**.

Reflect

WAYS OF WORKING Whole class

IN FOCUS Children need to create a word problem with an answer of 24. Ask them to share their problems with the class and get the other children to check if the answer is 24. What numbers were used? Did children use the same numbers? If children are unsure which multiplications make 24, then ask them to draw a bar model and see how they might be able to divide it up. They can then work out the correct multiplication to make 24.

ASSESSMENT CHECKPOINT Check if children can make up a word problem that requires them to multiply.

ANSWERS Answers for the **Reflect** part of the lesson appear in the separate **Practice and Reflect answer guide**.

After the lesson ⏸

- Can children solve a range of multiplication and division word problems?
- Can children identify key information that is important to the problem?
- Can they draw a bar model to represent a multiplication problem and the two types of division problem?

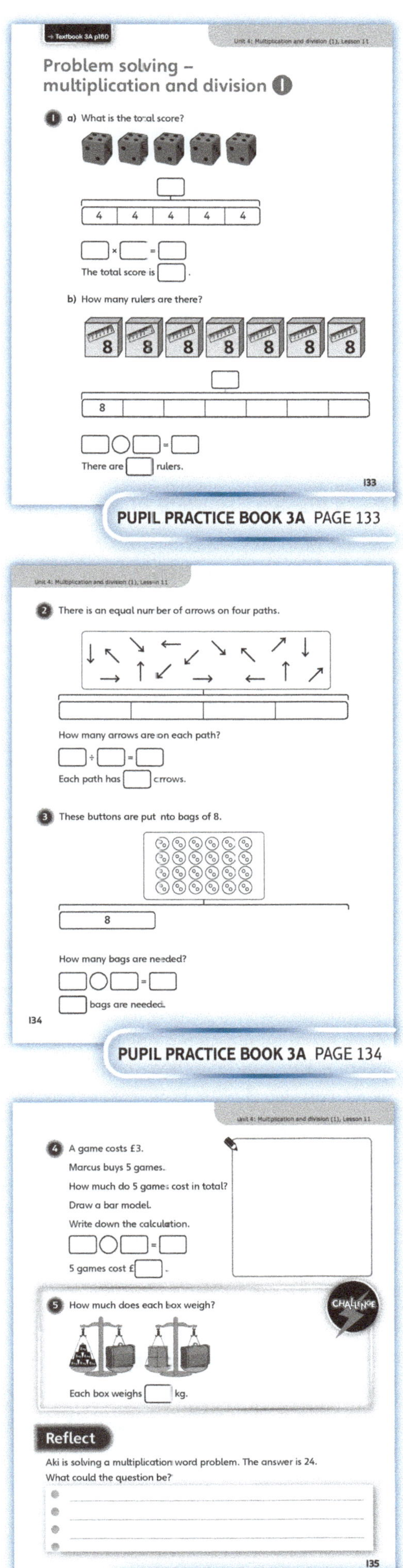

PUPIL PRACTICE BOOK 3A PAGE 133

PUPIL PRACTICE BOOK 3A PAGE 134

PUPIL PRACTICE BOOK 3A PAGE 135

Problem solving – multiplication and division ②

Learning focus

In this lesson, children will solve simple one-step multiplication problems. They will also begin to tackle solving simple two- and three-step multiplication and division problems that may involve an addition or a subtraction.

Small steps

→ Previous step: Problem solving – multiplication and division (1)
→ **This step: Problem solving – multiplication and division (2)**
→ Next step: Understanding divisibility (1)

NATIONAL CURRICULUM LINKS

Year 3 Number – Multiplication and Division

- Solve problems, including missing number problems, involving multiplication and division, including positive integer scaling problems and correspondence problems in which *n* objects are connected to *m* objects.
- Write and calculate mathematical statements for multiplication and division using the multiplication tables that they know, including for two-digit numbers times one-digit numbers, using mental and progressing to formal written methods.
- Recall and use multiplication and division facts for the 3, 4 and 8 multiplication tables.

ASSESSING MASTERY

Children can solve a range of multiplication and division problems that may involve more than one step. They are able to use key information from a question to draw a bar model and work out whether they need to multiply or divide first.

COMMON MISCONCEPTIONS

Children may not be able to work out whether they need to multiply or divide. Drawing a bar model to represent the situation may help with this. Ask:

- *Is this multiplication or division? How do you know? What clues are there in the question?*

STRENGTHENING UNDERSTANDING

To support children in understanding what they need to do, read through the questions line-by-line, asking children to identify the key information. If necessary, get them to represent the objects with counters or cubes. Making towers of cubes (or arrays) for multi-step problems will help them see that they have to carry out two separate multiplications as opposed to just one. This may help them to understand the more abstract questions better. Encourage children to draw a bar model to help them see whether they need to multiply or divide.

GOING DEEPER

Encourage children to make up their own one-step and multi-step problems. Working out the answers is not as important as designing the question. If children can recognise when a problem is one-step or two-step, this will help them gain a deeper understanding of problem-solving questions.

KEY LANGUAGE

In lesson: multiplication, multiply (×), division, total, bar model, pattern

Other language to be used by the teacher: method, times-table, divide (÷), multiplication statement, recall

STRUCTURES AND REPRESENTATIONS

Bar model

RESOURCES

Optional: cubes, counters, number lines, wooden blocks

 In the eTextbook of this lesson, you will find interactive links to a selection of teaching tools.

Before you teach 

- Do children know the 2, 3, 4, 5, 8 and 10 times-table facts?
- Can children draw a bar model for addition?
- Can they work out the answer to simple one-step multiplication and division word problems?

Discover

WAYS OF WORKING Pair work

ASK

- Question ❶ a): *What 3D shape is the block? What is the length of the block? What about the width and the height of the block? What does it mean when the blocks are said to be put 'end to end'? How can you work out the length of the new shape? Is it a multiplication or division? What shape is the new shape?*
- Question ❶ b): *How does this question differ from the last one? What calculation can you do to work out the answer? Is there more than one possible answer?*

IN FOCUS Ensure that children fully understand the context, as this is the first time they have been presented with dimensions of cuboids. Encourage drawing around the shapes if this helps children to work out the total length. Question ❶ a) asks children to work out the length of 3 blocks that have been put together. Some children may want to add three 8s together; they should realise that this is a multiplication. Children could be encouraged to build on their work on bar models from the previous lesson. Question ❶ b) is the opposite and gives children the total length, which should help them realise they need to do a division. Some children may count up in 8s until they get to 32. Encourage them to see the link with the division. Ask children to show how they can lay the blocks in a different way to make a different shape with a length of 32 cm. Draw out the fact that the blocks can be laid end to end in different orientations.

PRACTICAL TIPS Using wooden building blocks and blocks of cubes will be useful in this context.

ANSWERS

Question ❶ a): $3 \times 8 = 24$ cm

Question ❶ b): $32 \div 8 = 4$ blocks, or $32 \div 4 = 8$ blocks.

Share

WAYS OF WORKING Whole class teacher led

ASK

- Question ❶ a): *How has Astrid answered the question? Why has she used a bar model? What does each part of the bar model represent? How can you use the bar model to work out if you need to do a multiplication or an addition?*
- Question ❶ b): *What is the difference between the two methods? Is there more than one way to lay the blocks end to end? How does the answer change?*

IN FOCUS Discuss as a class the different methods used to get the two answers. In question ❶ a) Astrid shows how you can use a bar model to represent the situation. Ensure children know what each part of the bar model represents and how it leads to a multiplication. In question ❶ b), discuss the difference between the two methods used. Explain to children that eventually they need to try to do the division like Flo, but that Dexter's method explains exactly what is going on. Discuss that there are two solutions to question ❶ b), as the blocks can be laid in different orientations.

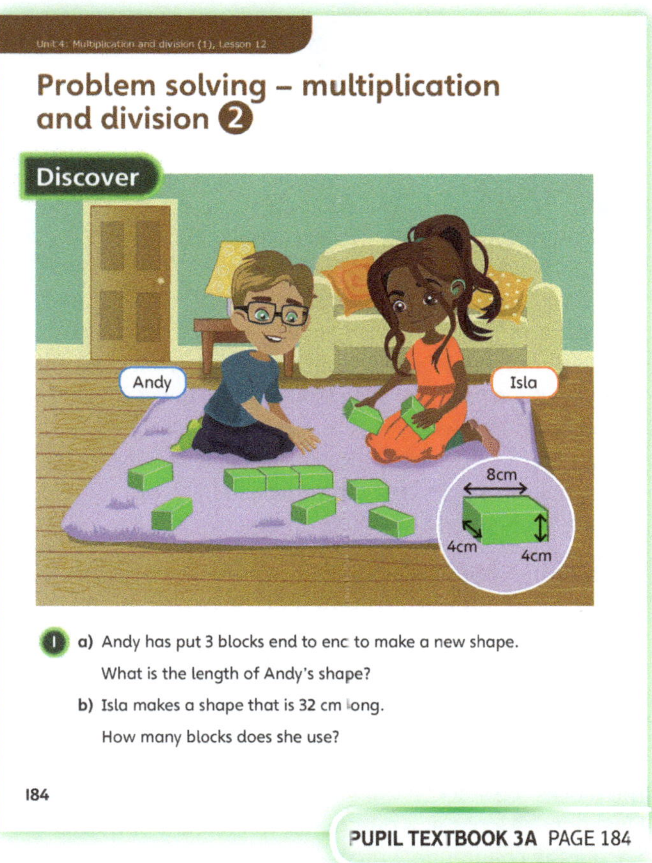

PUPIL TEXTBOOK 3A PAGE 184

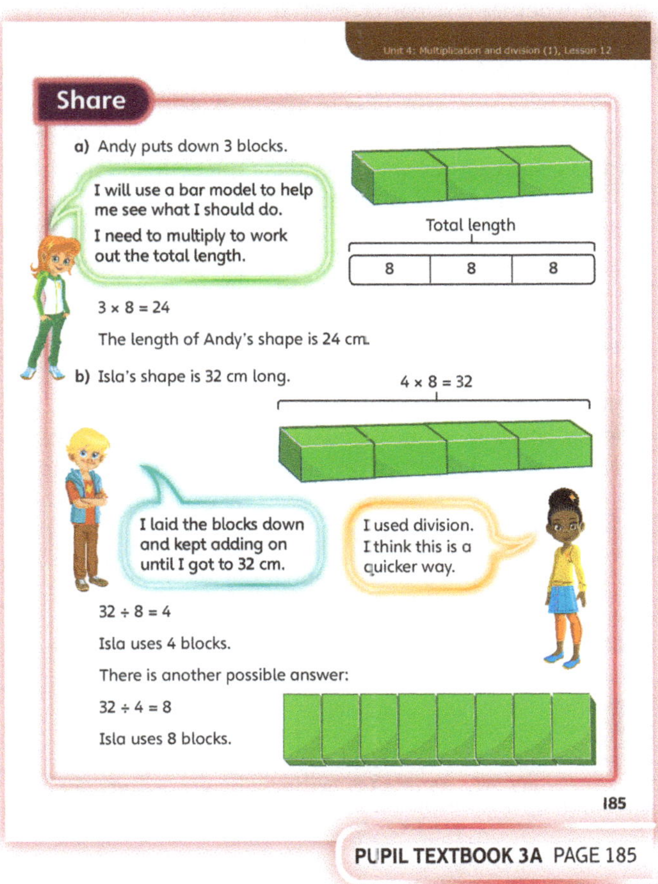

PUPIL TEXTBOOK 3A PAGE 185

Think together

WAYS OF WORKING Whole class teacher led (I do, We do, You do)

ASK

- Question **1** : *How many blocks are there? How long is each block? Why is it a multiplication? How does the bar model help you understand it is a multiplication?*
- Question **2** : *What is the height of each block in each tower? How many blocks are in each tower? How can you work out the height of each tower? Which tower is taller? How do you know?*
- Question **3** : *Can you make this using your own blocks? Why does Astrid suggest that you need to do two multiplications and add them together?*

IN FOCUS Question **1** builds on the **Discover** task by giving children seven blocks to find the length of. The bar model should help them understand why it is a multiplication. You may also want to use blocks to help children see this. Question **2** looks at the height of some towers with the blocks stacked in different ways. This is the first time that children will need to realise that the way the blocks are stacked will affect the height of the tower. Question **3** looks at a different arrangement of blocks. Encourage children to write two multiplications as opposed to a single addition sentence. For each multiplication, ask children if they understand what length it finds and why they need to add the two together.

STRENGTHEN Encourage children to work through each problem line by line. Ask children to highlight what the key information is in the question (each question in the **Practice Book** for this lesson is a multiplication). Ask children to draw or use a bar model to explain each of the steps. For question **1**, they will need to draw 7 bars, each with 4 in. Ask children what the bars represent and how they can work out the total length. If children find the shapes in question **2** hard to visualise, they can use wooden blocks or cubes to help them. Children need to see how stacking the blocks in different ways will give different heights.

DEEPEN For question **2**, probe the question further and ask children what would happen if the tower was 32 cm high. Ask: *How many blocks might you use? Are there two different answers?* In question **3**, can children write their answer as a single multiplication? Why not? Can they find the distance all the way around the shape?

ASSESSMENT CHECKPOINT Children can solve simple multiplication word problems and represent a simple problem as a bar model to help them see whether they need to do a multiplication or division. They recall times-table facts to work out the answers.

ANSWERS

Question **1** : 7 × 4 = 28 The shape is 28 cm long.

Question **2** : Tower A: 5 × 4 = 20 cm tall. Tower B: 2 × 8 = 16 cm tall. Tower A is the taller tower. It is 4 cm taller.

Question **3** : 3 × 4 = 12 cm, 2 × 8 = 16 cm, 12 + 16 = 28 cm The pattern is 28 cm long.

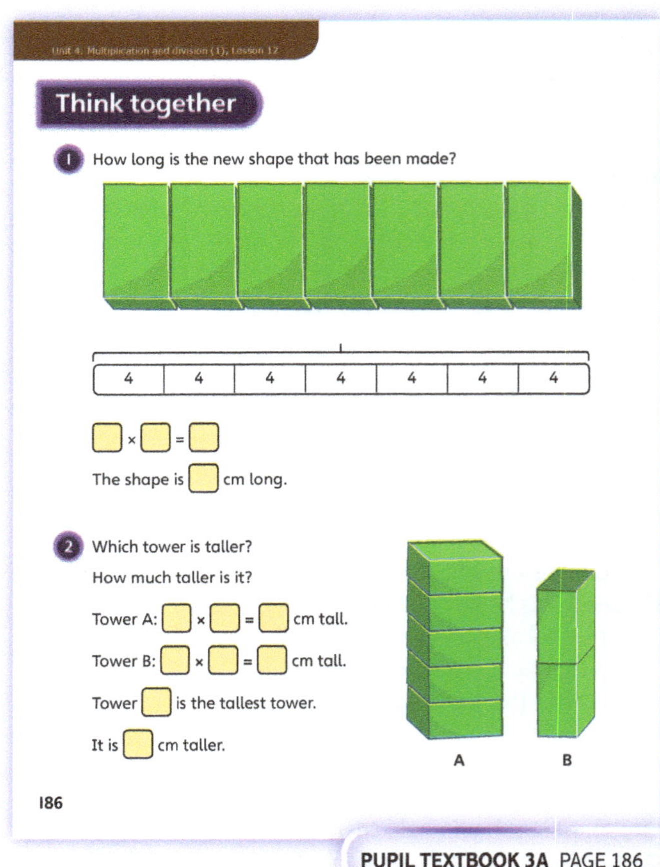

PUPIL TEXTBOOK 3A PAGE 186

PUPIL TEXTBOOK 3A PAGE 187

Practice

WAYS OF WORKING Independent thinking

IN FOCUS In questions ❶ and ❷, children have to do two multiplications and then add them together to find the total. Using a single bar model may help children understand the steps that they need to take. They will see that there are two different-sized bars in the model and they need to work out the separate lengths and then add them together. Question ❺ presents children with a common type of repacking problem. They first use multiplication to work out the total and then do the division to work out how many packs there will be if they are repacked into different-sized packs. Due to the numbers involved, children may do this in just one step.

STRENGTHEN To support understanding, use counters and cubes to represent the different groups of objects; this will help children see that in questions ❶ and ❷ they first need to do two multiplications and then an addition. For question ❸, children may find that drawing a simple array helps them compare the two groups of objects. For two-step problems, it is helpful to see a visualisation of why they have to work out one part before the other. Encourage children to draw a bar model, which may help them to see whether they need to multiply or divide.

DEEPEN In question ❹, children can make up their own multi-step problems. For example: how much do five buckets and spades and two beach balls cost? How much change will you get from £50?

ASSESSMENT CHECKPOINT Can children convert a word problem into a bar model? Can they use the bar model to work out whether they need to do a multiplication or division? Can children solve two-step problems across all four operations?

ANSWERS Answers for the **Practice** part of the lesson appear in the separate **Practice and Reflect answer guide**.

Reflect

WAYS OF WORKING Independent thinking

IN FOCUS There are two types of problem that children may come up with: either a simple one-step problem working out a total cost, or a two-step problem (such as finding a total cost and then calculating change). Some children may want to use the structure of the earlier practice and devise a problem where they have to work out the result of two multiplications and add or subtract them. Encourage children to make up several problems to demonstrate the difference between one-step and multi-step problems. Children can share their problems for friends to work out.

ASSESSMENT CHECKPOINT Check whether children can make up a word problem that requires either division or multiplication. Do children write a simple one-step problem or problems where multiple steps are involved?

ANSWERS Answers for the **Reflect** part of the lesson appear in the separate **Practice and Reflect answer guide**.

After the lesson ⏸

- Can children solve a range of multiplication and division multiple-step word problems?
- Can children identify key information that is important to a problem and realise what type of calculation they have to do?
- Can they draw bar models to represent a multiplication problem and the two types of division problem?

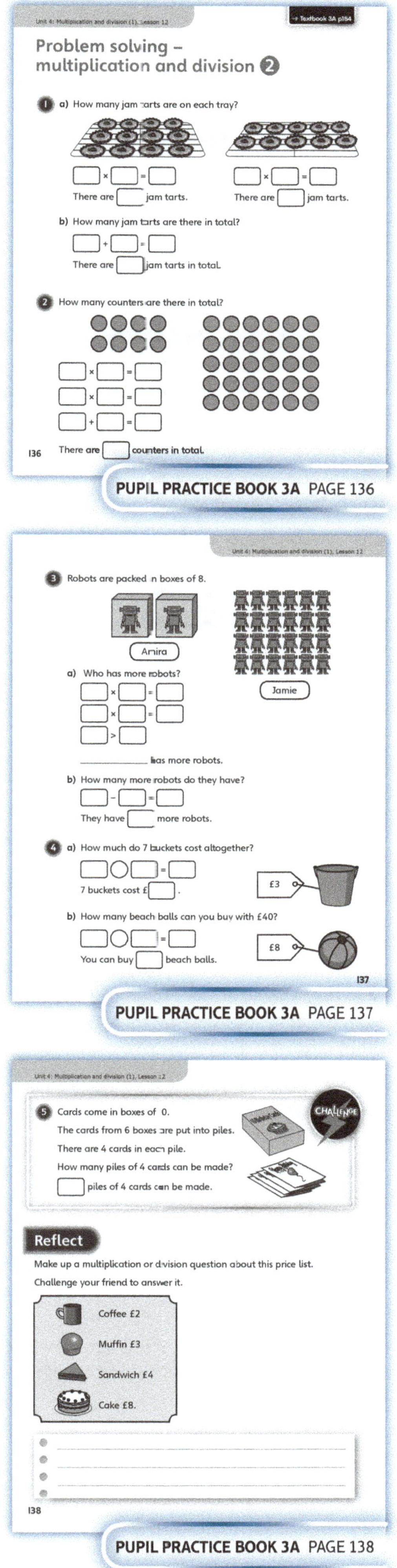

PUPIL PRACTICE BOOK 3A PAGE 136

PUPIL PRACTICE BOOK 3A PAGE 137

PUPIL PRACTICE BOOK 3A PAGE 138

Understanding divisibility ❶

Learning focus

In this lesson, children will realise that some division problems leave a remainder. They will also realise that the greatest possible remainder is 1 less than the number they divide by.

Small steps

→ Previous step: Problem solving – multiplication and division (2)
→ **This step: Understanding divisibility (1)**
→ Next step: Understanding divisibility (2)

NATIONAL CURRICULUM LINKS

Year 3 Number – Multiplication and Division

Solve problems, including missing number problems, involving multiplication and division, including positive integer scaling problems and correspondence problems in which *n* objects are connected to *m* objects.

ASSESSING MASTERY

Children can realise that some divisions are not always exact and can leave a remainder. Children start to understand that the greatest possible remainder is 1 less than the number they are dividing by.

COMMON MISCONCEPTIONS

Children may not recognise that what they are doing in this lesson is division. To support this, remind children that division can be seen as either grouping or sharing. Explain that making squares is like taking a group of 4 lollipop sticks from the starting amount. Children should continue to do this until they have some left over. Ask:
• *What does this remind you of? Is it grouping or sharing?*

STRENGTHENING UNDERSTANDING

Children should have access to lollipop sticks or strips of paper to make squares, triangles and other shapes. Work with children to present the results of their divisions in the form of a table. Encourage them to record all their working, including diagrams of what they made. Some children may find it easier to begin again each time by taking the correct number of lollipop sticks and making the squares. Eventually, children may see that they can just add 1 more lollipop stick each time.

GOING DEEPER

Children could do the reverse to work out the number of lollipop sticks that are needed to make 5 squares and a remainder of 3. They should also try to predict amounts of lollipop sticks that have a remainder of 1 when they are making squares. For example, they may say that there is a remainder of 1 for numbers that are 1 more than a multiple of 4. They should solve other problems like this, such as: when does a division have a remainder of 0?

KEY LANGUAGE

In lesson: left over, remainder, pattern

Other language to be used by the teacher: divide, grouping, complete, whole

RESOURCES

Mandatory: lollipop sticks or strips of paper to make shapes

Optional: cubes

 In the eTextbook of this lesson, you will find interactive links to a selection of teaching tools.

Before you teach ⓫

• Do children understand simple division by grouping?
• Do they know the names of basic shapes and the number of sides they have?

Discover

 Pair work

ASK

- Question ❶ a): *How many lollipop sticks do they have in total? How many whole squares can they make? How many are left over? Why is this a division problem? Why are you dividing by 4?*
- Question ❶ b): *What happens to your answers as you increase the number of lollipop sticks? Is this still dividing by 4? Why? What is the smallest number of lollipop sticks that are left over? What is the greatest? Why? Do you notice a pattern in the number of lollipop sticks left over?*

IN FOCUS In question ❶ a), children have to deal with an amount left over in division, and are introduced to the term 'remainder' for the first time. In question ❶ b), children start to investigate further by looking at how the answer changes if they have 14, 15, 16 or 17 sticks. They may start to realise that what they are doing is a division (equal grouping). In this case, they are dividing by 4, because they are making squares. The squares represent the whole and the ones left over represent the remainder(s). Children may start to notice a pattern in the number of whole squares and remainders.

PRACTICAL TIPS Provide children with lollipop sticks or strips of paper to make the squares.

ANSWERS

Question ❶ a): 13 sticks – 3 whole squares and 1 left over

Question ❶ b): 14 sticks – 3 whole squares and 2 left over
15 sticks – 3 whole squares and 3 left over
16 sticks – 4 whole squares and 0 left over
17 sticks – 4 whole squares and 1 left over

Share

 Whole class teacher led

ASK

- Question ❶ a): *How many squares have been made? How many are left over? What name do we give to the amount left over? What calculation is represented here?*
- Question ❶ b): *How has Flo organised her work? Why do you think this is a good way? Did you need to start again each time? What has Flo put into her table? Do you notice any patterns in the table? Why will the remainder never be greater than 3?*

IN FOCUS Question ❶ a) is an opportunity to look at wholes and remainders. Explain to children that what they have done is the calculation 13 divided by 4. This will be introduced formally in the next lesson, though children do need to understand that they are doing a division. Explain that they had 13 lollipop sticks to start with and were grouping into 4s; making squares helps children see the number of wholes. Sparks explains that the amount left over is called the remainder.

PUPIL TEXTBOOK 3A PAGE 188

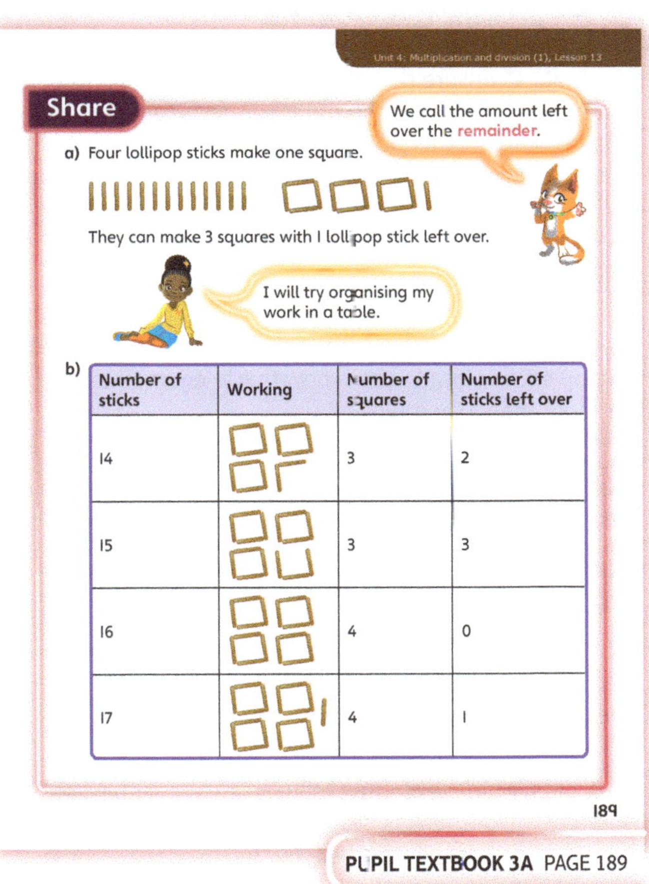

b)

Number of sticks	Working	Number of squares	Number of sticks left over
14		3	2
15		3	3
16		4	0
17		4	1

PUPIL TEXTBOOK 3A PAGE 189

Think together

WAYS OF WORKING Whole class teacher led (I do, We do, You do)

ASK

- Question ❶ : *How many lollipop sticks are we going to use this time? What do you need to record in the table? Do you need to start again each time? Can you do this without making the squares?*
- Question ❷ : *What patterns can you see? Why do you think that this is the case? What is the most the remainder can be? Can you explain why?*
- Question ❸ : *What shape are we making this time? What are we dividing by this time? Can you explain why? Can you predict the most you can have left over? What are the reasons behind your prediction?*

IN FOCUS In question ❶, children extend the table from earlier. They should not need to start again each time and should now be able to build on the squares they already have, adding 1 each time. They should now be looking for patterns. Remind children that they are dividing by 4. Can they explain why? This leads into question ❷, where children are asked to spot patterns. Ask children what patterns they notice. For example, why does the number of whole squares stay the same for a while? Children should be able to explain the pattern of remainders going 0, 1, 2, 3. They should also reason that the remainder will never be 4 or above as this would make another square. Question ❸ is a similar investigation, but this time looking at making triangles with the lollipop sticks. Children should see this as dividing by 3 this time, because they are making triangles.

STRENGTHEN Ensure that children have lollipop sticks or strips of paper to make the squares. They should be able to record the number of whole squares and the number of lollipop sticks left over. Some children may find it easier to do the first few from the start again to see how many squares they can make.

DEEPEN Deepen understanding of this being a division task by seeing if children can write some of the answers as divisions. They could predict the remainder if they had 41 lollipop sticks. How do they know that the remainder is 1? (It is 1 more than a multiple of 4.)

ASSESSMENT CHECKPOINT Children recognise that they have been grouping objects. They see that the number of squares is the whole and the number left over is the remainder.

ANSWERS

Question ❶ : 18 sticks – 4 squares and 2 left over
19 sticks – 4 squares and 3 left over
20 sticks – 5 squares and 0 left over

Question ❷ a): The amount left over goes 0, 1, 2, 3 and then repeats itself.

Question ❷ b): The amount left over can never be 4 or more as otherwise they would be able to make another square.

Question ❸ : 11 sticks – 3 triangles and 2 left over
12 sticks – 4 triangles and 0 left over
13 sticks – 4 triangles and 1 left over
14 sticks – 4 triangles and 2 left over
15 sticks – 5 triangles and 0 left over

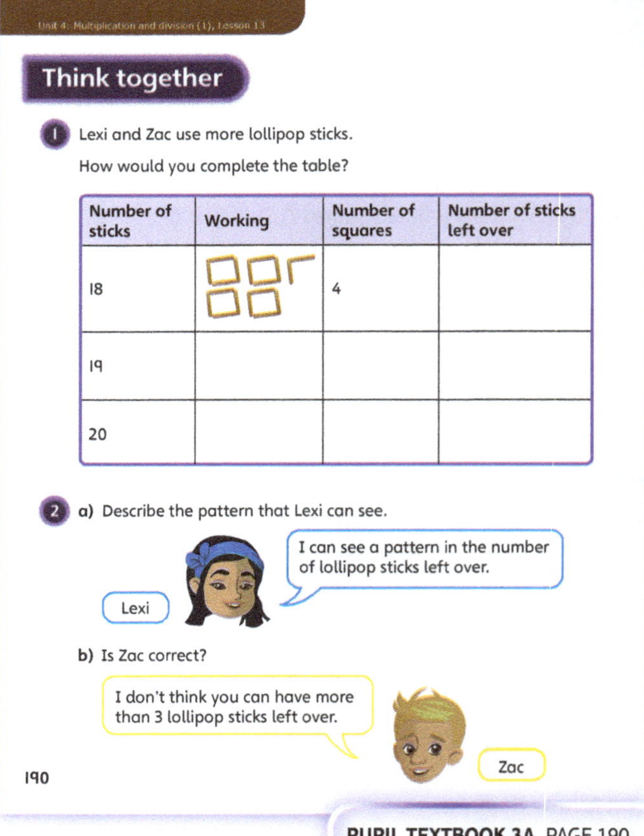

PUPIL TEXTBOOK 3A PAGE 190

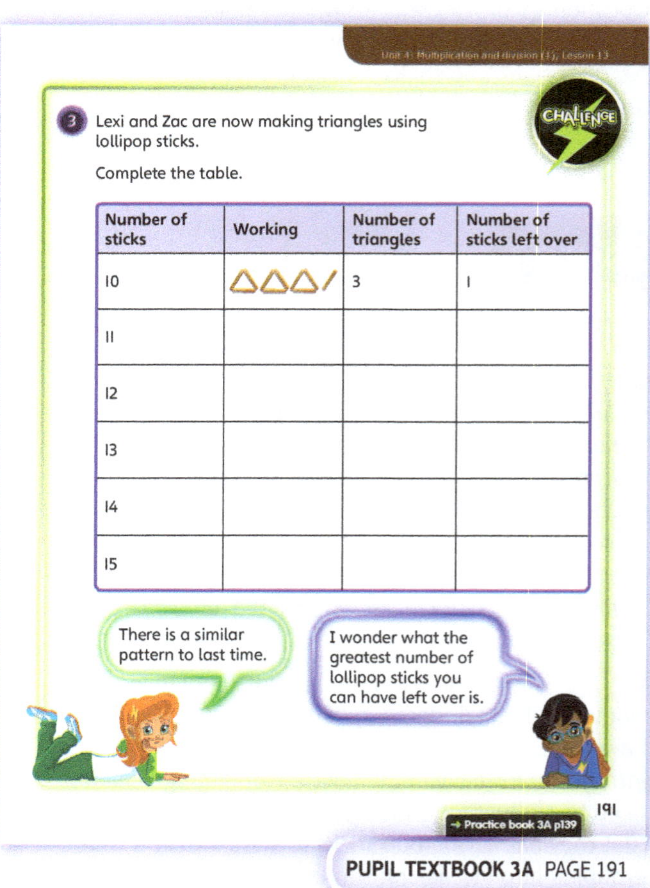

PUPIL TEXTBOOK 3A PAGE 191

Practice

WAYS OF WORKING Independent thinking

IN FOCUS Question ① builds on the **Textbook** by recapping making squares and triangles. Children must record their working out as well as the number of whole shapes made and the remainder. The word 'remainder' is now used instead of 'left over', to encourage correct mathematical language. In question ②, children extend their thinking by making regular pentagons. They should see that they are now dividing by 5, and should be able to predict the pattern of whole pentagons and remainders.

STRENGTHEN Some children will be able to tackle this without having lollipop sticks or strips of paper to help them. Children who are still struggling to work out the wholes by drawing can use lollipop sticks to do this, which will help them see the number of wholes and the remainder.

DEEPEN In question ③, children work from the wholes and the remainder to work out the number of cubes that Max started with. They may need to draw the blocks to help them. Ask children to explain their reasoning. Give them other wholes and remainders and see if they can find a method to work out the original starting number of cubes. To further deepen understanding, ask children to predict the remainder if you had 45 cubes, or 82 cubes, by thinking of the numbers in the 4 times-table.

ASSESSMENT CHECKPOINT Children recognise that they have been doing division throughout and understand the terms 'whole' and 'remainder'. Children can see a pattern in the number of wholes and the remainder.

ANSWERS Answers for the **Practice** part of the lesson appear in the separate **Practice and Reflect answer guide**.

Reflect

WAYS OF WORKING Pair work

IN FOCUS Children have been doing lots of repetitive practice and should realise that making squares is dividing by 4 and making triangles is dividing by 3. When dividing by 5, children should be able to communicate through words or diagrams that they are making pentagons. Children should be able to reason that the greatest remainder they can have is 4 because if they had 5, they would be able to make another pentagon.

ASSESSMENT CHECKPOINT Check whether children know that, when dividing by 5, the greatest possible remainder is 4.

ANSWERS Answers for the **Reflect** part of the lesson appear in the separate **Practice and Reflect answer guide**.

After the lesson ⏸

- Do children know that some situations will leave a remainder?
- Can they explain why the greatest possible remainder when dividing by 4 is 3, and so on?

PUPIL PRACTICE BOOK 3A PAGE 139

PUPIL PRACTICE BOOK 3A PAGE 140

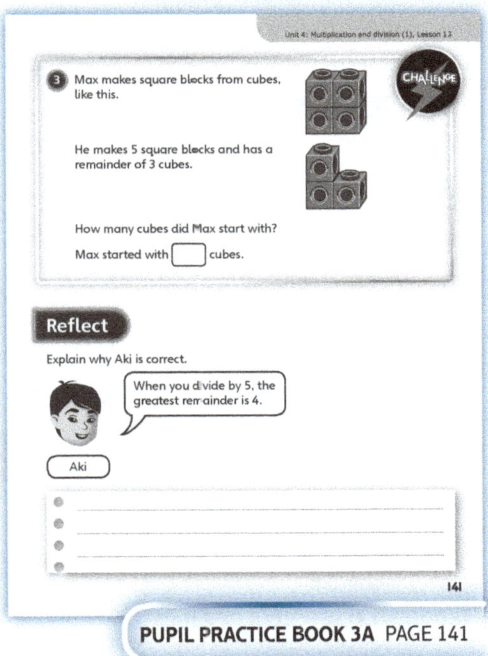

PUPIL PRACTICE BOOK 3A PAGE 141

223

Understanding divisibility ❷

Learning focus

In this lesson, children will continue to realise that some division problems leave a remainder. Children will learn to write the result of a division with a remainder more formally.

Small steps

➔ Previous step: Understanding divisibility (1)
➔ **This step: Understanding divisibility (2)**
➔ Next step: Related facts – multiplication and division

NATIONAL CURRICULUM LINKS

Year 3 Number – Multiplication and Division

- Solve problems, including missing number problems, involving multiplication and division, including positive integer scaling problems and correspondence problems in which *n* objects are connected to *m* objects.
- Write and calculate mathematical statements for multiplication and division using the multiplication tables that they know, including for two-digit numbers times one-digit numbers, using mental and progressing to formal written methods.
- Recall and use multiplication and division facts for the 3, 4 and 8 multiplication tables.

ASSESSING MASTERY

Children can understand that some divisions may leave a remainder and should start to recognise (using their knowledge of times-tables) when a division may lead to a remainder. Children understand that a remainder may result from both equal sharing and equal grouping and can find the result of a division with a remainder and write it in the form '*a* remainder *b*'.

COMMON MISCONCEPTIONS

Children may lack understanding of what the remainder means. They need to understand that the remainder is the number left over at the end (because there are either not enough to group or not enough to share). Ask:
- *Do you have enough to make another whole? Why? Why not? What is that called?*

Children may mix up the whole and the remainder when writing out answers formally. To help overcome this, ask children to link the final answer with the context in the question: for example, 13 apples shared between 5 people; each person receives 2 apples and the remainder is 3 apples. Remind children that we always write the remainder at the end. Showing this as an array may also help. Ask:
- *Can every person have the same amount? How many do they each have? How many are left?*

STRENGTHENING UNDERSTANDING

Ensure that children have access to concrete objects and demonstrate how to carry out both grouping and sharing procedures for division. Children should realise that the answer remains the same regardless. Ask children to carry out the grouping or sharing step by step, explicitly linking it to the concrete objects. They may find putting objects into arrays helpful for understanding the whole and the remainder.

GOING DEEPER

Children work back and solve missing number problems such as: ___ ÷ 4 = 5 remainder 3, and 15 ÷ ◯ = 7 remainder 1. Children may then tackle problems such as: ___ ÷ ___ = 2 remainder 3. Can they find several answers to this problem? Can they reason why this cannot be a division by 2 or 3?

KEY LANGUAGE

In lesson: left over, division, divide (÷), remainder, shared equally, array, group

Other language to be used by the teacher: whole, complete

STRUCTURES AND REPRESENTATIONS

Number lines, arrays

RESOURCES

Mandatory: counters or cubes

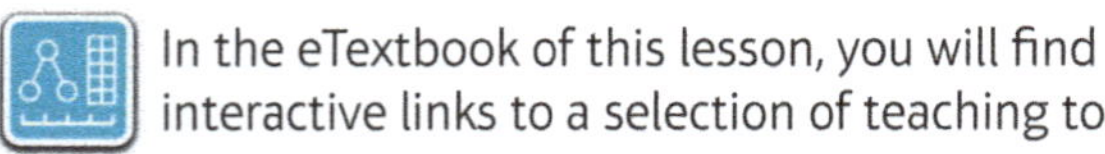 In the eTextbook of this lesson, you will find interactive links to a selection of teaching tools.

Before you teach 

- Do children understand the difference between equal grouping and equal sharing?
- Do children know that the remainder must be less than the number you are dividing by?
- Do children know the 2, 3, 4, 5, 8 and 10 times-table facts?

Discover

 Pair work

ASK

- Question **1** a): *How many apples are there? How many should you put into a group? Do you think you will have any left? Can you form an array to help you? How many do you have left each time? How many do you have left at the end? What is the amount left called?*
- Question **1** b): *Why can what you have done be written as a division? Is it grouping or sharing? What number are you dividing by what? How many whole lines or groups have you formed? What is the remainder?*

IN FOCUS This lesson formalises writing division with a remainder. In question **1** a) children start with 22 counters (representing apples) and have to group them into 5s. Before doing the grouping, children should consider if they are likely to have any apples left at the end. They may be able to reason that they will because it is not a multiple of 5. Children may find it easier to form an array made up of rows of 5 counters. This will help them see how many complete rows are formed and how many are left over. Question **1** b) focuses on how children may record their answer. Children need to understand that they are carrying out a division, as they are grouping, and then they need to realise that the answer is not an exact whole and that there will be a remainder.

PRACTICAL TIPS Ensure that there are enough counters or cubes for children to be able to group the apples into 5s.

ANSWERS

Question **1** a): There are 4 full bags and 2 apples left over.

Question **1** b): 22 ÷ 5 = 4 remainder 2

Share

 Whole class teacher led

ASK

- Question **1** a): *Can you explain what Flo is doing in each step? Is this an example of grouping or sharing? Could you have predicted that there were going to be some left over at the end? If so, how did you know? How does the array help you find the number of groups and the remainder?*
- Question **1** b): *Can you explain why it is a division? Why is this grouping and not sharing? Could we have done it as sharing? If so, what would the remainder have been?*

IN FOCUS In question **1** a) talk through the approach to dividing 22 by 5. 5 apples are grouped in a bag and there are 17 left, and so on, until there are 4 bags and 2 apples left. Model this with cubes, counters or other objects. Also show the method where children form an array. The array helps children to keep track of the number of groups and ensures that they have the same number in each group. Arrays can be used for grouping and for sharing. Question **1** b) shows how this can be written and recorded more formally. Children need to understand that this is a division, as they are grouping objects. Show that the answer would have been the same if they were sharing 22 apples between 5 people: each person would have 4 apples with 2 left over.

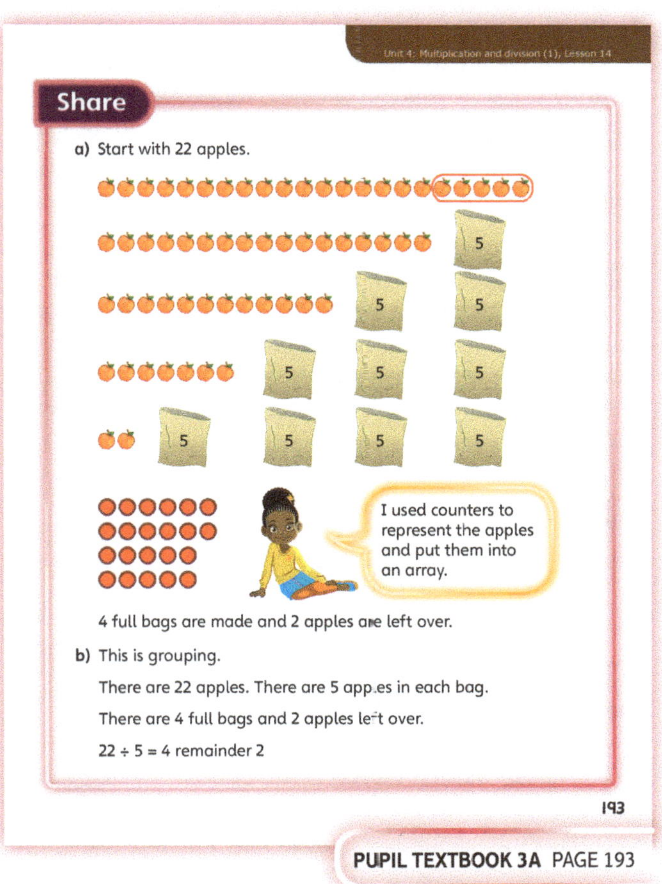

PUPIL TEXTBOOK 3A PAGE 193

Think together

WAYS OF WORKING Whole class teacher led (I do, We do, You do)

ASK

- Question **1** : *Is this equal grouping or sharing? How many pieces of fruit will go in each bowl? How do you know at the start that there are going to be some left over?*
- Question **3** : *Can you tell what the remainder is going to be at the start? Can you do any of these without using equipment? Did you use a sharing or a grouping method? What do you notice about numbers that do not have a remainder when you divide by 3? Where have you seen these numbers before?*

IN FOCUS Question **1** is a sharing example as opposed to a grouping example. Children should see that the remainder is the amount left over that cannot be shared equally. In question **3**, children can choose to use either a grouping or a sharing method to work out the whole and the remainder. Can children work out any of the remainders without doing the full calculation? Question **3** b) asks children to find a number that can be divided by 3 with no remainder. Children may generalise that these are numbers in the 3 times-table.

STRENGTHEN To support understanding of remainder, children should use concrete objects to carry out both grouping and sharing procedures. They should realise that the answer remains the same regardless. Ask children to carry out the grouping or sharing step by step and, at the end, count the number of groups formed (or the number in each group for sharing) and the amount remaining. Show children how the answer can then be written. Children may find that making arrays helps them see the answers more clearly: the number of rows (or number in each row) will help them see the whole, and the number left over not in the array is the remainder.

DEEPEN Build on question **3** by asking: *When you divide by 10, what other numbers have a remainder that is also 1? What numbers have a remainder of 2 when you divide by 10? Can you spot a pattern?*

ASSESSMENT CHECKPOINT Children use a grouping or sharing strategy to work out the answers to a simple division that includes a remainder. Children know how to record their answers in the form '☐ remainder ☐'. They should be able to talk about what each number in the answer means (such as the number of pieces of fruit in each bowl and the amount left over).

ANSWERS

Question **1** : There are 3 oranges in each bowl.
There is 1 orange left over.
7 ÷ 2 = 3 remainder 1

Question **2** : 14 ÷ 3 = 4 remainder 2

Question **3** a): 12 ÷ 5 = 2 remainder 2
17 ÷ 4 = 4 remainder 1
13 ÷ 8 = 1 remainder 5
51 ÷ 10 = 5 remainder 1

Question **3** b): Numbers in the 3 times-table have a remainder of 0.

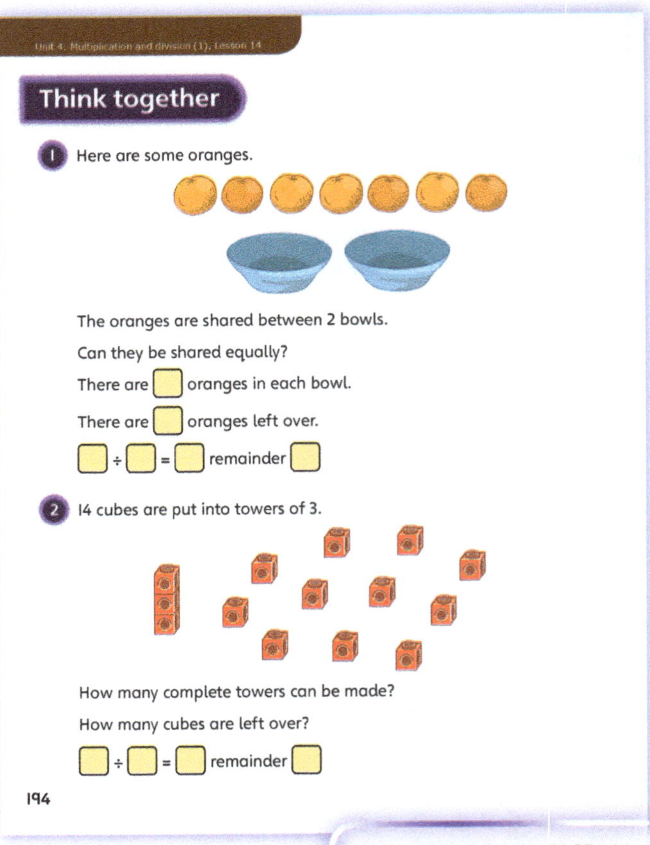

PUPIL TEXTBOOK 3A PAGE 194

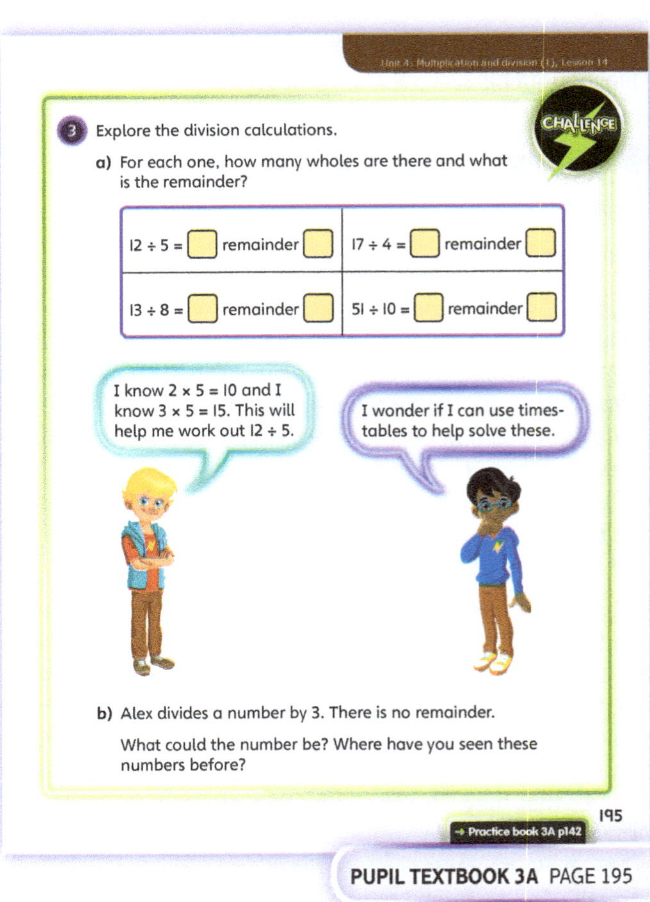

PUPIL TEXTBOOK 3A PAGE 195

Practice

WAYS OF WORKING Independent thinking

IN FOCUS Question ④ keeps the amount children are dividing the same and changes the number they are dividing by. This helps children focus on the process. Question ⑤ provides abstract practice. Here, children need to work out which ones will have a remainder: they should realise that the questions without a remainder are those where the number they are dividing is in the relevant times-table. You should ensure that children know what each number in a division and the answer mean.

STRENGTHEN To support understanding of remainder, children should use concrete objects and carry out both grouping and sharing procedures for division. In questions ④ and ⑤, children may need support to realise how many counters they need to take and the number of groups they need to make (or the number in each group). Children need to see the clear link between the numbers in the question and the objects.

DEEPEN To build on question ④, ask children to come up with a division where the remainder is 5. Ask: *What would you divide by?* Ask children to create their own problems similar to the ones in questions ⑥ and ⑦, where they have to work out the number being divided. You can also ask children to find examples, such as where ___ ÷ ___ = 2 remainder 1.

THINK DIFFERENTLY Question ⑥ a) requires children to think about the numbers that will give a remainder of 1 when divided by 4. Children should notice that the numbers that have a remainder of 1 are all numbers that are 1 greater than a number in the 4 times-table.
Question ⑥ b) asks children to reason that the remainder cannot be bigger than the number you are dividing by.

ASSESSMENT CHECKPOINT Children know how to find the answer to a division that includes a remainder, and how to write their answers formally.

ANSWERS Answers for the **Practice** part of the lesson appear in the separate **Practice and Reflect answer guide**.

Reflect

WAYS OF WORKING Pair work

IN FOCUS This focuses on numbers that have a remainder of 0 when you divide by 3. Children should notice through discussion that numbers in the 3 times-table have no remainder when they divide by 3. They then attempt to find numbers that have a remainder of 1. They may use a trial and improvement strategy to realise that numbers that are 1 more than numbers in the 3 times-table have a remainder of 1.

ASSESSMENT CHECKPOINT Children know that numbers in the 3 times-table have no remainder when you divide by 3. They know that if a number is 1 more than a number in the 3 times-table, then it has a remainder of 1.

ANSWERS Answers for the **Reflect** part of the lesson appear in the separate **Practice and Reflect answer guide**.

After the lesson 

- Can children complete a simple division that has a remainder, using either a grouping or a sharing strategy?
- Do children understand why sometimes a division has a remainder and can they predict when there will be a remainder?
- Can children record their answers for division formally, such as: 13 ÷ 5 = 2 remainder 3?

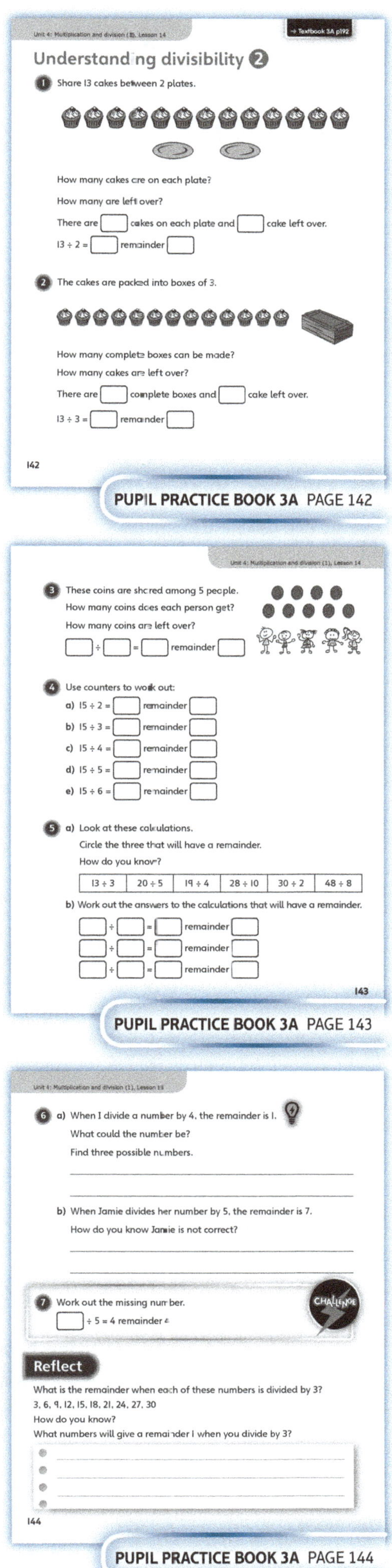

PUPIL PRACTICE BOOK 3A PAGE 142

PUPIL PRACTICE BOOK 3A PAGE 143

PUPIL PRACTICE BOOK 3A PAGE 144

Related facts – multiplication and division

Learning focus

In this lesson, children will learn that an array can be used to determine at least two multiplication facts and two related division facts. Children should also be able to find other related facts, when just given one fact.

Small steps

→ Previous step: Understanding divisibility (2)
→ **This step: Related facts – multiplication and division**

NATIONAL CURRICULUM LINKS

Year 3 Number – Multiplication and Division

- Write and calculate mathematical statements for multiplication and division using the multiplication tables that they know, including for two-digit numbers times one-digit numbers, using mental and progressing to formal written methods.
- Recall and use multiplication and division facts for the 3, 4 and 8 multiplication tables.
- Solve problems, including missing number problems, involving multiplication and division, including positive integer scaling problems and correspondence problems in which *n* objects are connected to *m* objects.

ASSESSING MASTERY

Children can, given an array or similar, find the set of multiplication and division facts that are associated with it and can also communicate what each fact tells them about the array. Children should be able to use a given fact to work out the other related calculations.

COMMON MISCONCEPTIONS

Children may write $4 \div 24 = 6$ instead of $24 \div 4 = 6$. This is a common mistake where children think that division can be commutative, like multiplication. Ask:
- *What was the whole that you started with? What are you dividing it into?*

STRENGTHENING UNDERSTANDING

To support understanding of the concept of related facts, use an array and explicitly link each number in the array with the relevant number in the calculation. For example, show children a 5×3 array and ask them how they can work out the total number of counters. They could count in 5s or 3s. Show them that it could be 3 groups of 5, or 5 groups of 3. Make it clear that the multiplication works out the total and the division works out either the number of rows or columns (or the number in each row or column).

GOING DEEPER

Given a calculation such as $15 \div 3$, can children draw the array and work out the other related facts? Can they challenge a friend to work out the related facts?

KEY LANGUAGE

In lesson: left over, division, remainder, group, array, shared equally, share, whole, group, multiplication fact, **division fact**, multiply (×), divide (÷), array

Other language to be used by the teacher: related fact, division statement, multiplication statement, commutativity

STRUCTURES AND REPRESENTATIONS

Arrays

RESOURCES

Mandatory: counters or cubes

 In the eTextbook of this lesson, you will find interactive links to a selection of teaching tools.

Before you teach 

- Do children know the 2, 3, 4, 5, 8 and 10 times-table facts?
- Do children know how to use $6 \times 5 = 30$ to work out $30 \div 5$?
- Do children know that multiplication is commutative?

Discover

 Pair work

- Question **1** a): *How are the flamingos arranged? How many rows and columns are in the array? What array is this? What multiplication facts can you see? What division facts can be derived from this? How many can you find?*
- Question **1** b): *Can you represent this fact as an array? Could the array go a different way? What facts can you find from the array?*

 Children look at related facts that can be determined from arrays. Children see that a fact such as 4 × 5 = 20 can lead to a family of facts such as 5 × 4 = 20, 20 = 5 × 4 and 20 = 4 × 5 and then related division facts, such as 20 ÷ 4 = 5, 20 ÷ 5 = 4, 5 = 20 ÷ 4 and 4 = 20 ÷ 5. Children have already used multiplication facts to determine answers to related to division problems. This lesson is about children trying find the full family of facts. Question **1** a) shows the flamingos presented as an array. Get children to use the array to generate a family of facts. Children should realise that they need to consider the number of rows and columns to derive the facts. Question **1** b) presents a multiplication fact and children make an array to help them derive the other facts.

 Ensure that there are enough counters or cubes for children to be able to make the arrays.

Question **1** a): 4 × 5 = 20 and 5 × 4 = 20, 20 ÷ 4 = 5 and 20 ÷ 5 = 4

Question **1** b): 3 × 8 = 24 and 8 × 3 = 24, 24 ÷ 3 = 8 and 24 ÷ 8 = 3

Share

 Whole class teacher led

- Question **1** a): *Can you make the array with counters? Does every array have four facts? What do you think Ash means when he says there are more than four facts?*
- Question **1** b): *What does the array look like for 3 × 8 = 24? What is the 3, what is the 8 and what is the 24? What three other facts can you determine from this?*

 Explain that an array can show two division facts and two multiplication facts. Talk children through the multiplications that are formed by the rows multiplied by the columns and show that the division is the total divided by the number of rows or columns. In question **1** a) discuss Ash's comment about there being four more facts. They are the ones where the equals sign is towards the start of the statement (for example, 20 = 5 × 4). Remind children that '=' means 'equivalent to'. Question **1** b) is very similar to question **1** a), except that this time children are given the multiplication fact and they have to think about what the array looks like. Ensure that at each stage you make it clear what each number corresponds to (for example, the total number of counters, the number of columns or the number of rows).

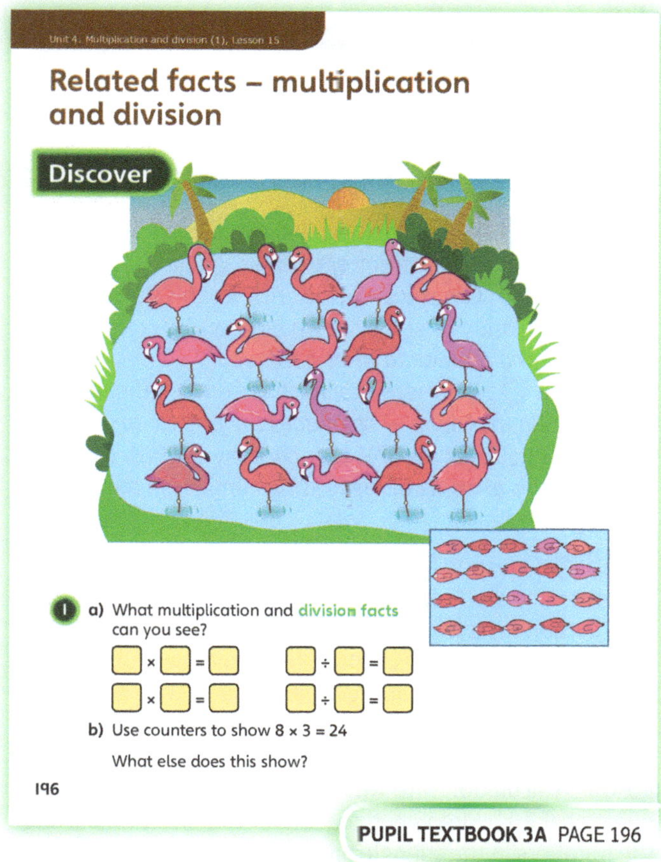

PUPIL TEXTBOOK 3A PAGE 196

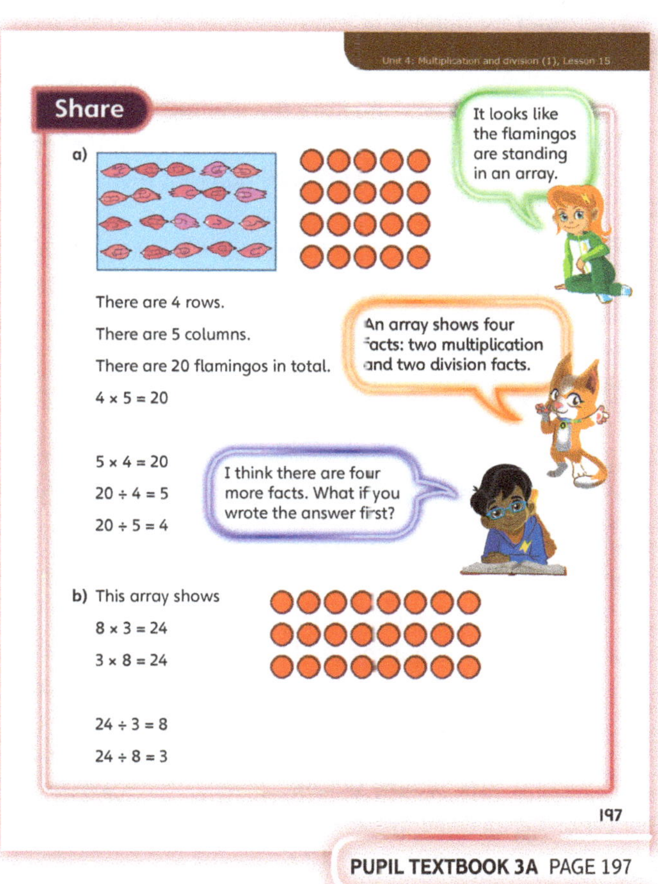

PUPIL TEXTBOOK 3A PAGE 197

Think together

WAYS OF WORKING Whole class teacher led (I do, We do, You do)

ASK

- Question **1** : *What multiplication and division sentences can you find from the array of trees?*
- Question **2** : *Can you show the array for this situation? What calculations can you see from the array? Can you match the calculation to what it tells you?*
- Question **3** : *What is the connection between the three numbers in the triangle? What does the 2 relate to in the array? What about the 5? What about the 10? How can you use this to work out the missing numbers in the other triangles?*

IN FOCUS Question **1** is another example like that of the **Discover** activity. Children must find the four multiplication and division facts that the array shows. Question **2** is similar except this time children need to understand the meaning of each calculation; for example, if they do 8 × 4, they need to understand that this tells them the total number of cakes. It is important that children can link the abstract calculation with what it is used to work out. In question **3**, children need to look for a connection between the numbers in the triangle and the array that is generated. Children should start to see that the two numbers at the bottom multiply to make the top number. Children need to work out the missing number(s) in the other trios. Some children may need to use counters to help them with this.

STRENGTHEN To support understanding of the related facts, use the array and explicitly link each number in the array with the relevant number in the calculation. For example 3 × 5 = 15 shows that 3 and 5 are the number of rows and columns and 15 is the total number in the array. You might need to show through repeated addition that there are 15 counters in the array.

DEEPEN After working through question **3**, ask children to make up their own trio for other children to solve. Give them a multiplication or division fact and ask them to write down the family of facts linked to this calculation.

ASSESSMENT CHECKPOINT Children should be able to find four facts from an array. In addition, given a multiplication or division fact children need to be able to show the array and/or work out the other related multiplication and division facts.

ANSWERS

Question **1** : 3 × 10 = 30, 10 × 3 = 30, 30 ÷ 3 = 10, 30 ÷ 10 = 3

Question **2** : 4 × 8 and 8 × 4 match to 'This calculation works out the total number of cakes.'
32 ÷ 4 matches to 'This calculation works out how many cakes are in a row.'
32 ÷ 8 matches to 'This calculation works out the number of rows.'

Question **3** a): They show the relationship between 2, 5 and 10, such as: 2 × 5 = 10.

Question **3** b): Triangle 1: 27
Triangle 2: 9
Triangle 3: 8 and 3, 4 and 6, 2 and 12, or 1 and 24.

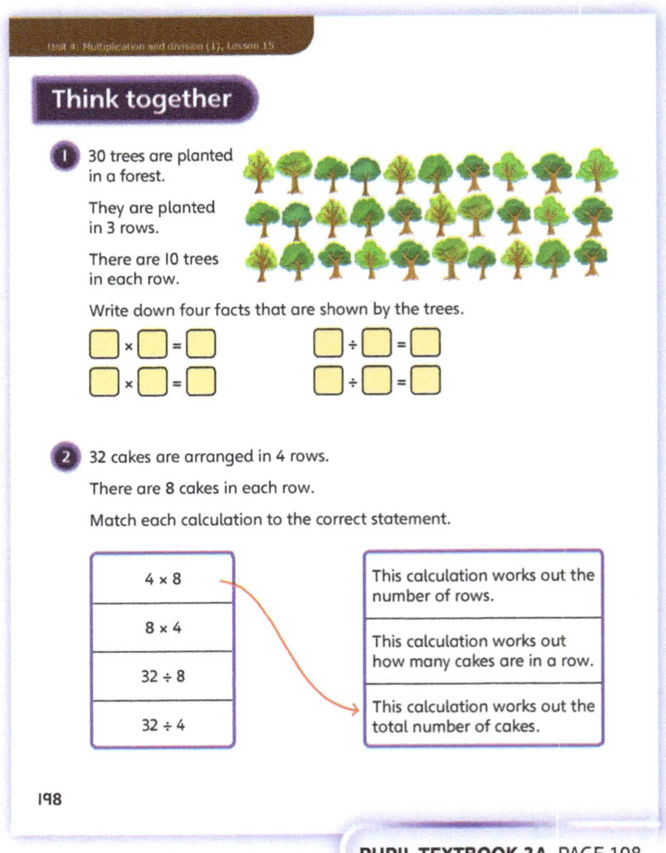

PUPIL TEXTBOOK 3A PAGE 198

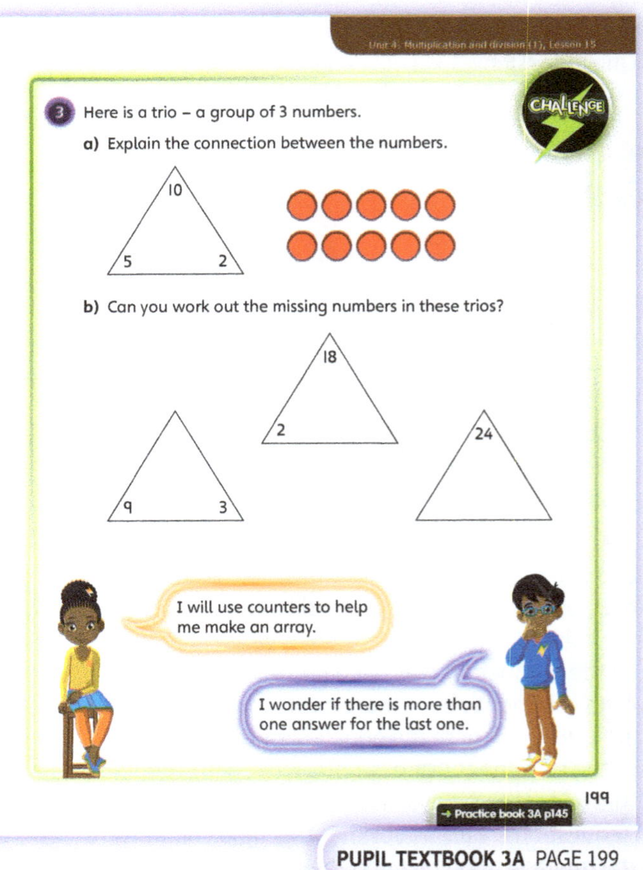

PUPIL TEXTBOOK 3A PAGE 199

Practice

WAYS OF WORKING Independent thinking

IN FOCUS Questions ① and ② develop fluency in finding four related facts from an array and other mathematical equipment, such as bead strings and towers of cubes. Question ③ requires children to realise that there are eight possible facts that children can find, as discussed in **Share**. Question ⑥ looks at whether children can apply the related facts concept to numbers they will not have met in times-tables. This will test whether children have an understanding of the concept of related facts.

STRENGTHEN To support understanding the concept of related facts, use an array and explicitly link each number in the array with the relevant number in the calculation.

DEEPEN Move on from question ⑥ by giving children a calculation such as 15 ÷ 3. Can they draw the array and work out the other related facts?

THINK DIFFERENTLY Question ⑤ requires children to use their knowledge of multiplication and division facts to understand the patterns in the trios. They can use either multiplication or division to find the value of each symbol.

ASSESSMENT CHECKPOINT Children should show that they know how to find at least two multiplication facts and two related division facts from an array. Children know what each fact tells them about the array. Children should be confident in applying their knowledge of related facts to finding answers to other calculations.

ANSWERS Answers for the **Practice** part of the lesson appear in the separate **Practice and Reflect answer guide**.

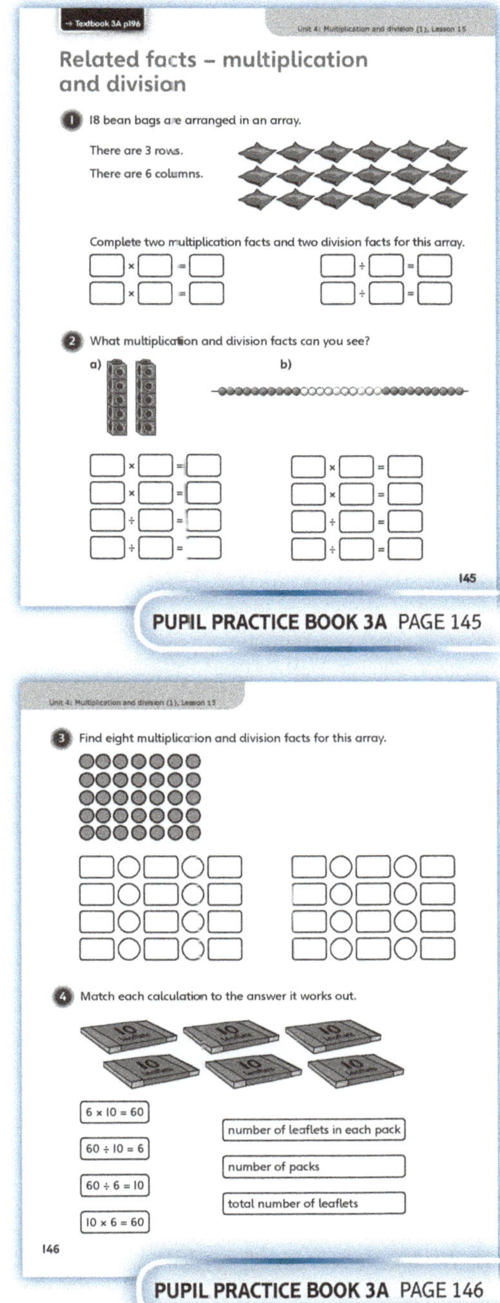

PUPIL PRACTICE BOOK 3A PAGE 145

PUPIL PRACTICE BOOK 3A PAGE 146

Reflect

WAYS OF WORKING Independent thinking

IN FOCUS Children show that they understand that an array shows at least four related multiplication and division facts. Some children may be able to show all eight facts. Ask children to share the answers that they have with their friends and check that they understand the meaning of each of the calculations and can relate them back to the original array.

ASSESSMENT CHECKPOINT Children know how to find four or eight related multiplication and division facts from an array.

ANSWERS Answers for the **Reflect** part of the lesson appear in the separate **Practice and Reflect answer guide**.

After the lesson ⏸

- Can children determine two multiplication facts and two division facts from an array?
- Do children understand what each fact tells them in relation to an array?
- Are children able to find the answer to a calculation (such as 24 ÷ 6) if they know a related fact (such as 4 × 6 = 24)?

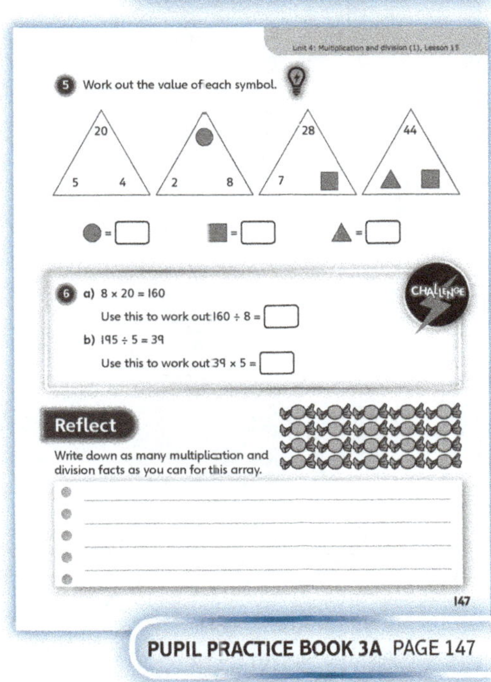

PUPIL PRACTICE BOOK 3A PAGE 147

End of unit check

Don't forget the *Power Maths* unit assessment grid on p26.

WAYS OF WORKING Group work adult led

IN FOCUS These questions cover the multiplication and division methods from this unit. They are designed to draw out particular misconceptions or misunderstandings.

Questions **3** and **4** look at what strategies children adopt. In question **4**, for example, do children just work out all the answers or do they reason correctly why b) must be correct (12 is double 6, and 4 is half of 8, so 12 × 4 must be the same).

For question **7**, ask children to check that both sides are balanced once they have worked out the answer.

ANSWERS AND COMMENTARY Children who have mastered this unit will start to know their 3, 4, and 8 times-tables off by heart and be able to confidently apply them to calculations. Children can explain when to multiply and when to divide. They will know that some divisions do not always give a whole answer and can have remainder.

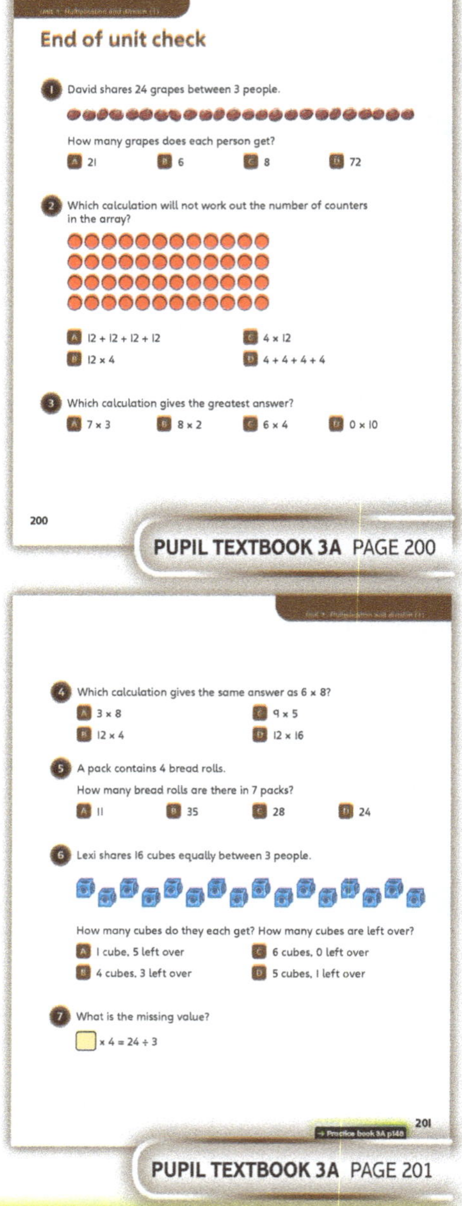

PUPIL TEXTBOOK 3A PAGE 200

PUPIL TEXTBOOK 3A PAGE 201

Q	A	WRONG ANSWERS AND MISCONCEPTIONS	STRENGTHENING UNDERSTANDING
1	C	A and D indicate children do not realise it is a division, B reflects a common misconception that $24 \div 3 = 6$	For question 1, use counters to represent the grapes and share them into 3 groups.
2	D	D shows children recognise there are four 12s being added together, not four 4s.	For question 4, children can rearrange an array to show why b) is the correct answer.
3	C	B shows they think the greater the number in the calculation, the greater the answer.	In question 5, children can use a bar model to represent the situation and it will show it is a multiplication.
4	B	A, C or D shows they have not noticed both numbers have halved or doubled.	For question 6, children can use a sharing model (with counters).
5	C	The other answers show they have not understood they need to multiply.	In question 7, both sides of the statement must be balanced.
6	D	A shows they have got the whole and the remainder confused. B or C shows they have not understood what a remainder is, or have divided incorrectly.	
7	2	Incorrect answers may suggest that they have misunderstood the problem and solve $24 \div 3$ or create a multiplication with an answer of 24.	

My journal

WAYS OF WORKING Independent thinking

ANSWERS AND COMMENTARY

This activity draws together learning about multiplication and division times-table facts.

Look for the different strategies that children use.

a) 30, 40, 50 … Look out for children who realise that this has to be a multiple of 10 above 25.
b) 24, 48 … Look out for children who realise that they need to focus on the 3 and 8 times-tables, as a number in the 8 times-table will definitely be in the 4 times-table.
c) 40 or 0. Children often overlook 0 in the times-tables. Remind them that 0 is in every times-table.
d) 0 or 60. Children may notice that 60 is 3 × 4 × 5 multiplied together because these numbers have no common factors except 1.
e) 0 or 120.

Power check

WAYS OF WORKING Independent thinking

ASK

- *How many times-table facts do you think you know off by heart?*
- *Do you feel confident about telling when a division is equal grouping and when it is equal sharing?*
- *Do you know how to solve word problems? Can you work out if it is multiplication or division?*

Power play

WAYS OF WORKING Independent thinking

IN FOCUS Use this Power play to test fluency with times-tables. Fluency and rapid recall of times-table facts are essential for problem solving and free up valuable working memory; regular practice will help. Support children in understanding what times-table facts mean; for example, 2 × 3 can be visualised as 2 groups of 3 or as a 2 × 3 array.

ANSWERS AND COMMENTARY These wheels test times-table fluency.

×3: 21, 6, 15, 18, 30, 36, 3, 0, 12, 9   ×5: 35, 2, 40, 12, 25, 9, 15, 6, 55

×4: 16, 24, 36, 48, 0, 4, 32, 12, 20, 44   ×8: 8, 1, 2, 3, 4, 0, 10, 11, 7, 5

The second set of wheels tests multiple times-tables. The inner number multiplied by the middle number gives the outer number.

Top left: 60, 20, 32, 8, 8, 18, 80, 48

Top right: inner: 4, 11, 12, 4, 8; middle: 8, 7; outer: 20, 21, 18

Bottom left: This wheel can have different answers as children need to find two numbers that make the outer number when multiplied together.

After the unit ⏸

- Do children know the 2, 3, 4, 5, 8 and 10 times-tables off by heart?
- Do children know how to use these times-tables to solve multiplication and division problems?
- Do they know that a division may have a remainder?

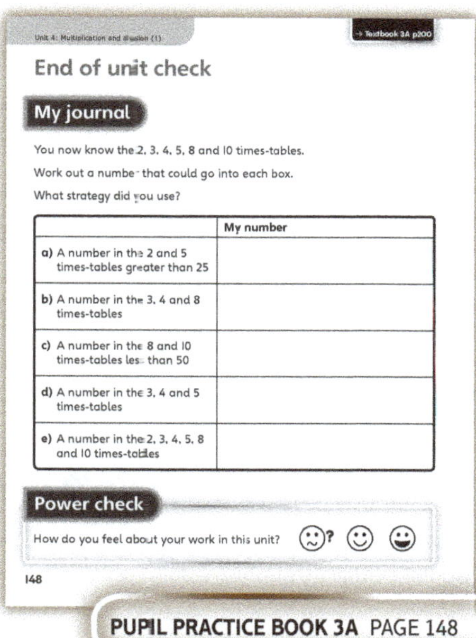

PUPIL PRACTICE BOOK 3A PAGE 148

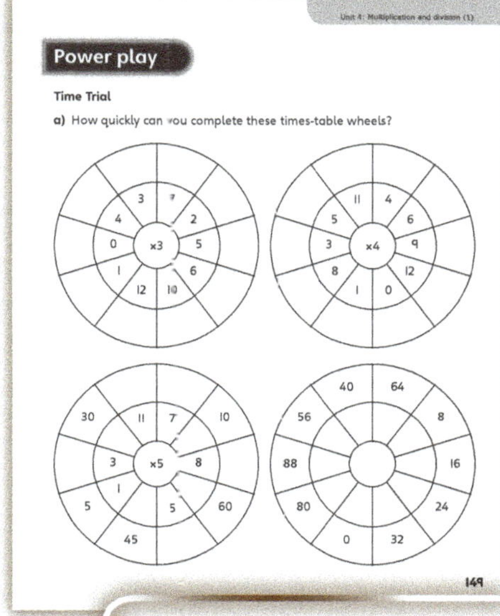

PUPIL PRACTICE BOOK 3A PAGE 149

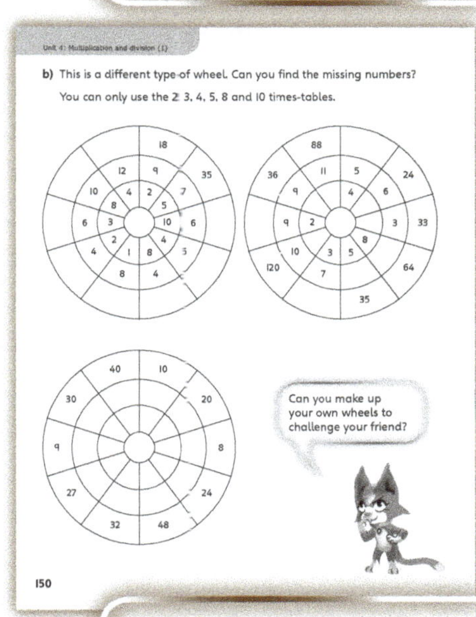

PUPIL PRACTICE BOOK 3A PAGE 150

Strengthen and **Deepen** activities for this unit can be found in the *Power Maths* online subscription.

Published by Pearson Education Limited, 80 Strand, London, WC2R 0RL.

www.pearsonschools.co.uk

Text © Pearson Education Limited 2018
Edited by Pearson, Little Grey Cells Publishing Services and Haremi Ltd
Designed and typeset by Kamae Design
Original illustrations © Pearson Education Limited 2018
Illustrated by Nigel Dobbyn, Virginia Fontanabona, Paul Moran, Nadene Naude and Emily Skinner at Beehive Illustration; and Kamae Design.
Cover design by Pearson Education Ltd
Back cover illustration © Diago Diaz and Nadene Naude at Beehive Illustration.
Coins © Crown copyright pages 189, 194

Series Editor: Tony Staneff
Consultants: Professor Liu Jian and Professor Zhang Dan

The rights of Tony Staneff and Josh Lury to be identified as authors of this work have been asserted by them in accordance with the Copyright, Designs and Patents Act 1988.

First published 2018

21 20 19
10 9 8 7 6 5 4 3

British Library Cataloguing in Publication Data
A catalogue record for this book is available from the British Library

ISBN 978 0 435 18997 6

Printed in Great Britain by Ashford Colour Press Ltd.

www.activelearnprimary.co.uk

Note from the publisher
Pearson has robust editorial processes, including answer and fact checks, to ensure the accuracy of the content in this publication, and every effort is made to ensure this publication is free of errors. We are, however, only human, and occasionally errors do occur. Pearson is not liable for any misunderstandings that arise as a result of errors in this publication, but it is our priority to ensure that the content is accurate. If you spot an error, please do contact us at resourcescorrections@pearson.com so we can make sure it is corrected.